Infrared and
Raman Spectroscopy

PRACTICAL SPECTROSCOPY

A SERIES

Edited by Edward G. Brame, Jr.

Elastomer Chemicals Department
Experimental Station
E. I. duPont de Nemours and Co., Inc.
Wilmington, Delaware

Volume I: Infrared and Raman Spectroscopy (in three parts)
edited by Edward G. Brame, Jr. and Jeanette G. Grasselli

ADDITIONAL VOLUMES IN PREPARATION

Infrared and Raman Spectroscopy

(in three parts)

PART A

Edited by

EDWARD G. BRAME, Jr.

Elastomer Chemicals Department
Experimental Station
E. I. duPont de Nemours and Co., Inc.
Wilmington, Delaware

JEANETTE G. GRASSELLI

The Standard Oil Company
Research and Engineering Department
Cleveland, Ohio

MARCEL DEKKER, INC.   New York and Basel

MARCEL DEKKER, INC.

270 Madison Avenue, New York, New York 10016

LIBRARY OF CONGRESS CATALOG CARD NUMBER: 75-32391

ISBN: 0-8247-6392-0

Current printing (last digit):
10 9 8 7 6 5 4 3 2 1

PRINTED IN THE UNITED STATES OF AMERICA

This book is the first one of a new series on spectroscopy. It is Part A of Volume I which is devoted entirely to the field of infrared and Raman spectroscopy. Volume I will also consist of Parts B and C which will be published subsequently.

The aim and objective of this new series is to cover all possible applications of the different fields of spectroscopy in order to show the usefulness of these different fields in such diverse areas as inorganic chemistry, polymer chemistry, biological science, environmental science, food, textiles, etc. Through these illustrations better appreciation and understanding of the different uses to which spectroscopy has been applied will become apparent. Also, it is hoped that this series will not only provide a resource but also provide a cross fertilization of applications between and among the different fields of biology, chemistry, and physics.

Volume I, Part A contains four chapters. Appropriately, it contains the introductory chapter for the entire volume. This chapter, which is authored by Drs. Bryce Crawford, Jr. and Douglas Swanson, discusses the nature of molecular vibrations, their appearance in the infrared spectrum and the Raman spectrum, and some of the experimental aspects of measuring them. A short historical section is also provided in this chapter.

Chapter 2 covers the application of infrared and Raman spectroscopy to inorganic materials. It is authored by Dr. Robert L. Carter. Included is a discussion on the theoretical principles, the experimental measurements and useful techniques, and a listing of the character and correlation tables.

Chapters 3 and 4 cover the applications of molecular spectroscopy to organometallic materials. They are both authored by

Dr. Walter F. Edgell. The first one is devoted to the vibrational analysis of these materials, while the latter one is devoted to the analysis of their ionic solutions.

Since this book initiates a brand new series of books on spectroscopy under the heading of *Practical Spectroscopy Series,* it serves the dual purpose of showing what the future books are like and how the material will be covered. Subsequent volumes will cover x-ray spectroscopy, mass spectroscopy, nuclear magnetic resonance spectroscopy, ultraviolet-visible spectroscopy, emission spectroscopy, etc. They will be published as the material is collected, edited, and prepared for publication on a sequential basis. There will not be a rigid schedule for publication as some of the books may even be published simultaneously, but the attempt will be to publish these books and volumes on this basis.

One final objective is to bring to the practicing spectroscopist a consistent and practical approach in describing the many and varied uses of spectroscopy.

E. G. Brame, Jr.

CONTRIBUTORS TO PART A

Robert L. Carter, Department of Chemistry, University of
Massachusetts, Boston, Massachusetts

Bryce Crawford, Jr., Department of Chemistry, University of
Minnesota, Minneapolis, Minnesota

Walter F. Edgell, Department of Chemistry, Purdue University,
West Lafayette, Indiana

Douglas Swanson, Department of Chemistry, University of Minnesota,
Minneapolis, Minnesota

CONTENTS

Cumulative Index appears in Part C

CONTENTS FOR OTHER PARTS

Part B

Computer Systems, R. P. Young

Organic Materials, R. A. Nyquist and R. O. Kagel

Environmental Science, Donald S. Lavery

Food Industry, A. Eskamani

Petroleum, P. B. Tooke

Textiles, Gultekin Celikiz

Part C

Biological Science, George J. Thomas, Jr. and Yashimasa Kyogoku

Polymers, Sandra C. Brown and Albert B. Harvey

Surfaces, Clara D. Craver

Infrared and
Raman Spectroscopy

Chapter 1

AN INTRODUCTION TO MOLECULAR VIBRATIONS

Bryce Crawford, Jr.
and
Douglas Swanson

Department of Chemistry
University of Minnesota
Minneapolis, Minnesota

I. INTRODUCTION

When beginning a book on "infrared (and Raman) spectroscopy in the
real world" it seems appropriate that we recall how early the newly
discovered infrared radiation was used to look at very practical
samples. A century and three-quarters ago, Sir William Herschel,
who started as an organist and later became an astronomer, reported
to the Royal Society [1] his discovery of thermal radiation beyond
the red end of a dispersed spectrum from sunlight. Using a thermom-
eter as his detector, he was able to obtain crude measurements of
the intensity as a function of the distance beyond the red, and he
thus recorded the first infrared spectrum (Fig. 1). Moreover, he
went beyond this to measure the absorption of his infrared radiation by various substances, beginning with distilled water, the first
substance to have its infrared (IR) absorption spectrum taken (Fig.
2). He then went on to those very practical liquids seawater, gin,
and brandy.

At that time neither the nature of light nor the structure of
molecules were understood. However, by the end of the nineteenth
century quite a bit was known about the former and chemists had at
least developed the ideas of chemical bonds holding atoms together
in definite geometrical arrangements. Various scientists, mostly
physicists, were engaged in observing infrared molecular spectra.
At the beginning of the twentieth century a Cornell graduate student
named William W. Coblentz undertook a systematic study of infrared
absorption, and continued his investigation as a research associate
of the Carnegie Institution of Washington, publishing in 1905 a collection of remarkably accurate IR spectra for 131 substances [2].
Even though this was the start of so called "diagnostic" IR spectroscopy, applied in a purely empirical manner for chemical purposes,
Coblentz took little interest in the dynamics of molecular vibrations.
From the beginning he was alert to the consistent appearance of certain absorption peaks in spectra of compounds of a given chemical
class.

The blossoming of diagnostic infrared did not get under way,
however, until the second quarter of the twentieth century. Progress

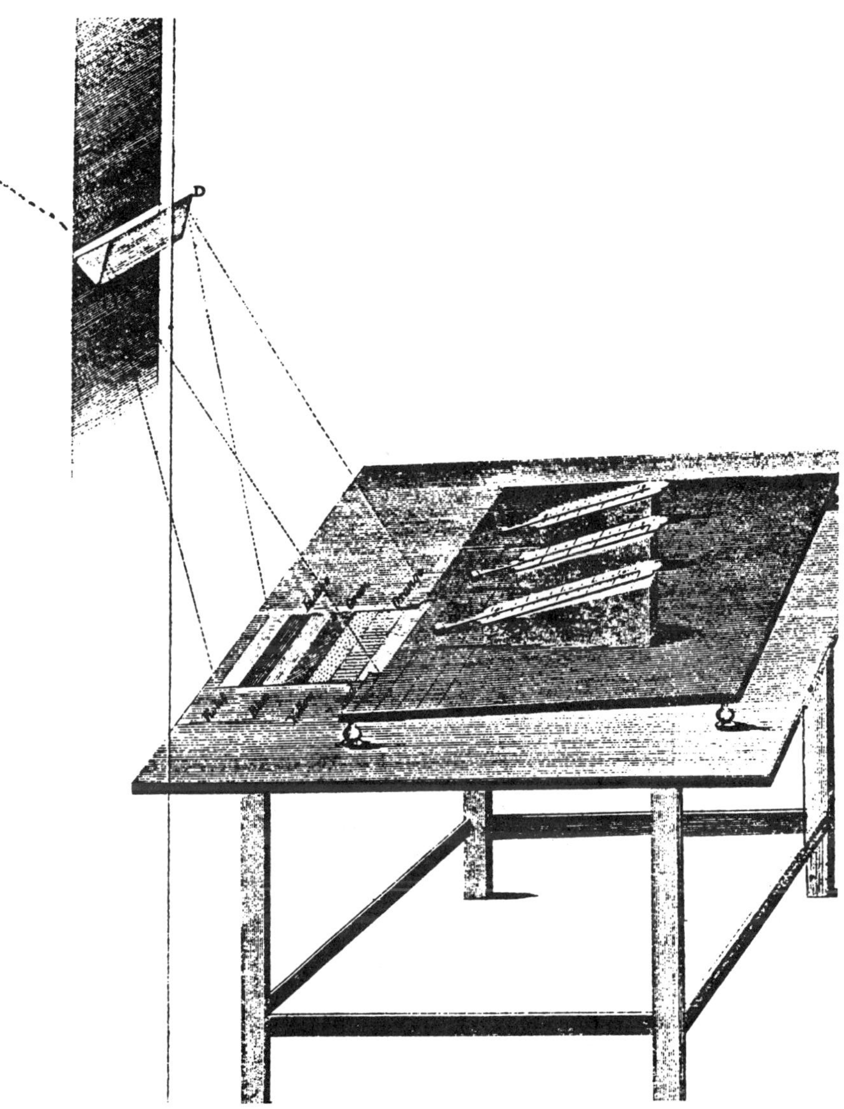

FIG. 1. Herschel's observations. (From Ref. 1, p. 292. Courtesy of The Royal Society.)

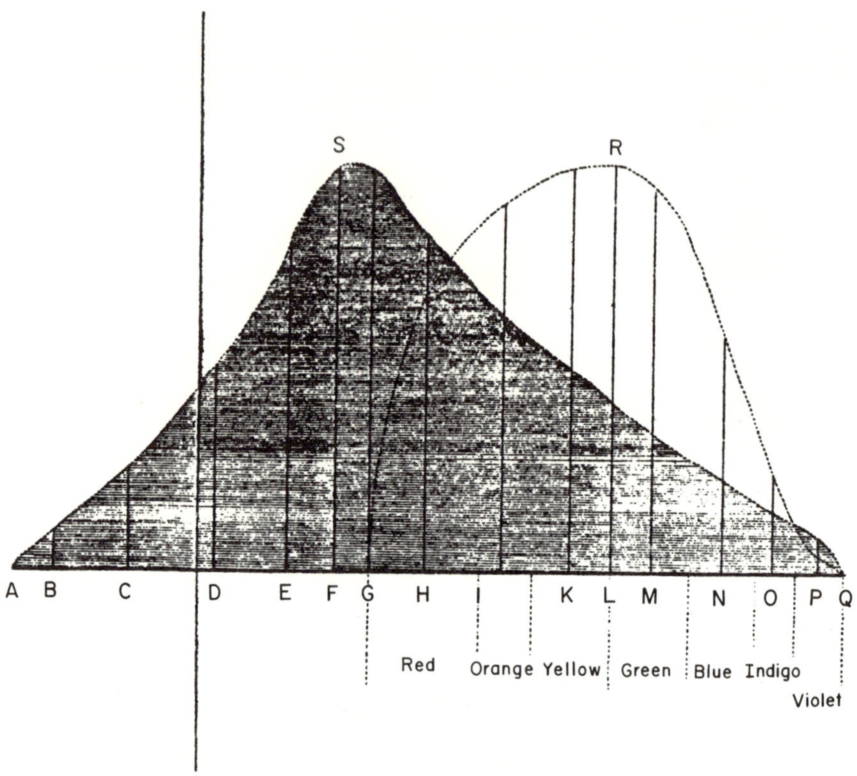

FIG. 2. The first infrared spectrum. (From Ref. 1, p. 538. Courtesy of The Royal Society.)

toward the understanding of molecular structure and molecular spectra was made, of course, especially with the coming of quantum mechanics; but these applications were made largely by physicists. They had little impact on chemists. But here and there studies with chemical applications were made, and of particular importance was the systematic use of diagnostic IR begun in the mid-thirties by three different workers at three industrial locations in the United States: Bowling Barnes at American Cyanamid, Norman Wright at Dow Chemical, and Robert Brattain at Shell Development. By the time World War II came upon us, with its enormous demands on our chemical technology and industry, the usefulness of IR had been established in these locations; moreover, the concept of chemists using "black-box" instruments making physical observations had become somewhat accepted. The designs of the instruments which had been homemade in the industrial laboratories for their own use were made available to two instrument manufacturers: Beckman on one coast and Perkin-Elmer on the other. Hence, the era of the commercial IR spectrometer was born.

Similar things were happening, of course, in other countries. Lecomte in France, by the time of the thirties, had built up a good deal of information on characteristic frequencies. And the impact of the war in Britain brought the quite specific problem of aviation fuel before certain groups of infrared spectroscopists. The usefulness of the "fingerprint region" in the analysis of hydrocarbons was established by Delia Agar (then Delia Simpson) in the laboratories at Cambridge University -- a study in which the early spectra of Coblentz demonstrated their quality. In Britain, as in the United States, there arose enthusiasm for the diagnostic and analytical use of infrared spectroscopy in the "real world" of chemistry, and the IR spectrometer entered the chemical laboratory as a working tool.

II. THE CHEMICAL BOND AND MOLECULAR VIBRATIONS

It will serve us well if, before we consider the several areas of chemical application of vibrational spectroscopy, we become acquainted with what is happening to the molecule so that we understand the reasons why infrared is so peculiarly "chemical"; we shall then have greater insight regarding the problems where infrared observation will be most likely to help us. Coblentz and the pioneers of the early century had to work empirically, but we now have the fruits of the basic investigations of that period and we now know which structural and dynamic factors control the absorption of infrared radiation by molecules.

Our experiment is the classical spectrometric one in which our material sample (a solution) is placed in a beam of radiation and the ratio of the intensity of radiation transmitted through the sample to that incident on the sample is measured as a function of either the wavelength or frequency of the light. We find that, to a good approximation, the Beer-Lambert-Bouguer law holds, and this ratio is related to sample properties by the equation

$$\log \frac{I}{I_0} = \log T = -A$$

$$= -abc \tag{1}$$

where

$$I = \text{intensity of transmitted beam}$$
$$I_0 = \text{intensity of incident beam}$$
$$T = \text{transmittance}$$
$$A = \text{absorbance}$$
$$a = \text{absorptivity of solute}$$
$$c = \text{concentration of solute}$$
$$b = \text{path length through solution}$$

The quantities T, A, and a will vary with radiation frequency. In the chemical infrared we are dealing with the wavelength range from 2 to 25 μm in general; we can more usefully consider the wave number of the radiation, which then runs from 5000 down to 400 cm^{-1}. The advantage of thinking in terms of wave number is that this measure is proportional to the frequency of vibration of the light wave ($\nu = c\omega$ where c is the velocity of light) and hence, by the basic quantum law, to the energy of the quantum of this radiation which a molecule will absorb ($\Delta E = h\nu = hc\omega$). We shall see that this also means that ω is proportional to the frequency at which a vibration is occurring in the molecule.

It turns out that the infrared region corresponds to energy quanta which involve change in the vibrational excitation of the molecules. This has significant implications for the ease of interpretation of infrared spectra. The very existence of vibrational frequencies and the possibility of separation of vibrational energies from rotational and from electronic implies that molecules must have a semirigid structure; and this is not an inherent property of all mechanical systems of particles which are held together. The nucleons within a nucleus, for example, do not exhibit such nice behavior: their motion cannot be approximated by small vibrations, and hence intranuclear energy spectra are much more difficult. We chemists owe our good fortune to the fact that the forces between intramolecular particles (electrons and nuclei) are of Coulomb nature, and that the natural constants the proton mass M, the electron mass m, the electronic charge e, and Planck's constant h have the values they do.* It is illuminating to follow through the reasoning which leads both to "small vibrations" of nuclei in molecules and simultaneously to the justification of the chemical bond, showing how vibrational energies are especially closely linked to the bond concept.

First, we begin with the fact that forces between particles are of the same magnitude, e^2/r^2, whether the particles are two nuclei, two electrons, or one of each, it follows that the nuclear and electronic frequencies of motion, and likewise their amplitudes, will vary inversely with a power of their mass ratio; and M/m is about 2000. This means that we can consider the nuclear motion as taking place in the averaged field of the electrons -- the "electron cloud" concept or the Born-Oppenheimer separation as discussed in any sound quantum-chemistry text. Second, we can find the size of the electron cloud by applying the simplest quantum ideas [3] to find that the basic distance will be $h^2/4\pi^2me^2$, the Bohr distance, which works out to 0.53 Å; hence the fact that bond distances will run to a few angstroms is a consequence of the magnitudes of e, h, and m. Incidentally, the same constants in the same quantum considerations lead to the conclusion that electronic energy spacings will be of the order of $2\pi^2me^4/h^2$ or some 14 eV (or 112,000 cm^{-1}), which for our present interest merely assures us that infrared quanta will not cause electronic transitions or break chemical bonds.

With regard to nuclear motions, if they are to move under coulomb forces applied at distances of approximately a Bohr unit, then

---

*We are indebted to Professor David Dennison for sharing with us his insight concerning these causes and effects.

there will be a sharp change of force with motion. The second derivative of the coulomb potential, the force constant, will be
$2e^2/r^3$, and with r put at two or three Bohr units this works out to
2 or 3 x 10^5 dyn/cm, which means, for a proton mass, a vibrational
motion of frequency approximately 2000 cm^{-1}, with a vibrational
amplitude of a few tenths of an angstrom. All of which means that
the nuclei in a molecule will be held in a semirigid configuration,
executing fairly small excursions around their equilibrium positions.
This not only coincides with and justifies the concept of chemical
bonds -- even to their rather constant bond distances -- but also
permits us to analyze nuclear motions in terms of "small vibrations"
or "normal modes of vibration," which we shall presently consider.
This analysis not only enables us to understand molecular vibrations,
but leads to the concept and the reality of group frequencies, which
can be observed in the infrared region, and which can also be related
to certain chemical bonds or bond patterns.

We can in fact back off now, with these basic points under control, and take as our model for considering intramolecular motion a
collection of "atoms," mass points endowed only with mass and perhaps
some small electrical charge, fastened to each other by bonds which
can be thought of as "springs," permitting only small excursions of
the atoms. Such a model clearly coincides almost exactly with the
chemist's structural model of the molecule with its chemical bonds.

To be sure, this model is an approximation. Even if we consider
the HCl molecule to be made of an H atom of mass 1 amu carrying a
small positive charge and a Cl atom of mass 35 amu carrying a small
negative charge, held together by an ideal spring which in accord
with Hooke's law, exerts a restoring force exactly proportional to
the deviation from the equilibrium internuclear distance, we cannot
explain every aspect of its IR spectrum. On occasion we shall need
to recall that the "springs" on the bonds do not follow Hooke's law
exactly; more correctly for the diatomic HCl we should write that
the potential energy controlling nuclear motion may be written as a
function of the internuclear distance r, or more conveniently the

displacement $\Delta r = r - r_{eq}$ where r_{eq} is the equilibrium internuclear distance:

$$V(\Delta r) = V_0 + V'(\Delta r) + \frac{1}{2} V''(\Delta r)^2 + \frac{1}{6} V'''(\Delta r)^3 + \cdot\ \cdot\ \cdot \tag{2}$$

We then obtain the harmonic or Hooke's law approximation by noting that (a) we may take $V_0 = 0$, since this merely shifts the arbitrary datum line for energy, (b) we may take $V' = 0$, since at the equilibrium distance the derivative of potential energy, or force exerted, must vanish, and (c) we may disregard all terms in $(\Delta r)^3$ and higher because the excursions will be small. Note how we make use of the fact that the excursions will be small, a benefit which comes from the Coulomb law and the magnitudes of M, m, e, and h. This gives us the harmonic potential

$$V(\Delta r) = \frac{1}{2} V''(\Delta r)^2 = \frac{1}{2} k(\Delta r)^2 \tag{3}$$

where we introduce the usual notation k for the force constant. In going from (2) to (3) only assertions (a) and (b), that $V_0 = V' = 0$, were exact. Our assertion (c) concerning small excursions is a very good approximation, but on occasion we may want to allow for correction, taking into account the small but finite magnitude of the cubic and even higher anharmonic terms in the potential function.

Likewise in considering the motion of electric charges, we may on occasion want to recall that we do not really have point charges riding on the H and Cl atoms. We may want to take into account the fact that the dipole moment p of the HCl molecule is not simply a linear function of the displacement Δr; we may want to consider the expansion

$$p(\Delta r) = p_0 + \frac{\partial p}{\partial r} \Delta r + \frac{1}{2} \frac{\partial^2 p}{\partial r^2} (\Delta r)^2 + \cdot\ \cdot\ \cdot \tag{4}$$

and go beyond the linear term. When we come to consider Raman spectra, we shall need to consider the electron cloud which surrounds, or rather constitutes, the bond, and its polarizability and mobility. For a good understanding of all the major facts of IR spectra, and

their chemical interpretation, we can stay with the simple model of
the "point-masses carrying change" and the "bond with a spring."

III. OSCILLATORS, COUPLED OSCILLATORS, AND NORMAL MODES

Let us apply this model to simple molecules, diatomic ones at first,
using our chemical intuition freely to discern broad qualitative
classifications. The diatomic molecule is clearly the place to start:
two atoms of masses m_1 and m_2, connected by one bond of force constant k, and involving one vibration. From physics we recall that,
if two masses m_1 and m_2 are connected by a spring of force constant
k, the vibration frequency is given in terms of k and the "reduced
mass" μ by

$$\nu = \frac{1}{2}\pi \sqrt{kG} = \frac{1}{2}\pi \sqrt{\frac{k}{\mu}}$$

$$G = \frac{1}{m_1} + \frac{1}{m_2} = \frac{1}{\mu}$$

For convenience we can measure the frequency in wave-number
units in cm^{-1}; force constants in mdyn/Å, and atomic masses in amu
or g/mole. And we can rewrite to obtain

$$\omega^2 = \frac{N_o}{4\pi^2 c^2} (10^8 \text{Å/cm}) (\text{dyn}/10^3 \text{ mdyn}) kG$$

or

$$\left(\frac{\omega}{1303}\right)^2 cm^2 = kG \text{ (amu-Å/mdyn)}$$

We note that the reciprocal-mass factor G will be dominated by
the effect of the "lighter" atom: indeed, if $m_1 < m_2$, then

$$\frac{1}{m_1} < G < \frac{2}{m_1}$$

And if we consider the bonds involving common atoms, G will be either
about 1.1 amu^{-1}, for H-X bonds, or much smaller, about 0.14 amu^{-1},

for all other X-Y bonds. With regard to k, chemical intuition suggests we try to systematize its values in terms of single, double,
and triple bonds.

For a few diatomic molecules (selected not altogether at random), Herzberg [4] gives these frequencies:

Type H-X		Type X≡Y		Type X=Y		Type X-Y	
HCl	2989	CN	2068	OO	1580	FF	892
CH	2824	CO	2168	CC	1642		
OH	3728	NN	2360				

Thus we arrive at a "zeroth approximation" in classifying bond
stretching frequencies:

Bond Type	Frequency, cm^{-1}	Range, cm^{-1}
H-X	3000	2600-3600
X≡Y	2200	2000-2400
X=Y	1600	1500-1800
X-Y	1000	800-1200

We can also go in the direction of deducing from diatomic spectra basic structural information concerning molecules, namely, the
actual values of the force constants k. We do indeed find that the
single-double-triple bond pattern holds and that for single bonds k
normally runs about 4 mdyn/Å, for double bonds 10 or 12, and for
triple bonds 15 or 16. But while the k value occasionally illumin-
ates a point of structure of a diatomic molecule, for example, con-
firming the fact that the bond in O_2 is a rather normal double bond
and that in CO a definite triple bond, we seldom gain new insights
from diatomic k values. It is when we come to polyatomic molecules,
and can look at the effect of neighboring interactions (e.g., conjugation) on bond force constants, that the study of these parameters
becomes visible.

For the present let us return to the qualitative characteriza-
tion of bond frequencies and the diagnostic approach to an IR spectrum. We see that we already have some ideas to try in understanding
a spectrum, and in relating the bands we find to the mechanical

structure of the molecule -- which as we have seen means its bond structure and hence its chemical nature. If we try to apply even this first step, say on a spectrum of CH_2=CH-C≡N, we find better success than we deserve: there are indeed frequencies ascribable to ν(CH), the stretching of the CH bond, and to ν(C=C) and ν(C≡N); and the polar CN bond looms very strong. We are encouraged to identify regions of the spectrum associated with hydrogen stretchings (2600-3600 cm^{-1}), triple bonds (2000-2400 cm^{-1}) and double bonds (1500-1800 cm^{-1}). However, we have a way to go since there are quite a few bands in the CH_2=CH-C≡N spectrum besides these. Indeed we realize the single=bond stretching frequencies seem not to be so constant but to move around from one polyatomic molecule to another, and of course, there will be bendings, along with other types of motion than just bond stretchings.

Let us move on now toward the "coupled oscillators" which we find in a polyatomic molecule a system that contains several atoms, several bonds, and several ways in which it can vibrate. We now want to consider a vibrating system which has more than one vibrational motion, but it would be nice if it had not too many -- two would be ideal. Let us step outside the molecular examples and consider a simple mechanical system which will have two basic vibrational motions, namely, a mass point suspended in the middle of an oval ring (Fig. 3). We might consider it an exoskeletal molecule with the atom in the middle vibrating against the outer ring. We assume also that the springs holding it in position are of an unequal force constant. Now, physically, we see that if we displaced the atom along the x axis, or along the y axis, it would be pulled straight back to the center. In either case it would then vibrate back and forth in a straight line. This would be a simple "mode" of vibration to think about. If we pulled it off center in a general direction, say about 45° northeast, the restoring forces due to the two springs would give components which would *not* pull the atom back directly to the center. Therefore, it would not vibrate in a straight line, but would execute one of those lovely but complicated Lissajous figures.

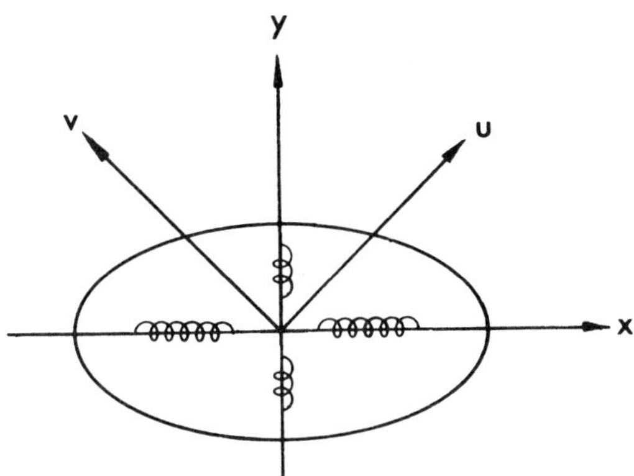

FIG. 3. An exoskeletal molecule.

We can, of course, describe any position, however off center,
in terms of x and y; similarly, we can describe any vibrational pattern, however complex, in terms of the two simple modes of straight-
line vibration. That is what we mean by analyzing vibrations in
terms of the normal modes. There will be as many normal modes as
there are vibrational degrees of freedom: two for our exoskeletal
molecule with its central atom moving in a plane. For an n-atomic
molecule there must be in all 3n degrees of freedom, and if we allow
for 3 translations and 3 rotations of the molecule as a whole, we
must have 3n - 6 normal modes. (A linear molecule, e.g., acetylene,
has only 2 rotations and hence has 3n - 5 normal modes.)

We might call x and y the "nearly normal coordinates." (There
is a mass factor we will care for shortly.) In this case it is easy
to see them and to measure them. In the general polyatomic molecule
it's a little different, and the name "normal coordinate" is not
really appropriate if we think of a coordinate in the naïve sense --
some sort of distance between witness marks that we could walk up to
with a ruler and measure. "Normal vibration variables" would perhaps
be better. What we really have are the normal modes, patterns of
vibration in which all the atoms move back and forth in phase and
in straight lines; the directions of the excursions we can indicate
pictorially with little arrows to obtain diagrams of their normal

modes. The extent of the excursion can be indicated by a variable,
which we can call the "normal coordinate." There will be 3n - 6
such variables.

Another classical physical characteristic of the normal modes
is that they are the resonance modes of the system. If the system
is forced to oscillate with a periodic applied force, and if the ap-
lied frequency matches the frequency of one of the normal modes,
excursions will be wildest and energy will be absorbed most effectively. Related to this is the outstanding quantum physical characteristic: that the energy levels of the system are given in terms
of the normal modes and normal coordinates.

Now how do we describe this in mathematical terms? Let us write
down, classically, the Hamiltonian of the oval system previously
mentioned, first in terms of the normal coordinates x and y:

$$2H = 2(T + V) = \mu(p_x^2 + p_y^2) + (k_x x^2 + k_y y^2) \tag{5}$$

Here we write μ for the reciprocal mass of our atom, and k_x and k_y
are the appropriate force constants. Now let us solve using Hamil-
ton's equations:

$$\dot{p} = - \frac{\partial H}{\partial q}$$

$$\dot{q} = \frac{\partial H}{\partial p}$$

$$\dot{p}_x = - \frac{\partial H}{\partial x} = -k_x x$$

$$\dot{x} = \frac{\partial H}{\partial p_x} = \mu p_x$$

$$\ddot{x} = -\mu k_x x$$

This differential equation is solved in every physics book: we get
the familiar sinusoidal simple harmonic motion, for x and y separately, with different frequencies and of course with different phases.

We mentioned the mass factor; to adjust our "nearly normal" x
and y we may divide by $\mu^{1/2}$ -- or what is the same thing multiply
by $m^{1/2}$ -- to obtain

$$Q_x = m^{1/2}x \qquad Q_y = m^{1/2}y \tag{6}$$

and if we follow through with the momenta

$$P_x = \mu^{1/2}\, p_x \qquad P_y = \mu^{1/2}\, p_y \tag{6'}$$

we may write the Hamiltonian as

$$2H = (P_x^2 + P_y^2) + (\Lambda_x Q_x^2 + \Lambda_y Q_y^2) \tag{7}$$

where in fact $\Lambda_x = \mu k_x$, $\Lambda_y = \mu k_y$, and these Λ's are proportional to the squares of the frequencies -- $\omega_x = 1303\, \Lambda_x^{1/2}$, and so on.

Suppose we had chosen "abnormal" coordinates such as u and v, inclined at 45° as shown. Then it is not hard to show that the transformed Hamiltonian will read

$$2H = \mu(p_u^2 + p_v^2) + k_u u^2 + k_v v^2 + 2k'uv \tag{8}$$

where

$$k_u = \frac{k_x + k_y}{2}$$

$$k_v = \frac{k_x + k_y}{2}$$

$$k' = \frac{k_y - k_x}{2}$$

We can easily see what the cross term does to our easy mathematics:

$$\dot{p}_u = -\frac{\partial H}{\partial u} = -k_u u - k'v \tag{9}$$

We shall get simultaneous differential equations with cross terms. Such things are fairly messy, and indeed the methods of solution are the same as choosing normal coordinates.

We can perhaps see that the mathematical description of normal coordinates is essentially that they permit both the potential energy and the kinetic energy to be expressed as a sum of squares, without cross terms.

The reader familiar with vector-matrix notation will see that
we may describe what we have done by saying that in terms of the abnormal coordinates u and v, if we collect them into a column vector
R with a column vector P_R for their conjugate momenta, we may write
the kinetic and potential energy expressions as

$$2T = P_R'GP_R \qquad 2V = R'FR \tag{10}$$

where the matrices G and F refer to the mass factors and the force
constants respectively, and neither is necessarily diagonal (although
in our simple exoskeletal molecule G happens to be). Mathematically,
the object of the transformation from "working" coordinates R to the
normal coordinates Q,

$$R = LQ \tag{11}$$

is to transform the kinetic and potential energy matrices into the
identity and diagonal form respectively; that is, Eq. (7) transcribes
as

$$2T = P_Q'EP_Q \qquad 2V = Q'\Lambda Q \tag{12}$$

These formal "normal-coordinate analyses" and the consequent determination of force-constant values can, as we mentioned before, be
very useful in deducing chemical information from vibrational spectra; however, it is not our purpose here to discuss these techniques.
The interested reader will find the Wilson FG-matrix technique well
described both in the canonical exposition of Wilson et al. [5] and
in the perhaps more elementary presentation of Brand and Speakman [6].

Our object here has been to elucidate the reality and the physical significance of the 3n - 6 normal modes which an n-atomic molecule has, and we may well stress their physical characteristics:

1. They are the resonance modes of the vibrating system.

2. In a normal mode the atoms all move back and forth in
phase, each moving in a straight line -- the basis of normal mode
diagrams.

3. Any pattern of vibration can be reproduced by combining
normal modes.

4. The energy levels of the system correspond to excitation of the normal modes with each mode separately quantized.

Let us then consider what frequencies we may expect in a poly-atomic molecule which is a system of coupled oscillators. Consider a linear triatomic molecule XYZ, with appropriate force constants in the XY diatomic part and the YZ diatomic part. Taken separately, we would expect the frequencies ω_{XY} and ω_{YZ}. But, since the two dia-tomics are "fused" at the atom Y, we must consider the complete XYZ system. As far as linear vibrations are concerned this system will have two normal modes, with two frequencies ω_+ and ω_-. To what extent will these frequencies of the complete-system normal modes depart from the "natural frequencies" or "group frequencies" of the uncoupled diatomics? An instructive example is the set of triatomics XCN, wherein we replace the H atom of $H-C\equiv N$ with heavier atoms; the series has been observed, and on the other hand we can easily calculate the expected change of frequencies as we change only the mass m_X, keep-ing the force constants fixed. The table shows the series of XCN molecules, giving for each different m_X the natural or uncoupled frequency ω_{CX} -- the other "bond frequency" of course stays as $\omega_{CN} = 2220$ cm^{-1} -- and showing the actual experimentally observed stretch-ing frequencies. Figure 4 shows these values, the circles being the experimental frequencies, and also shows solid curves not only for the calculated "bond frequency" ω_{CX} but also for the two calculated frequencies for the coupled system, ω_+ and ω_-.

	m_X	ω_{CX}, cm^{-1}	Observed Values, cm^{-1}	
HCN	1	3400	3312	2096
DCN	2	2490	2629	1928
TCN	3	2108	2460	1724
CH_3CN	15	1275	2283	918
ClCN	35	1100	2201	729
BrCN	80	1000	2187	580
ICN	127	950	2158	470

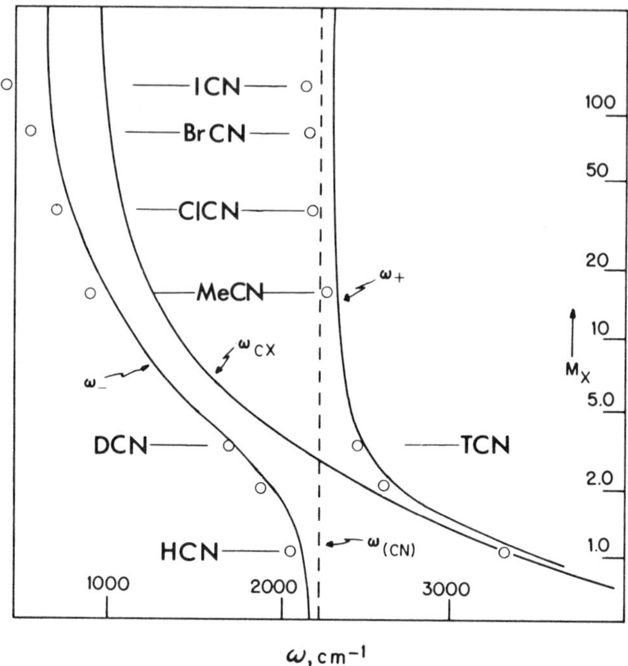

FIG. 4. Coupling of oscillators in XCN.

The example shows us that when two bond oscillators (1) are
tightly coupled and (2) have bond frequencies which are close in
value, we cannot expect the coupled system to show frequencies close
to the bond frequencies. Thus, for TCN, where the bonds are tightly
coupled, being fused at the C atom, and where the bond frequencies
are close (2220 and 2108 cm^{-1}) we get quite large shifts. Where
either condition fails to hold, we see that the bond frequency is
little shifted.

IV. GROUP FREQUENCIES
AND THE STANDARD GROUP MODES

This brings us to a position where, applying a bit of simplification
to the concepts regarding coupled oscillators and using a rather obvious "aufbauprinzip," we can quickly achieve an understanding of

group frequencies and can perceive what to expect from certain types
of commonly occurring groups of atoms -- the CH_3 group, and so on.
We can think of them as "vibrational groups" in the same way that,
in terms of chemical reactivity, we think of "functional groups."

Generalizing from the XCN example we shall consider that any
group of "stiff" bonds, (the H-X, the X=Y, and the X≡Y bonds) which
are fused together by common atoms will constitute a "vibrational
group." Putting it another way, a group of atoms linked together by
"stiff" bonds constitutes a vibrational group. We shall expect from
them not the individual bond frequencies, but group frequencies
which may be quite different from the bond values. Of course the
group will also show other frequencies which involve bending of bond
angles rather than bond stretchings.

If we consider an organic molecule we may predict its spectrum
by cutting it apart at the X-Y single bonds, into its vibrational
groups: thus propylene (CH_3-CH=CH_2) should show the group frequencies for the methyl (-CH_3) group and the vinyl (CH_2=CH) group; acryl-
onitrile (CH_2=CH-C≡N) should show those for the vinyl group, and the
nitrile (C≡N) group. This "zeroth approximation" would mean that the
vinyl group would always show its particular group modes, no matter
what was attached. Actually there will of course always be some
coupling and interaction, and what will happen is that some of the
group modes will change very little in frequency with different molecular surroundings -- these are the "good" group frequencies.
Others will be more sensitive. There are in fact perfectly sound
methods of distinguishing these stable frequencies in terms of the
force-constant analysis [7]; but for the diagnostic spectroscopist,
the empirical observations on which group frequencies are stable and
which are not provide a satisfactory guide. Thus in a "first approximation" we would recognize that some coupling will occur and some
group frequencies will be unreliable. In the next or "second approximation" we would include the significant coupling effects which
often reflect strong chemical interactions. It is in this approximation that we take note of the fact that the carbonyl group shows a

different bond-stretching frequency ν(C=O) depending on whether it
occurs in an ester, a ketone, or an acyl halide; the neighboring
oxygen or halide atom actually affects the effective force constant
of the C=O bond. Finally, in what we might call a "third approximation" we note that some groups interact so strongly as to form new
entities. That happens when the linking single bond of the X-Y type
is not quite an adequate description. The Kekule resonance in benzene is of course the classic example: the phenyl group is a single
vibration group and not a collection of separate -CH=CH- groups.
Other examples occur in the case of conjugation or in combining such
groups as the carbonyl and the -NH- groups in amides.

The usefulness of the group-frequency approach is precisely that
the zeroth or first approximation gives one so much; one can in fact
perceive quickly in an IR spectrum the presence of a vinyl group, or
a methyl group, or a carbonyl group. This degree of success is also
the danger, for too often a chemist will "interpret" a spectrum at
this level alone, using perhaps one of the "magic charts" which enable him to delude himself more conveniently. The wiser diagnostic
IR spectroscopist goes to the second and third approximation; he
checks to be sure that all of the reliable group frequencies for a
postulated group actually appear. He considers alternatives, against
a background of the study of the various groups and their vibrational
modes. There are good discussions of the group frequencies in general books on infrared [8]. However, the most complete and soundest
exploration of the empirical facts on group frequencies is available
in the works of Bellamy [9, 10].

Our purpose here is not to present such a survey, but to stress
that the use of IR (and Raman) will in many fields depend on the
chemist's familiarity with the wealth of information available on
the group frequencies -- including the second and third approximation.
In addition to the detailed consideration of a particular group, it
is helpful to have some grasp of what we might call the standard
group modes -- patterns of atomic grouping which frequently occur,
and whose vibrational modes are frequently under discussion. It is

helpful to have a clear idea of just how the atoms are moving in a given group mode. Moreover, it is useful to see how two or more identical smaller groups may be coupled, either through being fused at a common atom or by being linked with a stiff bond, to form a large group, whose standard modes can be understood through combining the modes of the component groups.

We shall proceed to describe briefly the more important of these standard group modes, giving their usual descriptive names and some examples and describing the atomic motions. We shall also give information as to which of these modes are particularly good group frequencies (in the sense of being stable).

Linear XY_2. Perhaps the simplest standard group may be made up by fusing two XY molecules at the X atom, in a linear configuration. We can of course combine the two bond stretches either in-phase, $\nu_s(XY)$, or out-of-phase, $\nu_a(XY)$, where the subscripts stand for symmetric and asymmetric. In addition we must add the bending mode which can be thought of as involving the rotation of the two XY groups against each other.

XY_2 mode	Components	CO_2	C=C=C
ν_s	$\nu(XY) + \nu(XY)$	1333	1070
ν_a	$\nu(XY) - \nu(XY)$	2350	1965
δ	$R(XY) - R(XY)$	667	852

The numbers given for the C=C=C system are those for $\nu_s(C=C)$ and $\nu_a(C=C)$ in allene (CH_2=C=CH_2). They carry over into substituted allenes rather well. This linear XY_2 model also gives us a start on ketenic (C=C=O), cyanate (C=N=O), and similar systems.

We might note that the great difference between the values for ν_s and ν_a in both instances exemplifies the strong-coupling splitting. Neither can be regarded as an XY bond frequency. We also note that the bending vibration δ can take place in any direction. There are in fact two degrees of freedom (Fig. 5). This example is our first one of a double degenerate frequency: there exist two normal

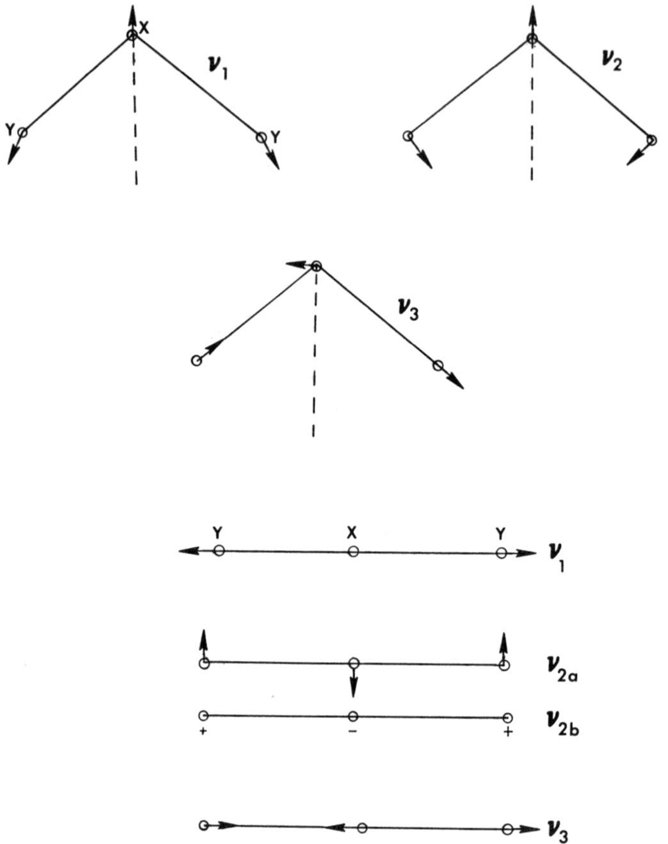

FIG. 5. Normal modes of linear and bent XY_2.

modes which have the same frequency by virtue of the inherent symmetry of this molecule. Thus the linear XY_2 molecule has four modes,
as it should $(3n - 5 = 4)$. Two of them, however, are degenerate so
that only three frequencies are observed.

Bent XY_2. We can also join the two XY groups not in linear
fashion but with some definite angle, usually something near the
tetrahedral angle of 109° (Fig. 6). This very important group also
has in-phase and out-of-phase stretches, and bending. Actually the
symmetric stretch has a little bend, and the bending mode a little

bond stretch; our labels are approximations. Usually ν_a lies higher than ν_s.

Mode	Components	H_2O	SO_2	NO_2
ν_s	$\nu(XY) + \nu(XY)$	3600	1152	1373
ν_a	$\nu(XY) - \nu(XY)$	3760	1361	1615
δ	$R(XY) - R(XY)$	1650	525	641

We give the frequencies for water, sulfur dioxide, and nitrogen dioxide. We will see later that ν_s and ν_a for the latter two give us

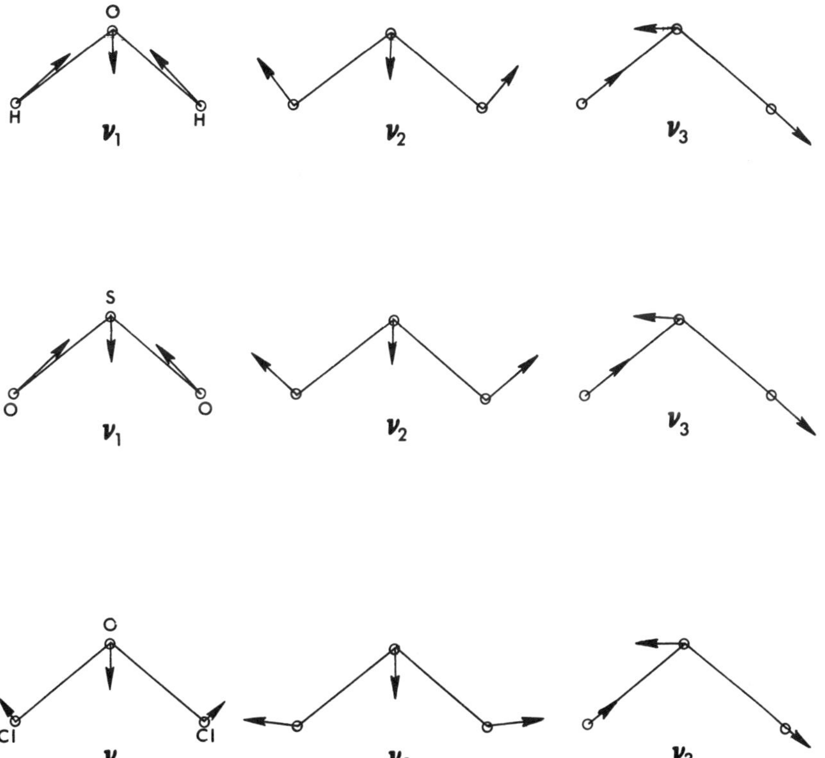

FIG. 6. Approximate form of normal modes for H_2O, SO_2, and Cl_2O. Only the relative amplitudes of the atomic excursions within a single mode are significant.

good group frequencies in the sulfones, sulfates, sulfonates, nitro compounds, and nitrates -- compounds wherein the NO_2 or SO_2 has other atoms linked by single bonds only. As we shall shortly set out specifically, the modes ν_s, ν_a, and δ are good group frequencies for the CH_2 group also.

Linked linear YX≡XY. If we join the two XY molecules by "linking" rather than "fusing," we can build up the linear X_2Y_2 molecule. We must now consider in-phase and out-of-phase coupling of the internal motions and also of the translations as well as rotations.

Components	Mode	C_2H_2
$\nu(XY)\ +\ \nu(XY)$	$\nu_s(XY)$	3374
$T_z\ -\ T_z$	$\nu(X{\equiv}X)$	1974
$\nu(XY)\ -\ \nu(XY)$	$\nu_a(XY)$	3287
$R_x\ +\ R_x$	δ_s	729
$R_x\ -\ R_x$	δ_a	612

Planar ZXY₂. If we return to our bent XY_2 and attach a fourth atom Z to the central X -- attaching it by a stiff bond so that we have a proper tetratomic vibrational group -- all in a plane with the ZX bond bisecting the XY_2 angle, then we shall have six modes. Three of them will be the XY_2 modes already examined. The other three are formed by combining translations and rotations of the Z atom and of the XY_2 group. We take the z direction along the ZX bond and the x direction perpendicular to it in the plane of the XY_2 group.

	Components	Mode	H_2CO	Cl_2CO
Symmetric	$\nu_s(XY)$	$\nu_s(XY)$	2780	575
Symmetric	$T_z\ -\ T_z$	$\nu\ (ZX)$	1744	1827
Symmetric	$\delta\ (XY_2)$	$\delta\ (XY_2)$	1503	297
Asymmetric	$\nu_a(XY)$	$\nu_a(XY)$	2874	849
Asymmetric	$T_x\ +\ R_y$	$\rho\ (XY_2)$	1280	240
Out-of-plane	$T_y\ +\ R_x$	$\omega\ (XY_2)$	1167	440

The symmetric δ is known as the "scissors" deformation, the in-plane ρ as the "rock," and the out-of-plane ω as the "wag." The Cl_2CO case is of course not a true example since the C-Cl bonds are not stiff; but the comparison with the truly coupled H_2CO is instructive; and comparisons with the trigonal CO_3^{2-} which we'll look at shortly are also instructive, especially as regards the $\nu(C=O)$ mode. As far as the two C-Cl bonds are concerned, they are indeed coupled -- ν_s and ν_a differ widely, and neither is close to an "unsplit" (C-Cl) as exhibited in CH_3Cl below.

XY_2 Fused into XY_2Z_2 (CH_2Cl_2). The XY_2 tied into a molecule with two single bonds at the central X atom is a particularly common group. The CH_2 example alone makes it worth our examination. The CH_2Cl_2 case is particularly instructive. Let us fuse a CH_2 group with a CCl_2 group at the C atom, with the CH_2 in the yz plane and the CCl_2 in the xz plane, the z axis bisecting both CH_2 and CCl_2 angles. Then we must couple the internal modes, and also rotations, about the fusing atom and note the symmetries of the various modes with respect to the two planes, xz (CCl_2 group plane) and yz (CH_2 group plane). The CCl_2 is so heavy that it serves almost as an anchor against which the light CH_2 group vibrates.

| Planes | | Components | | CH_2Cl_2 | CH_2Cl_2 | | |
xz(CCl_2)	yz(CH_2)	CH_2	CCl_2	Description	Number		
Sym.	Sym.	ν_s		ν_s(CH)	ν_1	2990	(1)
Sym.	Sym.		ν_s	ν_s(CCl)	ν_3	706	
Sym.	Sym.	δ		δ(CH_2)	ν_2	1424	(1)
Sym.	Sym.		δ	δ(CCl_2)	ν_4	286	
Sym.	Asym.		ν_a	ν_a(CCl)	ν_9	739	
Sym.	Asym.	R_y	R_y	ω(CH_2)	ν_8	1256	(2)
Asym.	Sym.	ν_a		ν_a(CH)	ν_6	3045	(1)
Asym.	Sym.	R_x	R_x	ρ(CH_2)	ν_7	897	(2)
Asym.	Asym.	R_z	R_z	τ(CH_2)	ν_5	1157	(2)

The three CH_2 frequencies marked (1) are very stable group frequencies indeed; the rock ρ, wag ω, and twist τ, marked (2) are more variable, though still sometimes useful. The numbers given in the sixth column are those assigned in structural studies. The normal modes of CH_2Cl_2 are shown in Fig. 7.

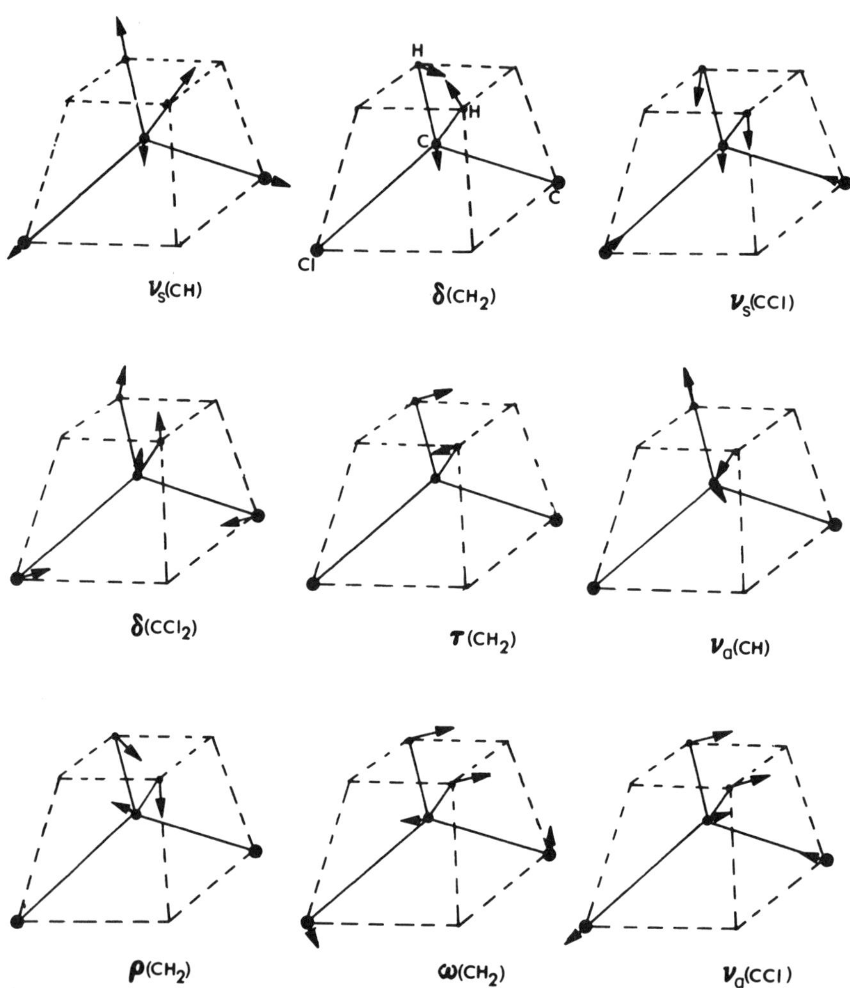

FIG. 7. Approximate modes of CH_2Cl_2.

Trigonal Patterns. A further set of groups involves three equivalent bonds with trigonal symmetry. We can approach their standard modes by changing the ZXY_2 case to let Z be the same as Y, observing the mixing of the modes. However, since a new factor comes in, namely, doubly degenerate modes, it is perhaps more enlightening to start with the simplest possible trigonal combination. It is useful to consider the hypothetical trigonal X_3 molecule. In this case the modes ν_s and δ of the X_2Y case mix in such a way that ν_s becomes a trigonally symmetric "breathing" mode, and δ becomes degenerate with the ν_a mode. We can then have a symmetrical ν_1, and a doubly degenerate ν_2, as shown in Fig. 8.

Planar XY_3. It's convenient to label motions as parallel or perpendicular to the trigonal axis: thus ν_1 is parallel and the degenerate ν_2 is perpendicular. The perpendicular vibrations will be degenerate. Combining the Y_3 with a single X atom in the center, all in the xy plane with the z axis on the trigonal axis, we have six modes with four frequencies. The actual form of the normal mode for BF_3 is shown in Fig. 9.

Y_3	X	XY_3	BF_3	SO_3	NO_3^-	CO_3^{2-}
ν_1		$\nu_\parallel$	888	1069	1050	1063
T_z	T_z	$\delta_\parallel$	719	652	831	879
ν_2		$\nu_\perp$	1503	1330	1390	1415
T_x	T_x	$\delta_\perp$	482	532	720	680

The comparison of the planar SO_3 molecule with SO_2, and of the planar nitrate ion with NO_2, is worth studying.

Pyramidal XY_3 (NH_3). These are much like the planar XY_3 modes, except that the dominant symmetry element is the trigonal axis, and the parallel-perpendicular classification becomes more important:

$\nu_\parallel$ (NH), breathing
$\delta_\parallel$ (NH$_3$), umbrella
$\nu_\perp$ (NH) $\Big\}$ each degenerate
$\delta_\perp$ (NH$_3$)

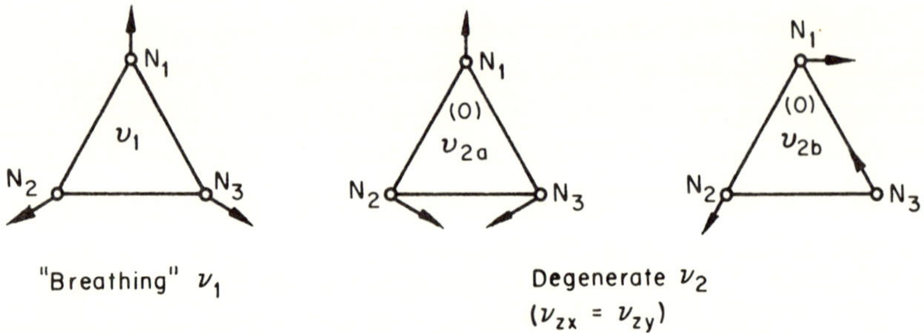

FIG. 8. Normal modes of a trigonal X_3 molecule.

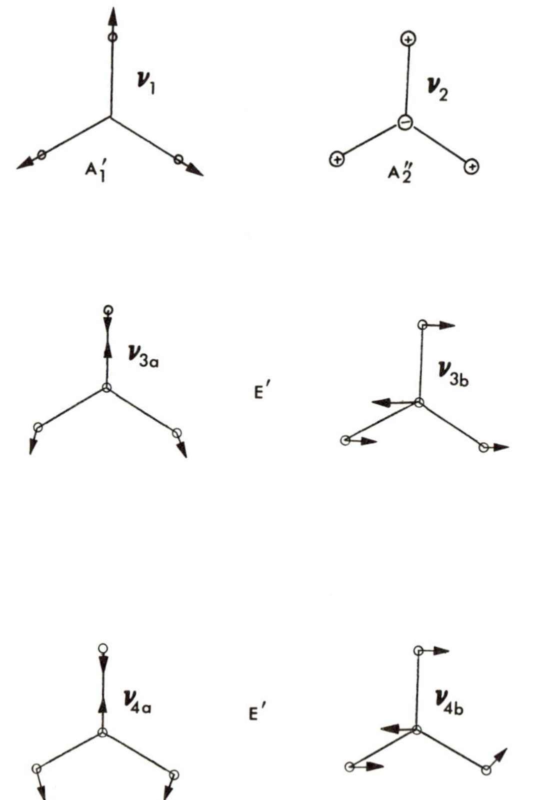

FIG. 9. Approximate form of normal modes of BF_3.

Pyramidal XY$_3$ linked to an atom (CH$_3$Cl) (Fig. 10).

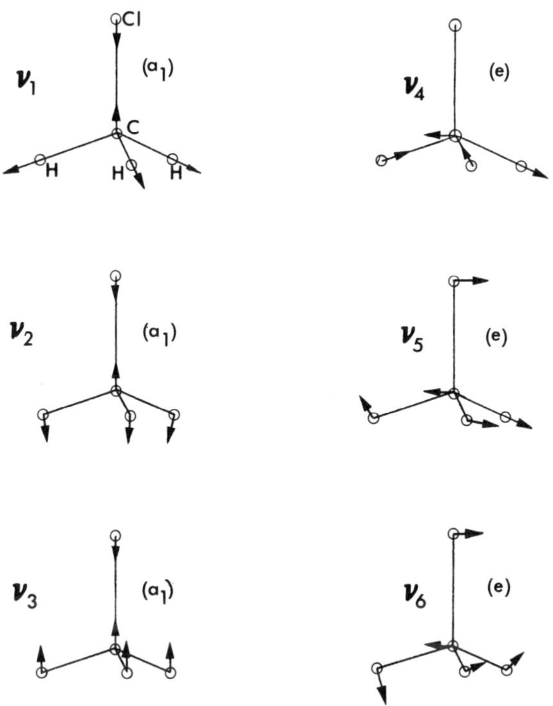

FIG. 10. Approximate modes of CH$_3$Cl (schematic).

Symmetry	CH$_3$	Cl	Molecule	Number	CH$_3$Cl Frequencies	"Group" Frequencies
	$\nu_{||}$		$\nu_{||}$(CH)	ν_1	2937	(2870)*
Parallel	$\delta_{||}$		$\delta_{||}$(CH$_3$)	ν_2	1355	(1380)*
	T_z	T_z	ν(CCl)	ν_3	732	
	$\nu_{\perp}$		$\nu_{\perp}$(CH)	ν_4	3039	(2960)*
Perpendicular	$\delta_{\perp}$		$\delta_{\perp}$(CH$_3$)	ν_5	1450	(1460)*
	R_x, R_y	T_y, T_x	ρ(CH$_3$)	ν_6	1017	

The frequencies marked with an asterisk (*) are the useful CH$_3$ frequencies as found in hydrocarbons (i.e., with the CH$_3$ attached not to a Cl atom but to another carbon). The $\delta_{||}$ is often called the "umbrella" mode, and the doubly degenerate ρ is the "rocking" mode.

XY_4, *tetrahedral.* Our final standard group will be the model which applies to CH_4, CCl_4, SO_4^{2-}, and many other molecules of interest. As when we moved to the trigonal symmetry XY_3, we ran into a considerable mixing of modes of components. For this XY_4 case to produce triply degenerate modes we can start with the ZXY_4 model and set Z equal to Y. The approach of considering the motion of a Y_4 group (or a P_4 molecule), with the atoms moving on the surface of a sphere, and then combining their modes with the three translations of the central X atom, can be used. Perhaps the most perspicuous approach is to begin with our CH_2Cl_2 case and allow the H and Cl atoms to become the same. There must of course be nine modes altogether, with four of them involving stretching of the XY bonds. Thus, five must involve bending or changes in the YXY bond angles, or in another view the sliding about of the four Y atoms on the sphere. The combinations which result, giving one double and two triple degeneracies so that nine modes involve only four separate frequencies, are shown in the table.

Components	Mode	CH_4	CCl_4	SO_4^{2-}
$\nu_s(CH) + \nu_s(CCl)$	ν_s	2917	459	981
$\tau(CH_2)$ $\delta(CH_2) + \delta(CCl_2)$	δ	1534	217	451
$\nu_s(CH) - \nu_s(CCl)$ $\nu_a(CH) + \nu_a(CCl)$ $\nu_a(CH) - \nu_a(CCl)$	ν_a	3019	776	1104
$\delta(CH_2) - \delta(CCl_2)$ $\omega(CH_2) + \rho(CH_2)$ $\omega(CH_2) - \rho(CH_2)$	δ_a	1309	314	613

The first three modes, involving the frequencies ν_s and δ, do not move the central atom at all. The first is the totally symmetric "breathing" mode, and the doubly degenerate pair represent two of the five modes of "skating on the sphere" of the Y_4 group. The other three "skating" modes are approximated by the degenerate triple with frequency δ_a. The motion involved is most easily seen by considering the $\delta(CH_2)-\delta(CCl_2)$ combination. These are shown in Fig. 11. The

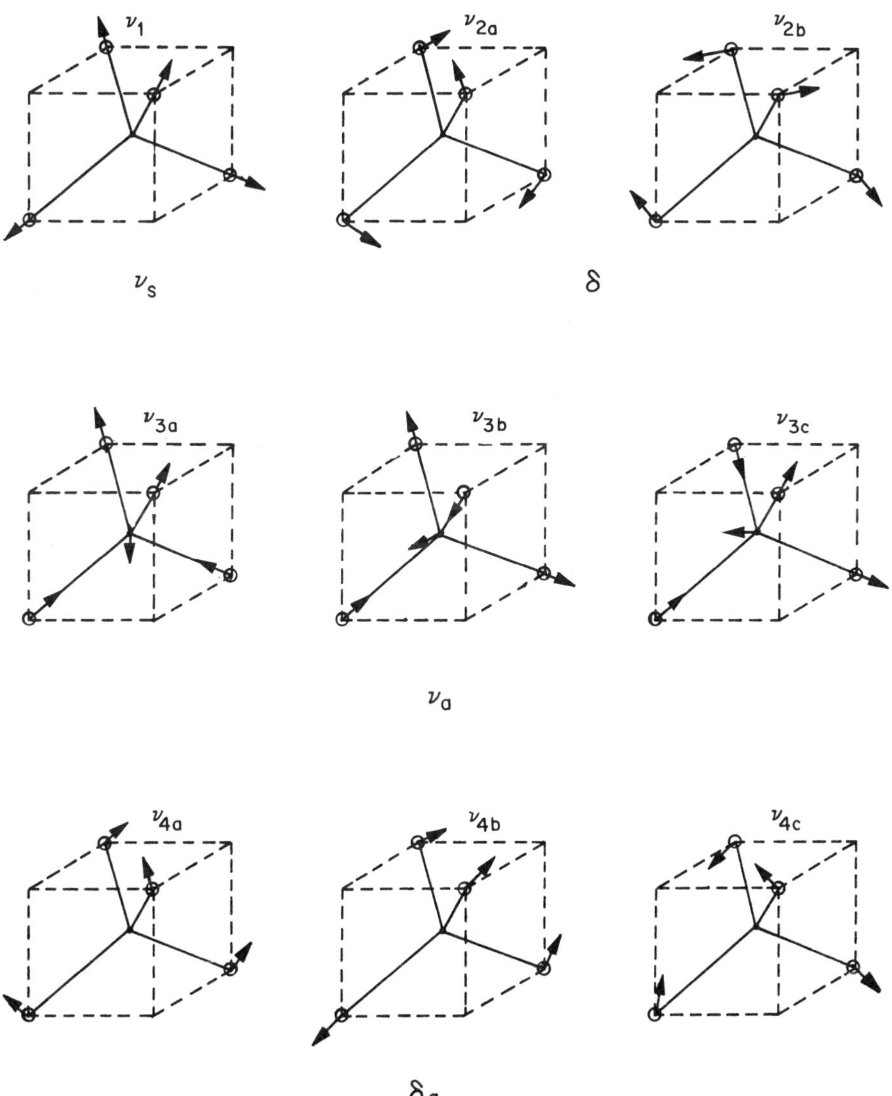

FIG. 11. Normal modes of a tetrahedral XY_4 molecule (schematic).

triple with frequency ν_a does involve more motion of the central X atom. In fact they can be thought of as arising from the three translations of that X atom against the Y_4 shell. Here the $\nu_s(CH)-\nu_s(CCl)$ combination gives the clearest picture of the motion.

There are other less important standard group modes, with symmetries such as the PCl_5 molecule, for example, or the square planar XY_4. Likewise, there are very important special vibrational groups, such as the vinyl group $CH_2=CH-$ already mentioned, or the two $-CH=CH-$ groups, cis and trans -- the two geometrical patterns of linking cause these two groups to have completely different modes and very different spectra -- but we have neither the space here nor the purpose to give a complete survey. The references cited will enable the reader to complete his thorough study of group modes and group frequencies. For a further study of the definitive guide to a basic understanding of molecular vibrational motions with detailed examination of many examples, the use of the book by Herzberg [11] will pay rich dividends.

V. INFRARED APPEARANCE: SELECTION RULES AND INTENSITIES

The purpose of infrared and Raman spectroscopy is to observe the vibrational modes we have discussed. We need to understand not only what vibrational modes a molecule may have, which reflect its chemical structure, but also which of those modes can be expected to appear and with what sort of intensity. We are aware of the fact that a molecule placed in a beam of infrared radiation can absorb energy and thus show an "absorption band." But what factors govern this process?

From the quantum viewpoint we know that the vibrational energy levels are quantized. We have observed that these levels are properly described in terms of the number of quanta of excitation in each of the normal modes. In fact each of these normal modes acts like an harmonic oscillator, with the energy content worked out in any quantum-chemistry book,

$$E_n = h\nu\left(n + \tfrac{1}{2}\right) = hc\omega\left(n + \tfrac{1}{2}\right)$$

where n quanta of excitation are involved, ν is the classical frequency of the particular oscillator, and ω is the corresponding wave
number. The quanta of light absorbed thus will have a frequency corresponding to the classical normal-mode frequency. An absorption band
at a frequency ν implies that the molecule has a normal mode of frequency ν. (This is true for harmonic oscillators; we shall presently mention the need for slight refinement to understand overtones
and combinations.) This will be true, of course, for both infrared
and Raman observations.

Fortunately the mode of interaction between the molecule and
the oscillating electromagnetic field is easily understood in classical terms. The oscillating field will push the positive charges in
the molecule alternately one way, and the negative charges similarly
but in opposite phase. Thus, if we think of an HCl molecule properly oriented we will perceive that at one point in time the field will
be pushing the positive H atom upward and the negative Cl atom down-
ward, stretching the bond, and half a period of oscillation later the
forces will be reversed and the HCl molecule will be under compres-
sion. If we think of a CF_4 molecule, the electromagnetic wave will
move the positive C atom against the negative F atoms in a fashion
which will tend to excite the triply degenerate ν_a mode already discussed. In short, an electric field tends to displace positive and
negative charges and thus change the dipole moment p of the molecule,
distorting it in such a fashion as to change p. The vibrational
modes which produce changes in the dipole moment will be excited.
If a molecule has symmetry, we can formalize this requirement into
"selection rules" which state that, since certain modes cannot produce a change in p, they cannot appear in the infrared; this use of
symmetry, most conveniently through the formalism of group theory,
has been a powerful tool in the basic interpretation of molecular
spectra [5]. Without group theory, we can see that the symmetric
breathing vibrations of the linear XY_2 case, the planar XY_3 case,
and certainly the tetrahedral XY_4 case, will be inactive in the infrared. In fact, by simply using the model of positive and negative

charges on the atoms, we can see that in the XY_4 case only the triply degenerate frequencies can be infrared active. Hence, the spectrum of the SO_4^{2-} ion may be expected to consist of two bands. (Modified of course to allow for splittings in a crystal, and for overtones; but essentially only two fundamentals.) The symmetry requirements work just the other way for Raman selection rules: the more symmetric modes give greater Raman intensity, so that a happy complementarity exists between the two techniques.

It is reasonable that the more a normal mode affects the dipole moment, the more intensely it will interact with radiation. Indeed, the quantitative expression for absorption intensity shows that this will vary with $(\partial p/\partial Q)^2$, where Q is the normal coordinate for the mode involved. It is not our purpose here to go into quantitative intensity measurements, nor their interpretation in terms of basic molecular parameters. Very good treatments in depth are in the literature [12, 13]. What we wish to do is to look at the qualitative factors to perceive what modes we may expect to see in the infrared and what may cause intensity changes. As a first approximation we may use the "point-charge riding the atoms" model, and thus understand why homonuclear diatomics such as N_2 are inactive in the infrared, why HCl absorbs more strongly than HI, or H_2O more strongly than H_2S. In fact, we see why in general fluorocarbons absorb much more strongly than the corresponding hydrocarbons, and why the $\nu(HS)$ bond vibration, though a good group mode, is not helpful since it normally isn't seen in the infrared at all.

We need to go a bit farther, however, and recall that the point-charge model is rather crude, and that as nuclei move the electron clouds distort and shift. Thus, carbon monoxide CO is hardly a polar molecule (its permanent dipole moment is only a few tenths of a debye) but its infrared absorption is strong. It is $(\partial p/\partial Q)^2$ that counts, and as the CO distance changes the electrons move about. Similarly, the out-of-plane dishing modes of ethylene and benzene are enormously intense in the infrared, because with the distortions of the nuclear framework the π-electron cloud moves vigorously.

The symmetry idea can also be applied in a usefully approximate manner, in terms not of group theory but of local symmetry. Clearly, by rigorous symmetry the stretching of the double bond, $\nu(C=C)$, in ethylene C_2H_4 cannot appear in the infrared. If we consider symmetrically substituted ethylene such as trans $C_2H_2Cl_2$, this will still be true rigorously. If we go to trans-pentene-2, where the groups on the corners of the double bond are not identical, the local symmetry of that double bond will still cause the $\nu(C=C)$ band to be inactive or so weak as to require a good (and perhaps imaginative) eye.

It is also worth bearing in mind that intensities are a bit more sensitive to molecular environment than are frequencies. This can be useful in getting the most out of a spectrum. However, one needs to have examined the situation for the particular group in mind [9, 10] to avoid being misled.

Finally, we must recall that although the harmonic-oscillator approximation serves us well, it is not rigorous. One of the results is that we shall observe more than just fundamentals, or absorption bands in which only one normal mode goes up one level. We can and do observe overtones, in which a normal mode goes up two levels, absorbing radiation at a frequency almost (but not necessarily exactly) twice the fundamental, and combinations, in which two normal modes each go up one or more levels. As a very rough general rule, we may expect overtones to have perhaps one-tenth the intensity of the fundamentals, or less. However, there are many exceptions. There are even instances in which an overtone or combination is more intense than any of the fundamentals involved. Some characteristic group frequencies have equally characteristic overtone and combination bands, which can be very useful. A good example is provided by the "fingerprint" region for aromatic compounds between 1660 and 2000 cm^{-1}, where the overtone pattern gives helpful information regarding the pattern of substitution on the aromatic ring. It is well to bear in mind always that, where there are very strong fundamentals, overtones, and combinations, it is well to think in wave-number terms.

A nitrate ester free of any hydrogen will still show a fairly good band in the ν(CH) region; for the $-O-NO_2$ group has two strong bands (ν_a and ν_s for NO_2) at 1630 and 1280 cm^{-1}, and the combination ($\nu_s + \nu_a$) appears at 2820 cm^{-1} quite nicely.

VI. RAMAN SCATTERING

As we have seen, molecular frequencies which involve a change in dipole moment may be investigated by use of infrared radiation. It is also possible to investigate certain molecular vibrations by use of a light scattering technique known as Raman scattering. Again, we will try to use a very simple, physical approach to explain Raman scattering, leaving more rigorous procedures to other texts [14-17]. Let us consider a molecule placed in an electric field. In this situation the electrons, or better the electron cloud, will be displaced relative to the nuclear framework. This distortion, or polarization, produces an induced dipole moment regardless of whether there is a permanent molecular dipole moment. This situation can be represented mathematically as follows:

$$\underline{P} = \alpha \underline{E} \tag{13}$$

where $\underline{P}$ is the induced electric moment, $\underline{E}$ is the electric field in which the molecule is placed, and α is called the polarizability. Since $\underline{P}$ and $\underline{E}$ are vectors, α takes the form of a tensor and Eq. (13) may be more completely expressed as

$$\begin{aligned}
P_x &= \alpha_{xx}E_x + \alpha_{xy}E_y + \alpha_{xz}E_z \\
P_y &= \alpha_{yx}E_x + \alpha_{yy}E_y + \alpha_{yz}E_z \\
P_z &= \alpha_{zx}E_x + \alpha_{zy}E_y + \alpha_{zz}E_z
\end{aligned} \tag{14}$$

where the term α_{ij} relates the polarization in the ith direction due to the field in the jth direction.

Let us now imagine that the electric field in which the molecule was placed is the oscillating electric field of a light beam. If the light is of frequency ν_o, we may express its associated electric field as

$$\underline{E} = \underline{E}_o \cos 2\pi\nu_o t \tag{15}$$

In this case, the molecular polarizability will fluctuate in time as

$$\underline{P} = \alpha\underline{E}_o \cos 2\pi\nu_o t \tag{16}$$

This oscillating induced dipole can now radiate electromagnetic radiation of frequency ν_o. This process is known as Rayleigh scattering. Since this process radiates light of the same frequency as the incident radiation, it is of little use in determining molecular structure. If, however, the molecular polarizability α in Eq. (16) was a function of time, some interesting effects could occur. A more detailed examination of the polarizability tensor is needed to determine when and how α may vary with time.

In order to more easily understand the concept of a molecular polarizatility, it may be helpful to recall that the units of polarizability are volume units [18]. Thus, the magnitude of the polarizability is related to the volume of the "bag of loose electrons" which are free to move or to be polarized. The greater the volume of "loose" electrons, the more the molecule can contribute to the scattering processes. Systems containing large polarizable atoms or groups will tend to be strong scatterers. Thus, molecules that contain I, S, double bonds, aromatic rings, etc., will be better scatterers than will systems made up largely of ionic bonds such as silica glasses.

To help visualize the polarizability tensor, it may be helpful to think of a polarizability ellipsoid. If arrows were drawn from a common origin, in the appropriate directions, having lengths proportional to the values of the α_{ij}'s in Eq. (14), the heads of these arrows would form an ellipsoid. It is the variation of this ellipsoid, which represents the molecular polarizability, that we are concerned with. Figure 12 illustrates polarizability ellipsoids for several simple molecules. Some molecular vibrations may cause the polarizability ellipsoid to change size or orientation. Figures 13 and 14 illustrate the way in which the polarizability of two simple molecules vary for various normal vibrations of the molecule. As

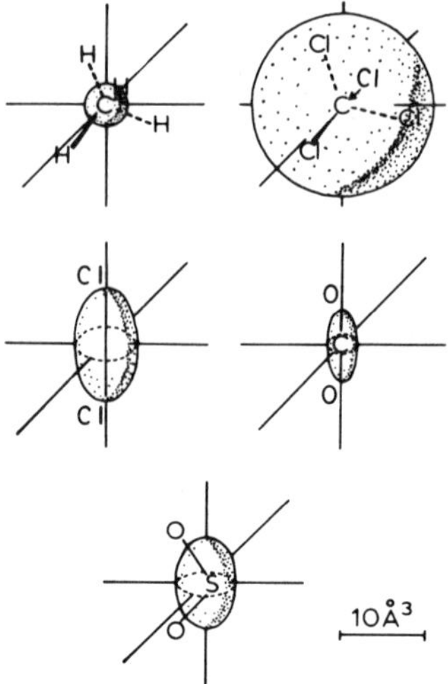

FIG. 12.   Polarizability ellipsoids for some simple molecules (drawn to scale).   (From Ref. 18, p. 6.   Courtesy of The American Chemical Society.)

can be seen from these figures, some vibrations cause the polarizability to vary at the frequency of molecular vibration. We can now examine this frequency dependence and its effect on scattering.

If we let Q_v represent some normal vibrational coordinate of the molecule we can expand the polarizability in terms of Q_v as

$$\alpha = \alpha_0 + \left(\frac{\partial \alpha}{\partial Q_v}\right)_0 Q_v \tag{17}$$

The coordinate of Q_v is a function of time and may be written as

$$Q_v = Q_v^0 \cos 2\pi \nu_v t \tag{18}$$

where ν_v is the frequency of the vth vibration. Upon substitution of Eqs. (17) and (18) into Eq. (16) we obtain

$$\underline{P} = \alpha_0 \underline{E}_0 \cos 2\pi \nu_0 t$$
$$+ \frac{1}{2} \underline{E}_0 Q_v^0 \left(\frac{\partial \alpha}{\partial Q_v}\right)_0 \{\cos 2\pi(\nu_0 + \nu_v)t + \cos 2\pi(\nu_0 - \nu_v)t\} \tag{19}$$

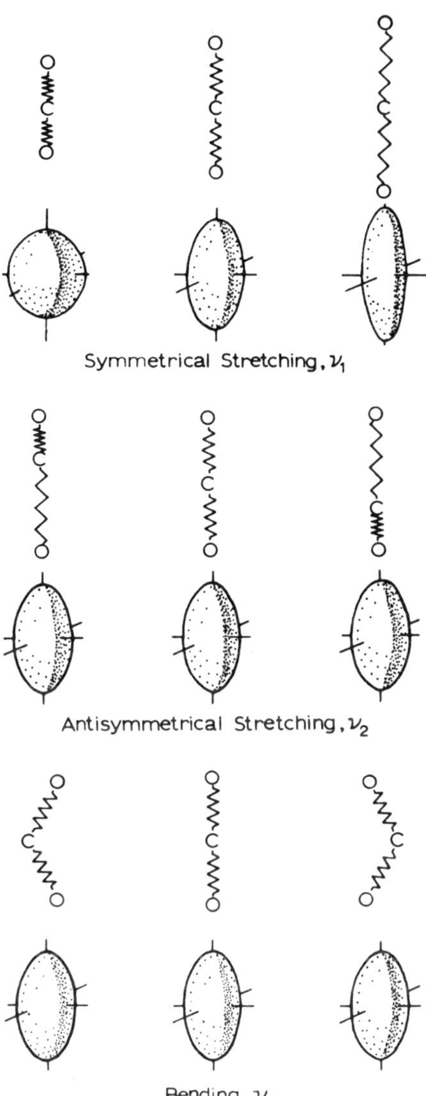

FIG. 13. Polarizability changes during the vibrations of carbon dioxide (exaggerated). (From Ref. 18, p. 6. Courtesy of The American Chemical Society.)

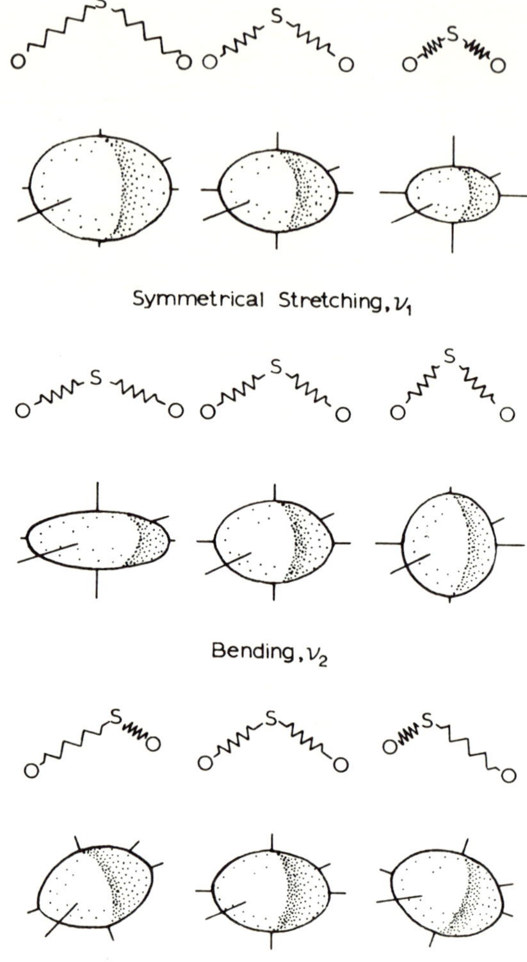

FIG. 14. Polarizability changes during the vibrations of sulfur dioxide (exaggerated). (From Ref. 18, p. 7. Courtesy of The American Chemical Society.)

The first term in Eq. (19) describes a classically oscillating dipole radiating frequency ν_o, i.e., Rayleigh scattering. The second term of Eq. (19) describes dipoles oscillating at frequencies $\nu_o + \nu_v$ and $\nu_o - \nu_v$ and thus will describe radiation of these frequencies.

It is the scattering of these frequencies that is known as Raman
scattering. Clearly, the radiated frequencies differ from the incident frequency by the frequency of molecular vibrations. It is in
this way that molecular group frequencies appear in a Raman spectrum.

It is seen from Eq. (19) that Raman scattering will not occur
unless the derivative term, $(\partial\alpha/\partial Q_v)_0$ is nonzero. In other words,
Raman scattering intensity depends on the change in polarizability
during molecular vibrations, as opposed to the change in dipole
moment that was found to be necessary for an infrared experiment.
It is the value of the derivative term that leads to selection rules
for Raman scattering, and any molecular vibration during which at
least one component of the polarizability tensor changes will be
Raman active and may appear in the Raman spectrum. Since we are not
concerned here with a rigorous group theoretical derivation of selection rules of Raman scattering, suffice it to say that all totally
symmetric vibrations are Raman active (e.g., the symmetric vibrations
shown in Figs. 12 and 13 show a change in the size of the polarizability tensor since the ellipsoid "breathes" with the vibration),
while vibrations which are not symmetric may or may not be Raman active. Thus the Raman spectrum furnishes information concerning the
magnitude of the polarizability change during vibrations which, when
coupled with group theoretical deductions, yields information about
molecular structure. The fact that the intensity of scattering depends on polarizability, and not on dipole moment, often makes Raman
scattering and infrared good complementary techniques. For many
group frequencies the Raman bands are strong and the associated infrared bands are weak, or vice versa.

Not only is the intensity of Raman lines important in elucida-
ting molecular structure but the study of the polarization of the
spectrum is found to be very important. Specifically, polarization
studies can be used to help determine the symmetries of the vibrations involved. In order to understand polarization effects it is
necessary to consider a so-called "derived" tensor of which the
components are defined by

$$\alpha'_{ij} = \left(\frac{\partial \alpha_{ij}}{\partial Q}\right)_0 \tag{20}$$

We expect the intensity of Raman scattering to be proportional to this tensor. Since, in a typical Raman experiment, the scattering molecule is rotating, the observed scattering will be an average over all orientations of the molecule. To express the scattering intensity in terms of the derived polarizability tensor, it is necessary to find quantities which are invariant under rotation. These invariants are found to be

$$\text{Mean value } \bar{\alpha}' = \frac{1}{3}(\alpha'_{xx} + \alpha'_{yy} + \alpha'_{zz}) \tag{21}$$

$$\text{Anisotropy } \gamma'^2 = \frac{1}{2}[(\alpha'_{xx} - \alpha'_{yy})^2 + (\alpha'_{yy} - \alpha'_{zz})^2$$
$$+ (\alpha'_{zz} - \alpha'_{xx})^2 + 6(\alpha'^2_{xy} + \alpha'^2_{yz} + \alpha'^2_{zx})] \tag{22}$$

Using these invariants to express the intensity of scattering we can now examine a specific scattering case. Consider incident unpolarized light propagating in the y direction and scattering in the x direction. This situation is portrayed in Fig. 15.

We will now define the degree of polarization as

$$\rho_n = \frac{I_y}{I_z} \tag{23}$$

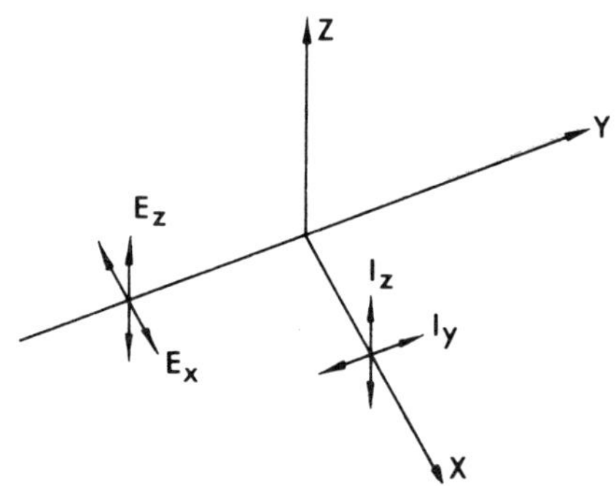

FIG. 15. Polarization of scattering for unpolarized incident light.

Averaging over all molecular orientations this depolarization ratio
is found to be

$$\rho_n = \frac{6(\gamma')^2}{45(\overline{\alpha'})^2 + 7(\gamma')^2} \tag{24}$$

It is well to note that this depolarization ratio is defined for
Raman scattering and not for Rayleigh scattering, which would be a
function of our original polarizability tensor components and not
of our "derived" tensor, which is the case for Raman scattering.
Using incident plane polarized light it is easily shown that

$$\rho_p = \frac{3(\gamma')^2}{45(\overline{\alpha'})^2 + 4(\gamma')^2} \tag{25}$$

where the subscript p denotes polarized incident light. If $\rho_p = 0.75$,
the line is said to be depolarized; if $\rho_p < 0.75$, the line is polarized; and if $\rho_p = 0$, the line is completely polarized.

The value of the depolarization ratio can be used to assist in
the determination of the symmetry of vibrations. This can be done
because the polarizability can be associated with an isotropic, or
spherical, part $\overline{\alpha}'$ and an anisotropic part γ'. The isotropic portion is related to the change in size of the polarizability ellip-
soid, while the anisotropic part is related to the change of orientation of the ellipsoid.

For the symmetric vibrations shown in Figs. 13 and 14 the size
of the ellipsoid changes but its orientation does not. Indeed, this
is found to be true for all symmetric vibrations. This means that
the diagonal elements of the polarizability ellipsoid change, imply-
ing that $\overline{\alpha}'$ changes Eq. (21). Therefore in Eq. (25) $\overline{\alpha}'$ is nonzero
and the line must be polarized, that is, $\rho_p < 0.75$. Antisymmetric
vibrations, on the other hand, do not lead to a change in the size
of the polarizability tensor and thus $\overline{\alpha}' = 0$. Hence, Raman lines
due to antisymmetric vibrations are depolarized. Information of
this type can be of great use in determining the symmetry of vibrations.

VII.   EXPERIMENTAL INFRARED SPECTROSCOPY

Spectroscopy in the infrared region is, after all, governed by the
same laws as spectroscopy in any frequency region, and most of the basic principles which govern sound design and operation of instruments hold in the infrared also [19]. There are a few special factors which differentiate this region in terms of experimental procedures. Some are profound and must be held very carefully in mind
in the design of the instruments to be used. Other factors are not
basic to design but are quite pertinent for effective use. In the
first category, the fact that sources in the infrared are not intense, and are complicated by large amounts of unwanted (stray) radiation will doubtless be overcome in the next few years or decades
by the reduction to routine practice of tunable infrared lasers.
These are not yet easily available and will bring certain problems
of their own when they do become a routine "black-box" component.
Meanwhile, this weakness of source combined with the lack of sensitivity, which characterizes the usual infrared radiation detectors,
results in a low signal-to-noise (S/N) ratio and causes infrared to
be an energy-limited field of spectrometric performance. The whole
question of the proper designs to effect the best compromises in
instrument design is tremendously fascinating, but it seems inappropriate for us to discuss here. We shall assume that the reader will
be using a commercially available infrared instrument, in fairly
standard fashion, and that he intends no special modification or
adaptation. If he does, then he should refer to the sources of
more basic and thorough discussion of instrumental factors involved --
and of course, it is a truism that even if one intends no modifications to an instrument, the more one understands the factors controlling its performance and the compromises built into its design,
the better one can make it perform [8, 19-21].

We shall couch our discussion with reference to the available
commercial instruments. A curious cycle has almost completed itself
in the thirty years or so since the first commercially built IR
spectrometer came on the market. Then, in the World War II period

and immediately after, the chemical spectroscopist was expected to
know enough about the basic physics involved in order to build his
own spectrometer. The first commercial instruments were something
of a complete "kit" but still required a good knowledge of IR physics to operate. Over the years there came, as was appropriate, the
development of better and more completely reliable instruments, approaching the "black-box" ideal, where the user needs to know nothing of the insides. Nevertheless, as progress was made from the use
of prisms for dispersion of the beam to gratings, and from single-
beam instruments to the more convenient double-beam type, there remained a "building-block" philosophy which, for some years, made it
possible for a reasonably well-informed spectroscopist to put together his own assemblage of components, particularly well adapted to a
special application. This is no longer the case. The "building
blocks" are not on the market. Commercially available infrared spectrometers are, save for one example, of the reliable compact type
which utilizes gratings for dispersion of the radiation, gives satisfactory resolution on the order of 0.5 cm^{-1}, records a satisfactory
S/N, and is admirably adapted to the routine scanning of the spectrum
of a sample which can be placed in the beam in typical absorption-
spectroscopy fashion. These instruments, at modest cost, do a splen-
did job for the vast majority of chemical IR applications. For certain applications, a "high-performance" instrument is justified,
providing usually somewhat better resolution, distinctly better S/N
and stability, and enormously improved flexibility with regard to
operating and recording parameters. This need, up until recently,
has traditionally been met by a high-performance grating spectrophotometer, and one, the Perkin-Elmer Model 180, retains a proud
position on the market. However, the trend in this field of high-
performance infrared spectrophotometry has been going more and more
to the use of an interferometer, coupled integrally to a computer
which will rapidly give the Fourier transform of the interferogram,
to produce the spectrum. The Fourier-transform instrument competes
well with the high-performance dispersion spectrophotometer, not so

much because the spectrum produced is superior -- it is altogether
comparable -- but because the inclusion of the computer and data
storage, required, of course, by the need to provide the Fourier
transform, gives the added benefit of enabling the user to store
spectra, to modify them, to compare them, to perform matching, to
subtract them, and to provide compensation comparisons which can be
helpful in unraveling the puzzles of an unknown sample by comparing
its spectrum to spectra of known materials or combinations of those
spectra.

It seems, therefore, inappropriate to give such detailed exposition of the basic factors involved with obtaining an infrared
spectrum as would guide the reader in building for any special instrumental applications or adaptations. The cycle has rolled around
so that anyone wishing to make use of infrared in a specialized fashion must once again, as before World War II, build his own instrument -- he cannot even find the useful "building blocks" or components which were so useful in the 1950s and 1960s. If a really specialized application is needed, then the reader is referred to the
basic references [19-21], or for the Fourier-transform technique
(which for some purposes offers very real advantages aside from the
inclusion of the data-handling computer) to such books as Bell [22].

We shall therefore assume that the application in mind involves
the type of measurement which can be made with the conventional absorption cells for the conventional instrument, which compares the
intensity of light transmitted through a reference cell and a sample
cell and records the ratio of these intensities as a function of wave
number (Fig. 16), producing the familiar infrared spectrum known to
all chemists. There are in addition specialized instruments for special purposes, and specialized adaptors for conventional instruments
permitting the use of hot cells, or long-path gas cells, or condenser
optics for micro samples, or the ATR mode of observation. Many
of these will be found discussed, and appropriately so, in the specific application chapters of this volume. However, all of these,
with the possible exception of the ATR adaptors, are merely optical

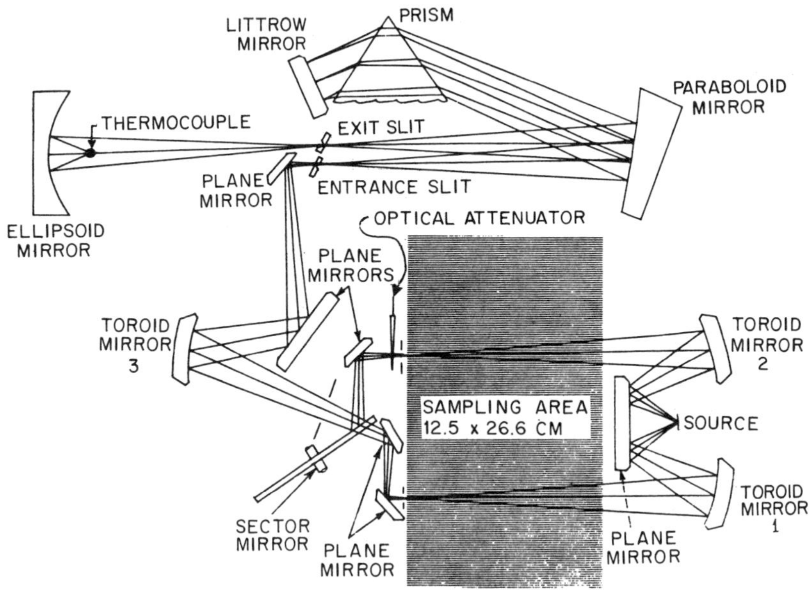

FIG. 16. Optical diagram of typical modest-cost spectrometer, the Perkin-Elmer Infracord Model 137. (From Ref. 8, p. 56. Courtesy of Academic Press and the Perkin-Elmer Company.)

folding and focussing devices to permit the sample beam (and often the reference beam as well) to be brought out of the confining "sample well" and passed in useful fashion through a sample of shape or geometry or surroundings different from the conventional absorption cell. In using any such device, the wise spectroscopist studies all the information available from the manufacturer, and then the controlling physical factors at a depth which cannot be provided in the space of this chapter. We make reference to the general sources already cited, and to the specific application references given in the several chapters which follow.

What we can quite usefully do in a brief introductory chapter is to consider the experimental situation from the viewpoint of the molecules in the sample being studied. The great usefulness of the infrared technique to the chemist lies not only in its peculiarly chemical character in being directly correlative with the bond struc-

ture of the molecules, as we have already noted, but also in the
amazing adaptability of sampling. Solids, liquids, gases; crystals,
oriented samples, slurries; powders, syrups, films -- all can be
examined by infrared with the resulting direct interpretability correlated to the chemical bonds present. If the spectroscopist can
somehow cause the beam to pass through the sample (or in some approaches reflect from it) information can be obtained. Although
nature strews many difficulties in the path of the infrared spectroscopist, one great favor is granted: the long wavelengths of IR
will tolerate optical inhomogeneities which at shorter wavelengths
would scatter so much that no observation could be made, so that
powders and mulls and dispersions can be studied quite well in the
IR spectrometer.

Against this great blessing the infrared spectroscopist has two
major handicaps in placing his sample in the beam (we have, as noted,
bypassed any discussion of the several curses which the infrared instrument designer finds afflicting this region of the spectrum): a
window material and a solvent. Sometimes of course he needs neither,
as when he can cast or press a film of some polymeric material he
wishes to study and handle, self-supporting on a frame, in the sample
well. Then his only care is to adjust properly the thickness, so
that the sample is neither so thin as not to register its absorption effectively, nor so thick as to blank out the beam completely.
A good rule of thumb for most organic materials is to have an effective path length of absorbing molecules some 10 μm thick -- thinner
for fluorocarbons, thicker for alkanes.

However, if we regard as the "ideal" method of interpreting
infrared spectra that based on the group-mode concept, then we must
take as the "ideal" sample situation the solution, with the sample
of interest dissolved in an inert but transparent solvent, at a concentration such that the solute molecules will not interact to any
great extent. If we can achieve this situation, and if we run all
our samples of knowns and unknowns under the same conditions, then
we can apply the group-mode idea with full confidence that frequencies and intensities will not be shifted or modified, either by

intermolecular effects between solute molecules or by solvent-field
effects. Then shifts in peak frequencies of even 1 or 2 cm^{-1} may be
significant and useful. Happy is the IR chemist who can consistently
examine his samples in a 10% solution (wt/vol) in CCl_4 for the region
from 4000 down to 1300 cm^{-1} and CS_2 for the longer wavelength region!
Indeed, there are a great number of materials which can be so dissolved and studied. Moreover, if many common solvents absorb heavily
in several regions of the infrared, it requires only ingenuity to ex-
plore solvents which will dissolve the material of interest and provide enough "windows" to permit observation of the particular group
modes needing observation. We recall a study on the kinetics of
thermal decomposition of polymeric nitrate esters in which adiponi-
trile was used as the solvent. The nitrile polarity gave good solvent power, the molecular weight permitted high-temperature reflux-
ing, the solvent permitted clear observation of the two nitrate
fundamentals, and the dibasic nitrile provided an olfactory accep-
tance which the otherwise satisfactory valeronitrile did not. In
some cases where a solvent does have a modest absorption in an un-
fortunate spectral location, compensation techniques will still permit its use -- however, care is required (see Ref. 20, pp. 92-106).
We stress again the point that, if solvents are changed between ex-
amining the "known" compound and the "unknown," group frequencies
and their intensities often can shift in very serious fasion.

Another solvent limitation comes from the fact that the solvent
must be at once a good solvent for the sample materials and a poor
solvent for the window materials; which leads us to the problem of
cell windows. The traditional window material is NaCl, rocksalt, or
one of the related alkali halides which transmits to longer wavelengths. They are indeed good, much used, and cheap materials, but
the list of window materials available today is sufficient to enable
almost any desired study to be carried out, certainly within the
wavelength limits of commercial instruments.

Another way in which many spectra are run is in the form of
"neat" liquid. This method is often convenient when the liquid is
not so strongly absorbing as to require too thin a cell for good

spectra. Except to mention the caution that the group modes observed will be affected by intermolecular forces between identical
molecules, and that there may be some risk in comparing the peak
values of group-mode bands with those usually quoted and derived
from careful solution studies, no objection can be made to this
method. Films, whose convenience has been mentioned, are clearly a
special case of the neat-liquid situation.

Properties of Some Infrared-Transmitting Materials[a]

Material	Practical wavelength limit[b] (μ)	Solubility in water (g/100g)
Glass	2.5	Insoluble
Vycor	3.5	Insoluble
Fused silica	4.5	Insoluble
Crystal quartz	4.5 and 50	Insoluble
Calcite	5.5	0.0014
Rutile	6.2	Insoluble
Sapphire	6.5	Insoluble
MgF_2 or Irtran 1	7.5	Insoluble
MgO or Irtran 5	8.5	Insoluble
LiF	9	0.27
CaF_2 or Irtran 3	12	0.0017
Arsenic trisulfide	13	Insoluble
ZnS or Irtran 2	14.5	Insoluble
BaF_2	15	0.17
NaF	15	4.22
Silicon	15	Insoluble
PbF_2	16	0.064
Selenium	20	Insoluble
Germanium	23	Insoluble
ZnSe or Irtran 4	24	Insoluble
NaCl	26	35.7
AgCl	28	Insoluble
KCl	30	34.7

Properties of Some Infrared-Transmitting Materials[a] (continued)

Material	Practical wavelength limit[b] (μ)	Solubility in water (g/100g)
CdTe or Irtran 6	30	Insoluble
KRS-6	35	0.32
KRS-5	40	0.05
KBr	40	53.5
KI	45	127.5
CsBr	55	124.3
CsI	80	44
Diamond	80	Insoluble

[a]Data by permission from J. Stewart, <u>Infrared Spectroscopy</u>, Marcel Dekker, Inc., New York, 1970, pp. 115-116.
[b]About the 10% T with 2-mm specimen.

When a sample is in fact not soluble in any useful solvent, then the typical resort is to a dispersing of the sample, finely ground, into an "inert" medium which will support the fine powder in the beam and, by virtue of the infrared's small intensity of scattering, permit a useful spectrum to be obtained. One variant is the mull. In this technique the crystalline sample is ground in a mortar with a minimum amount of Nujol to form a mull, which is then spread between two salt plates and placed in the spectrometer. Where the two or three bands of Nujol itself obscure important regions, recourse may be had to perhalocarbon oil (Flurolube) which obscures elsewhere but leaves the CH regions open. It takes some practice to learn the trick of preparing a good mull, but the spectra are very clean and useful. Alternately, the finely ground sample can be mixed with fine-ground KBr and pressed into a pellet, which is then placed in the IR beam. Here, of course, the sample is dispersed and suspended in a KBr "solvent." Both techniques are fully described in the standard references [20,21]. From the molecular viewpoint, we must point out that each offers some dangerous chances of distortion between the known sample and the unknown. In

the KBr pellet case, the "solvent" is not altogether inert: KBr
can exchange or react with samples, and has been known to do so. In
either case there is opportunity for the exertion of pressure --
straight pressure of pelletizing the KBr and the grinding forces in
preparing the fine-ground sample -- to cause a change in crystalline
phase. These factors can cause actual change of the nature of the
sample. In any case, one cannot without risk compare spectra run on
molecules in mulls or pellets with spectra run on molecules in solvents such as CCl_4 or CS_2. Where this form of comparison is the
only one available, the practical spectroscopist does the best he
can, but he is wise to realize the danger, to be cautious in his in-
terpretations, and to hedge his bets.

At the opposite extreme lie the gaseous samples, where one
might think that indeed intermolecular interaction would be at a
minimum and every molecule would be essentially alone. Unfortunately, from the present viewpoint, in the gas phase the rotational
structure of the bands comes into play. The liquid-phase band is a
continuous bell-shaped absorption over an appropriate bandwidth,
since the rotational levels are smeared into continuity by the liquid-phase collision environment; whereas in the gas phase the vibrational band is a picket fence made up of the individual sharp rotational lines. Even where the separate lines are not resolved, this
basic structure can affect the appearance of the band. The width of
the individual rotational lines is a function of the pressure, and
the slit average over the picket fence will vary in intensity with
the width of the individual fence pickets. Hence, there is a need,
in measuring gas-phase spectra, to observe pressure standardization
(see Ref. 20, pp. 122-130).

Perhaps the technique known as attenuated total reflection (ATR)
deserves a special comment. In this mode of observation the intensity of reflection at a given frequency is a sensitive function not
only of the absorption index of the sample being studied, but also
of its refractive index, and of the angle at which the internal reflection takes place. The method can in fact be used to make very

accurate measurements of both n and k, but not without special instrumentation and, in particular, far more precise goniometry than provided on any of the handy commercial "drop-in" attachments for conventional double-beam spectrometers. More commonly it can be used to study those types of samples which might be characterized as "one sided." The user is well advised to remember that the observation is not one of simple light absorption, and that the "band" observed will, in general, be somewhat different from the absorption band of the same sample both in shape and in peak frequency. Neither the distortion of shape nor the frequency shift prevents the ATR technique from being a most useful approach. A little time spent reading the references will help in understanding what ATR is capable of doing and how to keep from being misled [23,24].

VIII. EXPERIMENTAL RAMAN SPECTROSCOPY

A major problem in the collection of good Raman spectra is the extreme weakness of the effect. Illuminating a liquid sample with visible light, the total intensity of molecular scattering may be on the order of 10^{-5} of the incident intensity. This includes Rayleigh scattering which may comprise a vast majority of the scattered light. Recent advances in Raman instrumentation have greatly enhanced the experimentalists ability to record good quality Raman spectra. Monochromators and detection systems have been improved greatly and have eased some of the experimental problems, but it was primarily the advent of the laser that has brought Raman spectroscopy to its present prominent position. The laser supplied the Raman spectroscopist with a nearly ideal source of radiation. It is very intense, polarized, highly directional and almost perfectly monochromatic. Coupled with the presently available double monochromators and photoelectric detection systems, the laser has made high quality Raman spectra obtainable with comparative ease.

A typical Raman spectrometer consists of a source, a sample compartment, a double monochromator and a photoelectric detection system. The following are some important considerations concerning

a Raman spectrometer. Further information and more thorough discussions can be found in several recent books [15,17,25,30].

A. Source

Since mercury and other arc-lamp sources previously used have
been reduced to virtual extinction by laser sources, only laser excitation will be considered. A number of continuous-output lasers
are currently available as Raman sources. A list of the most common types along with their principal lines and typical power levels
is shown in the accompanying table.

Laser	Wavelength, Å	Output Power, watts*
Cadmium	4416	0.2 - 0.4
Krypton	5682	0.5 - 1.0
	6471	0.5 - 1.0
Argon	4880	0.5 - 2.0
	5145	0.5 - 2.0
He-Ne	6328	0.080

All of the lasers shown in the table are currently being used for
Raman spectroscopy. In choosing a laser for a given experiment
both wavelength and power requirements must be considered. He-Ne
and Argon ion lasers are the most commonly used. Each has advantages
and disadvantages. The He-Ne laser is very reliable. It has a long
operating lifetime, and its 6328 Å line is useful for colored samples.
However, the lower power of the He-Ne laser can at times limit the
experimentalist, and its frequency range is not conveniently suited
to most photocathode materials. The Argon system has advantages of
greater output power, multiline, and an inherently greater Raman

---

*Output power levels shown are typical but may vary for differ-
ent lasers.

intensity due to its wavelength. These advantages are somewhat offset by its cost and reliability, but the quality of Argon lasers is
rapidly improving.

B. Sample Compartment

The laser beam is generally directed into the sample compartment by use of dielectrically coated mirrors of very high reflectance. A simple lens is used to focus the beam on the sample. The
sample compartment itself depends on the type of sample being used
and may have provisions for low temperature, variable angle, or other modifications needed for the experiment. Liquid samples are generally contained in capillary tubes which may be mounted on various
types of adjustable stands to facilitate alignment. Very small sample volumes, on the order of 10^{-6} liter can be routinely used to obtain Raman spectra of modest resolution. For poor Raman scatterers or
dilute liquids, a multipass cell, one in which the beam traverses
the sample several times, may be used. Fig. 17 shows the schematic
of a typical multipass sample cell.

Fine crystal or powder samples may also be contained in capillary tubes for Raman experiments. Again, very small sample sizes
can be routinely used. Larger, intact samples may be examined by
placing the solid directly in the focus of the laser beam. Figure
18 shows the Raman spectrum of the clear, yellow, polymeric handle
of a screwdrive. In general, solids scatter more stray light which
may interfere with the Raman spectrum. When using the 5145 Å line
of an argon laser this problem may be alleviated by use of an iodine
filter.

Obtaining good Raman spectra of gases is, in general, more difficult than solids or liquids, because most vapors have an insuffi-
cient molecular density to yield good Raman spectra. Nevertheless,
by use of appropriate sample containers, good spectra of many gaseous samples can be obtained. In this case multipass cells are of
great importance in signal improvement.

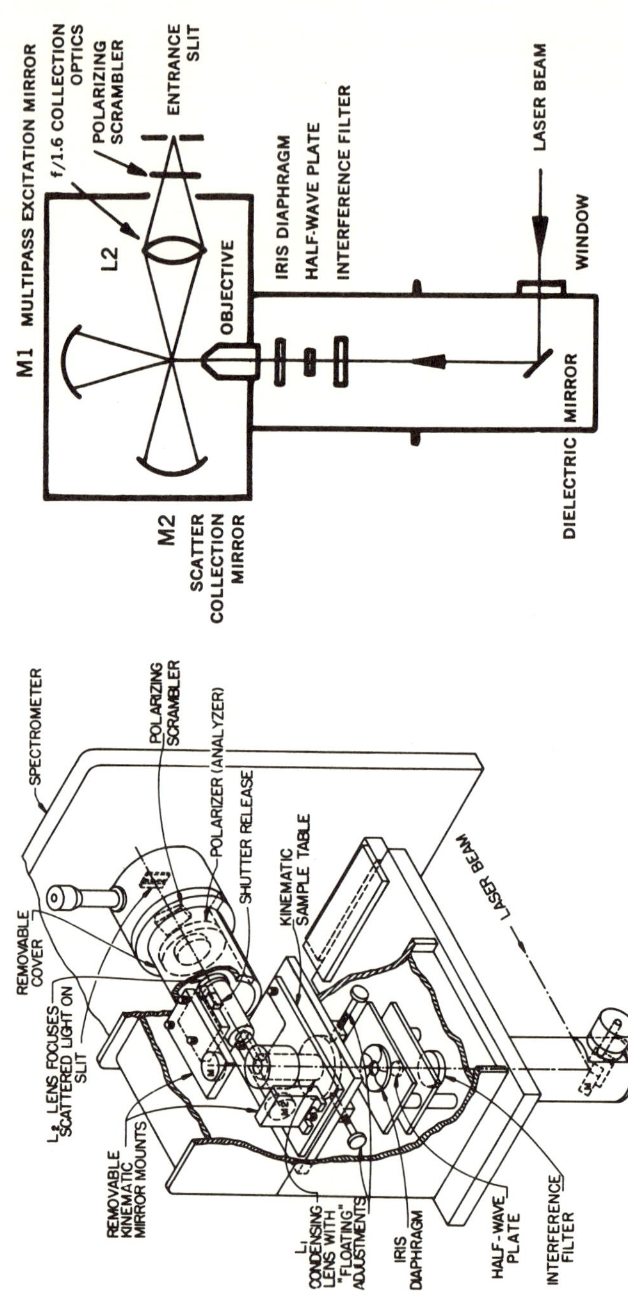

FIG. 17. A multipass sample illuminator. (From Ref. 25, p. 198. Courtesy of Marcel Dekker, Inc.)

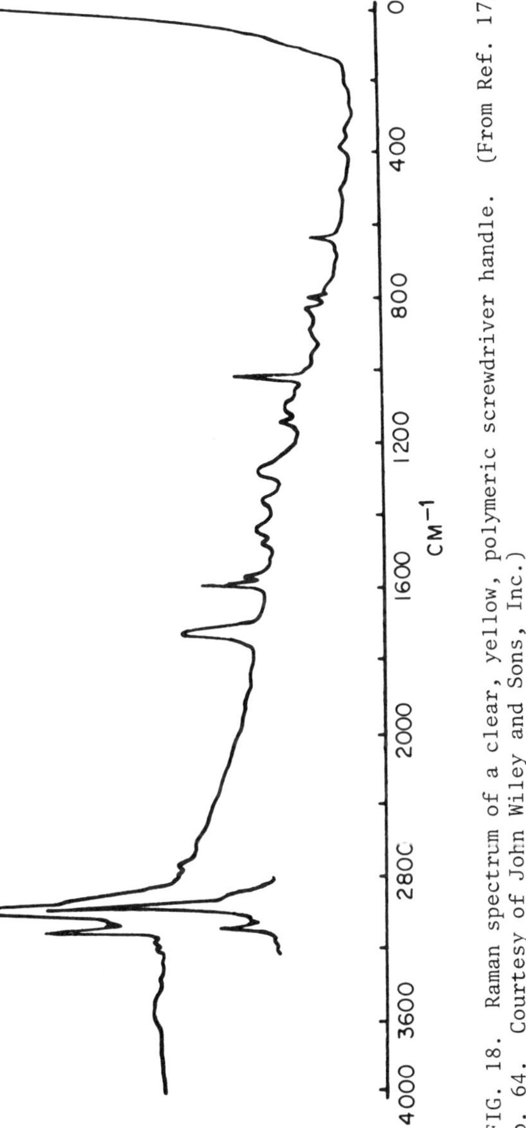

FIG. 18. Raman spectrum of a clear, yellow, polymeric screwdriver handle. (From Ref. 17, p. 64. Courtesy of John Wiley and Sons, Inc.)

C. Monochromators

The problem of stray light may seriously hamper the collection of Raman spectra making the rejection of stray light essential. Virtually all commercial Raman instruments incorporate double or triple monochromators in their systems, with the double being the most popular. Most instruments are capable of approximately 0.5 cm^{-1} resolution, although 2 to 4 cm^{-1} resolution is generally adequate for good survey spectra. A complete review of commercially available Raman spectrometers can be found elsewhere [15].

D. Detectors

The development of high-quality photomultiplier tubes for use as detectors has greatly aided the Raman spectroscopist. Two factors are of importance in the choice of a photomultiplier tube for the detection of weak signals. These factors are low dark current and high quantum efficiency. The dark current is the signal due to thermally excited electrons which leave the cathode of a phototube in the absence of light. Cooling the tube greatly reduces thermionic dark current. This practice is relatively common. Quantum efficiency is a measure of the ratio of the signal at the anode to the number of photons at the cathode. Many photocathode surfaces have a quantum efficiency of 10 to 20% in the blue-green portion of the spectrum while in the red portion quantum efficiency is on the order of 1%. This is a factor in the choice of a laser for a given system.

Signal processing of the output of a photomultiplier tube can be handled in any one of several ways. Direct-current amplification is the simplest method of processing output but is hindered by lack of sensitivity to very weak signals and drift present in direct-current amplifiers. Pulse-counting techniques, where one actually amplifies each pulse from the phototube individually, is another popular processing method. The advantages of this method include good sensitivity to weak signals and the presentation of data in digital form. The drawback of pulse counting is its failure to handle very large signals. However, with care this problem can be

avoided. Pulse counting is now the most popular form of signal
processing.

As previously mentioned, polarization effects and the measurement of the depolarization ratio are important in the analysis of
Raman spectra. The most common type of geometry used for the measurement of the depolarization ratio uses a polarizer between the
sample and the spectrometer slit to pass the desired component, parallel or perpendicular, of the scattered radiation to be used in the
calculation of the depolarization ratio. The 218 cm^{-1} band of CCl_4
is sometimes used as a check on the depolarization method and its
measured depolarization ratio should be 0.75 using a polarized laser
and a 90° scattering geometry. A number of problems must be considered in the measurement. Grating instruments do not pass different polarizations in the same way. Thus, these differences must be
corrected for. To do this, most Raman spectrometers have polarization scramblers at the entrance slit. Also, the depolarization ratio
is defined in terms of integrated intensities, although for usual
purposes the ratio of peak heights is adequate.

Polarization is very useful for Raman scattering from oriented
single crystals in that it allows one to measure the scattering due
to a given component of the scattering tensor. Porto et al. [26]
have initiated a notation describing polarization data obtained from
single crystals. Consider a case where the incident beam is propagated in the x direction with y polarization and the scattered beam
is propagated in the y direction with z polarization. This situation
is described by the notation x(yz)y, where the letters outside the
parentheses indicate propagation directions of the incident and scattered beams and the letters inside the parantheses indicate the polarization of the incident and scattered beams, respectively. The
coordinates inside the parentheses indicate the component of the
scattering tensor involved. By proper orientation of the crystal
and various combinations of polarization and propagation directions,
all components of the scattering tensor may be viewed separately.
This method is found to be of great assistance in the interpretation
of Raman spectra.

A problem often encountered in Raman spectroscopy is that of
fluorescence, either of the sample itself or of trace impurities.
The fluorescence may completely mask the Raman signal. There are,
however, a number of methods designed to help alleviate this problem. Since fluorescence spectra may differ for different excitation
frequencies, the use of a different laser line may sometimes help.
If it is trace impurities that are causing the problem, proper puri-
fication should help. In some cases fluorescence may be suppressed
by the addition of a quenching agent.

IX.   QUANTITATIVE ANALYSIS

In addition to being of use for molecular identification and structure determinations, spectroscopy is a useful tool for quantitative
analysis. This is particularly true for infrared spectroscopy due
to its almost universal applicability. Raman spectroscopy, on the
other hand, suffers from various experimental problems including reproducibility and calibration standards. Although some quantitative
analysis using Raman spectroscopy has been undertaken, in general,
this area is fraught with difficulties. Since chemists have only
recently had available Raman spectrometers capable of measuring subtle effects of polarization and intensity, most current work has
been concentrated on the areas of molecular structure determinations
and the study of crystal physics. Thus, the entire field of quantitative analysis in Raman spectroscopy has remained virtually un-
touched, although with the development of proper techniques this
area remains a bright possibility for future investigations. This
being the case, we will restrict our discussion of quantitative analysis by spectroscopy to the use of infrared, briefly outlining some
of the principles and problems therein.

In essence, quantitative analysis is accomplished by the comparison of the absorption strength of an infrared band of a compound in
the mixture to be analyzed with the corresponding measurement under
conditions of known concentration. In order to do this, the substance to be measured must exhibit one or more unique absorptions of
appropriate strength that are not overlapped by bands of other com-

pounds present. This procedure is found to be straightforward and reasonably rapid under restrictions of modest accuracy.

Consider a sample subjected to infrared radiation of radiant power I_o, where I_o is the number of quanta per second. If, after passing through the sample, the radiant power is I, the transmittance T of the sample is given by the equation

$$T = \frac{I}{I_o} \tag{26}$$

If the sample is in a cell of thickness b cm and is of concentration c g/liter, then the fundamental equation governing the absorption of radiation is found to be

$$\frac{I}{I_o} = 10^{-abc} \tag{27}$$

or

$$\log \frac{1}{T} = abc \tag{28}$$

where a is a constant characteristic of the sample and is called the absorptivity or absorption coefficient. Care must be taken in the comparison of coefficients since some ambiguity may arise due to the units used or the exact form of the basic equation. Defining A, the absorbance, as

$$A = \log \frac{1}{T} \tag{29}$$

one obtains

$$A = abc \tag{30}$$

which is one form of Beer's law of radiant absorption. It is convenient to use absorbance in quantitative analytical calculations in order to avoid mathematically unwieldy manipulations inherent in the logarithmic form of Beer's law.

Most quantitative analysis is done with measurements taken at absorption maxima, i.e., transmission minima. This is a reasonable point on the band profile to employ since it can be located easily

and accurately. This method is known as the method of "peak height,"
as opposed to the method of integrated absorption which is more difficult and will not be discussed here.

Two ways of making simple quantitative determinations are commonly used. They are the "cell-in--cell-out" method, performed at
fixed frequency, and the "base-line" method, performed with a scanned
spectrum. Each method including their respective advantages and disadvantages will be outlined briefly. More complete treatments of
these methods and other elements of quantitative analysis can be
found elsewhere [8,20].

A. "Cell-in--cell-out" Method

This method is performed with the spectrometer set at a fixed
frequency and fixed slit widths. The frequency is predetermined according to a well-defined absorption peak with slits set to give an
appropriate signal. The unknown to be determined is dissolved in a
suitable solvent and placed in a cell of known path length. Absorbance measurements involved include the absorbance of the unknown,
the absorbance of the same cell with pure solvent, and a determination of 0% transmission. The latter measurement can be taken by
placing an opaque object in the spectrometer sample compartment but
this may introduce errors due to scattered light and in practice one
must be aware of this problem and take appropriate steps to solve it.
From these measurements and ensuing manipulation of Eq. (30) for
each, one can quite easily determine the percentage of the measured
component in the unknown mixture.

This method has its advantages in that it gives very high precision and reproducibility because of fixed slit and wavelength.
It is rapid, since one need not scan the entire spectrum, as well as
simple and straight-forward. The primary disadvantage of the "cell-
in--cell-out" method is that, although it may be capable of high precision, it may be subject to serious systematic errors due to inaccurate measurement of the I_o value. This may result from scattering
effects, broad electronic absorption, or refractive index reflection
loss. Figure 19 represents a typical spectrum to which this method

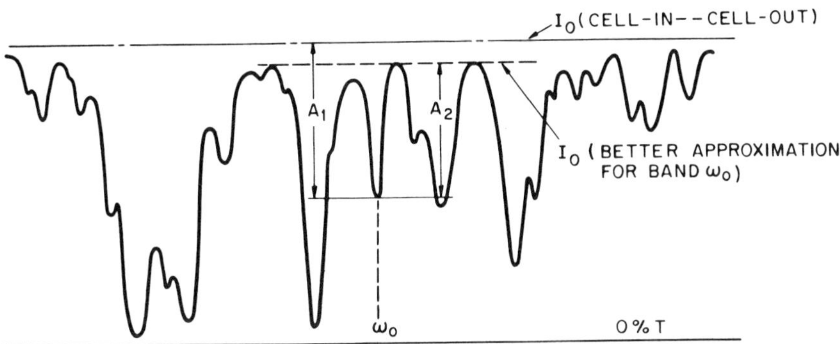

FIG. 19.   Error in I_0 measurement of "cell-in--cell-out" method. (From Ref. 20, p. 165.   Courtesy of John Wiley and Sons, Inc.)

could be applied. A_1 represents the absorbance measured in the cell-in--cell-out method which is not as good as A_2 which would require a scanned spectrum for the measurement.

B.   "Base-line" Method

This method results from an attempt to measure more accurately the I_0 value of an absorption band, i.e., determining the absorbance at a desired point if the absorption band at that point was absent. The dashed line in Fig. 19 is most probably a good place at which to measure I_0 and is called the base line.   There are various ways to construct the base line for a given band; some of these are shown in Fig. 20.   Each method has advantages for certain types of spectra and the decision of which procedure to use is determined by the factors affecting each specific experiment.   The actual analytical calculations for the base-line method are the same as those for the cell-in--cell-out method.

Although the base-line method gives somewhat poorer random error than the cell-in--cell-out method, it does give an excellent correction for systematic error.   It is especially useful for the analysis of complex mixtures, especially "dirty" ones, where the correction of base line is essential.

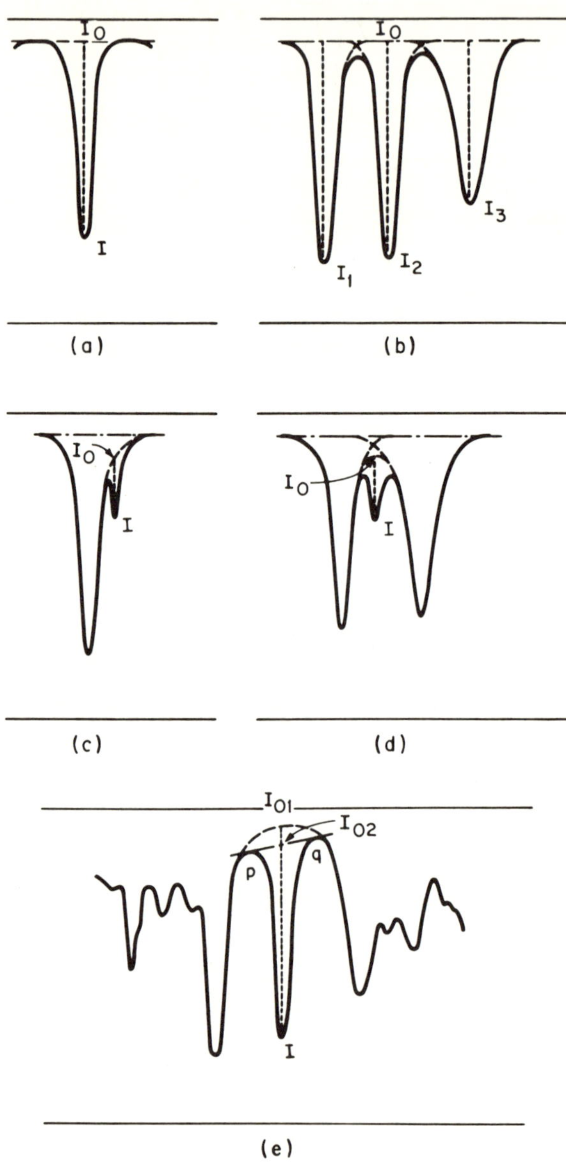

FIG. 20. Various methods of base-line construction. (From Ref. 20, pp. 166-167. Courtesy of John Wiley and Sons, Inc.)

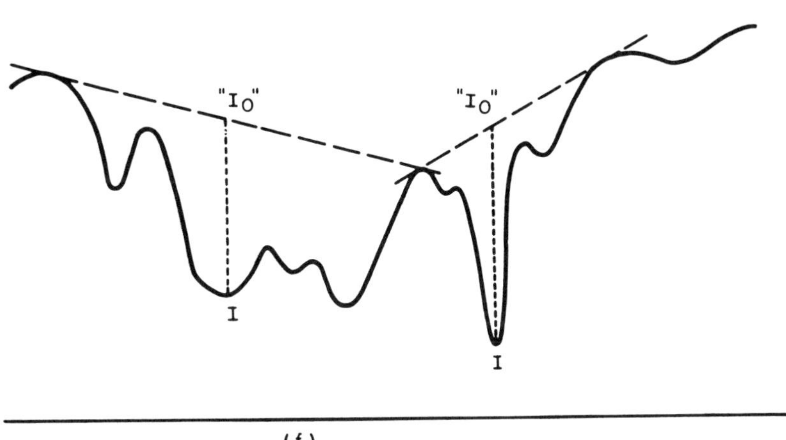

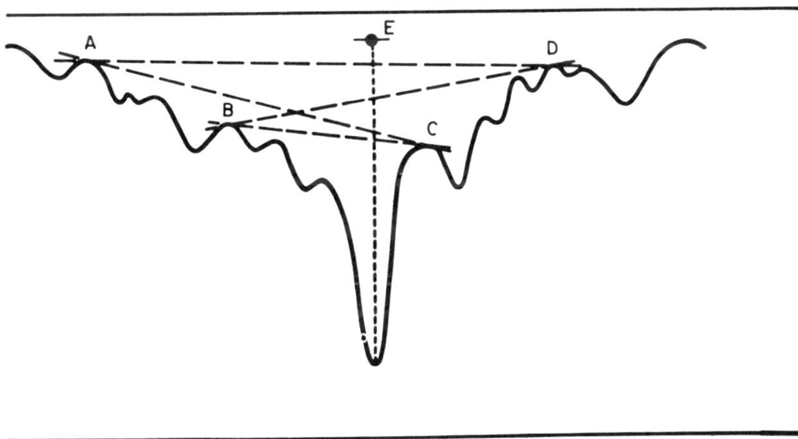

The previous discussions have assumed strict adherence to Beer's law, i.e., that absorbance is a linear function of the product of concentration and pathlength. In reality, some deviations do occur and result principally from two causes, these being chemical effects and spectrometer effects. Deviations arising from chemical effects are due to variations of absorption bands by association effects between solute molecules. These effects may change the frequency, shape, or intensity of an absorption band and may vary as the concentration is changed. A common example is that of intermolecular hydrogen bonding in alcohols and phenols. Some difficulties due to chemical effects can be eliminated by using lower concentrations in longer cells, appropriate solvents, or other methods. In some cases, chemical effects cannot be eliminated or corrected for. They destroy the possibility of quantitative analysis.

Spectrometer effects causing Beer's law deviations are of more concern since they generally occur, to some degree, in most absorption bands. These effects arise from the fact that Beer's law is not rigorously obeyed unless the distribution of radiation passed by the spectrometer exit slit is narrower than the absorption band being measured. This becomes more apparent when one realizes that any spectrometer has a finite spectral slit width which passes a range of frequencies which are seen by the detector when it is supposedly viewing a monochromatic signal. Without going into the geometric and diffraction considerations needed to explain the spectral slit width properly, let us just say that this effect causes problems which in some cases can be reduced, but that cannot be eliminated. Thus, a correction must be made in order to obtain good quantitative analytical results.

When confronted by spectrometer effects the experimentalist must reach a compromise between a small slit width, which reduces deviations from Beer's law and signal, and a large slit, which increases both signal and deviations. A good rule of thumb is that, generally, a larger slit width will cause fewer problems because the distribution of frequencies is proportional to the slit width, whereas

the signal-to-noise ratio, which is essential to good results, is
proportional to the square of the slit width.

X. THE PRACTICAL SPECTROSCOPIST
IN THE REAL WORLD

Finally in this introductory chapter let us point out that the applications of infrared and Raman spectra are constantly developing and
changing, and the need to keep abreast, important in any field of
scientific endeavor, is particularly acute. Neither this volume nor
any other can hope to bring the final or definitive word to the reader, though we hope both the general comments and the specific applications will prove illuminating and stimulating. We would of course
urge the reader to go beyond what can be included within these covers,
to the books we refer to here and in the chapters that follow, wherein greater depth of discussion will be found. Beyond that, we call
the reader's attention to certain journals, such as *Spectrochemica
Acta, Applied Spectroscopy,* and *Applied Spectroscopy Reviews,* which
focus on the field of "spectroscopy in the real world" and are in-
valuable to the practicing spectroscopist.

Even beyond the published journals, we urge the spectroscopist
to affiliate with his colleagues. If Raman spectroscopy is useful,
he should get in touch with the Raman Newsletter [27]; if Fourier-
transform spectroscopy is useful, then he should join the Fourier
Transform User's Group [28]. In any case, he will find it helpful
to affiliate with the Coblentz Socicty [29], if only to profit from
the maintenance of standards by that group for Class III and Class II
spectra, and the benefit all spectroscopists derive from that program.

Finally, the spectroscopist is well advised to participate in
one or more of the meetings each year, where the new techniques and
new instruments, and new wrinkles, are first exposed. Of particular
benefit are the Pittsburgh Symposium, now usually held in Cleveland,
and the Federation of Analytical Chemistry and Spectroscopy Societies
(FACSS), which is held each fall. Attending these gatherings, the
spectroscopist will not only pick up new ideas, and perhaps gather

friendly suggestions on solutions to his problems, but he will sense the activity, liveliness, and progress which, a century and three-quarters after Herschel's discovery, still pervade this remarkably applicable technique of investigating the nature of material samples.

REFERENCES

1. W. Herschel, *Philos. Trans. R. Soc., 90,* 284 (1800).

2. William W. Coblentz, "Investigations of Infrared Spectra," Carnegie Institution of Washington, Washington, 1905; reprinted by the Coblentz Society, 1962.

3. W. N. Lipscomb, *Atomic and Molecular Structure,* in "Comprehensive Inorganic Chemistry," (M. C. Sneed, J. L. Maynard, and R. C. Brasted, eds.), vol. 1, Van Nostrand, New York, 1953.

4. G. Herzberg, "Molecular Spectra and Molecular Structure: Diatomic Molecules" (2nd ed.), Van Nostrand, New York, 1950.

5. E. B. Wilson, Jr., J. C. Decius, and P. C. Cross, "Molecular Vibrations," McGraw-Hill, New York, 1955.

6. J. C. D. Brand and J. C. Speakman, "Molecular Structure," Chap. 6 and App. III, Arnold, London, 1960.

7. B. Crawford, Jr., and W. T. King, *J. Mol. Spectrosc., 5,* 421 (1960); *J. Mol. Spectrosc., 8,* 58 (1962).

8. N. B. Colthup, L. H. Daly, and S. E. Wiberley, "Introduction to Infrared and Raman Spectroscopy," Academic, New York, 1964.

9. L. J. Bellamy, "Infrared Spectra of Complex Molecules" (2nd ed.), Wiley, New York, 1958.

10. L. J. Bellamy, "Advances in Infrared Group Frequencies," Methner, London, 1968.

11. G. Herzberg, "Infrared and Raman Spectra," Van Nostrand, New York, 1945.

12. J. Overend, in "Infrared Spectroscopy and Molecular Structure" (M. Davies, ed.), Chap. 10, Elsevier, New York, 1963.

13. L. A. Gribov, "Intensity Theory for Infrared Spectra of Polyatomic Molecules," Consultants Bureau, New York, 1964.

14. G. W. Chantry, in "The Raman Effect," vol. 1 (A. Anderson, ed.), Marcel Dekker, Inc., New York, 1971.

15. M. C. Tobin, "Laser Raman Spectroscopy," Wiley, New York, 1971.

16. L. A. Woodward, in "Raman Spectroscopy," vol. 1 (H. A. Szymanski, ed.), Plenum, New York, 1967.

17. S. K. Freeman, "Applications of Laser Raman Spectroscopy," Wiley, New York, 1974.

18. R. S. Tobias, *J. Chem. Educ.*, *44*, 7 (1967).

19. R. A. Sawyer, "Experimental Spectroscopy" (3rd ed.), Dover,
     New York, 1963.

20. W. J. Potts, Jr., "Chemical Infrared Spectroscopy," vol. 1,
     Techniques, Wiley, New York, 1963.

21. J. E. Stewart, "Infrared Spectroscopy," Marcel Dekker, Inc.,
     New York, 1970.

22. R. J. Bell, "Introductory Fourier Transform Spectroscopy,"
     Academic, New York, 1972.

23. N. J. Harrick, "Internal Reflection Spectroscopy," Wiley (Inter-
     science), New York, 1967.

24. P. A. Wilks, Jr. and Tomas Hirschfeld, *Appl. Spectrosc. Rev.*, *1*,
     99 (1967).

25. A. Anderson (ed.), "The Raman Effect," vol. 1, Marcel Dekker,
     Inc., New York, 1971.

26. T. C. Damen, S. P. S. Porto, and B. Tell, *Phys. Rev.*, *142*, 570
     (1960); *Phys. Rev.*, *144*, 771 (1966).

27. Write to Miss P. R. Wakeling, Raman Newsletter, 1613 Nineteenth
     St. N.W., Washington, D.C. 20009.

28. Write to Dr. R. Kagel, Dow Chemical Company, Midland, Mich.
     48640.

29. Secretary, Dr. R. W. Hannah, c/o Perkin-Elmer Corporation,
     761 Main Ave., Norwalk, Conn. 06851.

30. C. E. Hathaway in "The Raman Effect," vol. 1 (A. Anderson, ed.),
     Marcel Dekker, Inc., New York, 1971, p. 198.

Chapter 2

INORGANIC MATERIALS

Robert L. Carter

Department of Chemistry
University of Massachusetts
Boston, Massachusetts

I. INTRODUCTION

If there ever was a distinct boundary between what is and what is
not inorganic chemistry, it has never been less obvious than it is
today. Of course, a similar statement could be made about any of
the traditional disciplines of chemistry, but this fact is nowhere
more evident than in the area of inorganic chemistry. The activities
of those individuals calling themselves inorganic chemists testify
to this, since their work so frequently takes them into what former-
ly could be perceived as the domains of organic, physical, and analytical chemistry. Indeed, this breadth of activity more than any-
thing else seems to be characteristic. This is quite natural. If
it can be said with some exaggeration that the organic chemist is
preoccupied with element number 6, then it can be said with equal
alacrity that the inorganic chemist is preoccupied with all the elements of the periodic table. Such a diversity of subjects leads to
a wide range of pursuits.

If the range of activities is vast in inorganic chemistry, so
too are the applications of Raman and infrared spectroscopy within
this field. These applications include qualitative analysis, quantitative analysis, determination of bond strengths, measurements of
thermodynamic parameters, and deduction of structure, i.e., everything for which vibrational spectroscopy is useful. Nonetheless,
the one area in which infrared and Raman spectroscopy seem to find
most frequent use in inorganic studies is the determination of molecular structure. It is the main purpose of this chapter to discuss
this important application in some detail.

It is well known that a molecule of n atoms has $3n - 6$ normal
modes of vibration ($3n - 5$ if linear). This useful rule, however,
says nothing about our ability to directly observe frequencies for
these normal modes by infrared or Raman spectroscopy. That ability
depends upon the spectroscopic selection rules that are operative.
The two primary selection rules of infrared and Raman spectroscopy
are familiar. A normal mode will be active in the infrared spectrum
(i.e., give rise to a transition which may be detected) if it causes

a change in the dipole moment of the molecule. A normal mode will
be active in the Raman spectrum if it causes a change in the polarizability of the molecule. In many cases it is a simple task to
describe the oscillating motions which define the several modes of
a molecule, but it is not always obvious whether or not a particular
mode will be infrared or Raman active. This determination is greatly facilitated by considering the equilibrium symmetry of the molecule of interest and applying the methods of group theory.

The results from group theory are exact in describing the symmetries of the normal modes and predicting their spectral activities
when the molecules are in a dilute gaseous state. The method can be
used with less ideal systems, such as liquids and solutions, but
some disagreement between prediction and observation may occur. Indeed, departures from the free-molecule predictions can be indica-
tive of environment-caused distortions or molecular associations.
The procedure used with free molecules usually shows poorest agreement with solid systems, but the methodology may be extended to determine the selection rules in crystals. In the solid state, the
actual symmetry of the molecule in the crystal (site symmetry) and
the symmetry of the crystal lattice itself (space group symmetry)
must be taken into consideration.

This ability to predict spectroscopic activities for a molecule
is the basis for using vibrational spectroscopy as a structural tool
in inorganic chemistry. In principle, the spectroscopist can determine the activities of normal modes for several probable structures
for a molecule, observe the infrared and Raman spectra, and deduce
the correct structure from the agreement between spectroscopic data
and the predictions for a particular model. This approach assumes
that each observed frequency can be associated with one normal mode,
or at least that all observed frequencies may be assigned to particular normal modes, unambiguously. A practical limit is reached with
large molecules, where several normal modes may have the same frequency (accidental degeneracy) or have frequencies so close together
that they cannot be resolved. In spite of this and other limitations
which can diminish the usefulness of this approach, the structures

of a surprisingly large number of molecules have been determined in
this manner. The reason for this lies in the tendency of inorganic
molecules to be relatively small by comparison to organic ones,
which owe their size and diversity to the tendency of carbon to
catenate. Even larger inorganic species, such as many transition
metal complexes or polymeric species, often can be treated as if
they were simpler systems. This generally permits observation of
distinct frequencies for the various normal modes, rather than the
characteristic group frequencies observed for more complicated molecules, and makes possible the correlation with the structure-based
selection rules derived by group theory.

Since the use of vibrational spectroscopy for structure studies
depends upon an understanding of symmetry and basic group theory,
an introduction to these topics will be presented. The level of
treatment is only sufficient for an understanding of the material
that follows, and this discussion makes no pretense of mathematical
sophistication. Those requiring a more rigorous treatment are referred to the many standard texts treating group theory [1-4].

The development of selection rules in this chapter begins with
the consideration of free molecules. The approach taken is largely
operational, rather than theoretical. The theoretical aspects of
vibrational spectroscopy of polyatomic molecules can be found in the
elegant expositions of Wilson et al. [3] and Herzberg [4]. The
emphasis has been placed on correlating the results of group theory
with the actual situations which are encountered in applying vibrational spectroscopy to structural problems. This includes a development of selection rules for solids and their potential for structural
analysis. A few of the problems and techniques which are uniquely
relevant to inorganic systems are presented in the final section.

II. SYMMETRY AND GROUP THEORY

A. Elements and Operations

Symmetry in an object is widely recognized as a harmonious bal-
ance of equal parts. Various types of relationships between the
equal parts can be recognized, and these relationships can be ex-

pressed in terms of symmetry elements and symmetry operations. A
symmetry element is a geometrical entity, such as a point, line, or
plane, with respect to which a symmetry operation is performed. A
symmetry operation is a movement about a symmetry element such that
the orientation and position of an object before and after the operation are indistinguishable. The terms element and operation are not
interchangeable. The element only serves as a reference object for
the movement of the operation. It is important to realize that the
operation does not necessarily carry each point of the object back
to its starting position, but rather to an equivalent position which
is indistinguishable. If we do not watch the operation from start
to finish, its effect on the object can not be perceived.

For the moment we will confine the discussion to the symmetry
of isolated molecules. The elements which describe the symmetry of
a molecule must intersect at least at a common point. Hence, the
symmetry of molecules is called point group symmetry. A requirement
of point group symmetry is that the elements must pass through the
molecule. This contrasts with the symmetry of crystals, where elements may lie parallel to one another and are not constrained to
pass through any chemically identifiable group. Four elements with
corresponding operations are possible within point symmetry. A fifth
element, identity, corresponds to the seemingly trivial operation of
doing nothing, and must be defined to satisfy the mathematical constraints of group theory. These elements and operations will now be
described.

1. *Rotational axis.*

A rotational axis, sometimes called a proper axis, is a line
running through a molecule about which the symmetry operation of
rotation occurs. The operation is given the notation C_n, where n
is the order of the axis. The order is the value of n in $2\pi/n$, where
$2\pi/n$ is the rotation in radians about the axis necessary to bring
the object to an equivalent configuration. Thus, if rotation of a
molecule through $2\pi/2$ ($180°$) results in a configuration which is indistinguishable from the starting position, this defines a C_2 or two-

fold axis. The value of n can be any integer up to infinity. Indeed, C_∞ is characteristic of all linear molecules, since any rotation by an infinitesimal amount or a series of infinitesimal amounts
about the linear axis results in an equivalent configuration.

A molecule may possess several rotational axes of various orders, and some of these axes may coincide. This can be illustrated
by considering the rotational axes of a square plane (Fig. 1), which
may be taken to represent any square planar species, such as $PtCl_4^{2-}$.
The highest order rotational axis, called the principal axis, lies
perpendicular to the plane in Fig. 1. This is a C_4 axis, since rotation by $2\pi/4$ (90°) results in an equivalent configuration. When
it becomes necessary to specify a coordinate system, the principal
axis is generally taken to be the z axis. If the operation C_4 is
carried out twice in succession about this axis, the effect on the
square is the same as the operation C_2. Thus we may write $C_4^2 = C_2$,
and the four fold axis is seen to be coincident with a twofold axis.
If three C_4 operations are performed in succession, the square is
carried into an equivalent configuration, and the operation is designated C_4^3. If we assume that the C_4^3 operation is performed in a
clockwise direction, then it is seen that the resulting configuration
is the same as a C_4 operation performed in a counterclockwise direction. An operation C_4^4 would carry the square to its starting position. This is the same as doing nothing, and the result is described
as identity, rather than C_4^4.

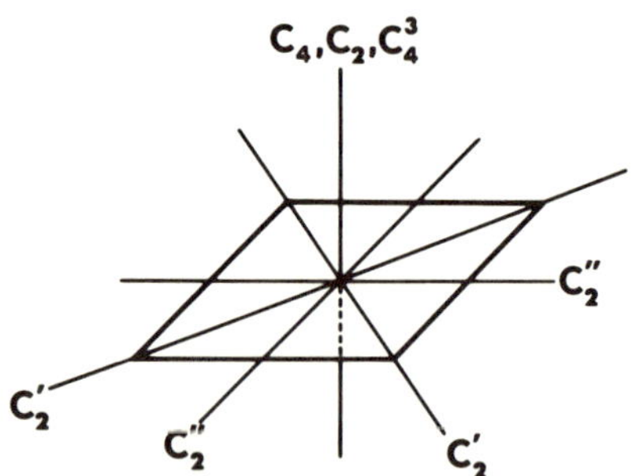

FIG. 1. Rotational axes of a square plane.

Four twofold axes lie perpendicular to the fourfold axis in Fig. 1. If we consider a C_2' axis, the presence of a fourfold axis perpendicular to it requires a second C_2' axis at 90° from the first. These two C_2' axes constitute a class. From a geometrical standpoint, operations belong to the same class when they are converted into one another by changing the reference coordinate system through some symmetry operation of the object. The two other twofold axes, designated C_2'', constitute a second class of C_2 operations. As with the two C_2' axes, the two C_2'' elements are related to one another by C_4. However, no symmetry operation of the square plane converts a C_2' into a C_2'', and these must belong to separate classes. The C_2' axes are customarily taken as the x and y axes of the coordinate system.

In general, if a molecule has a C_n axis, the operations C_n, C_n^2, ..., C_n^n will be present about that axis. The operation $C_1 = C_n^n$ is the same as identity, and usually not counted as a rotation. Any C_{2n}^2 operation is a C_n operation; for example, $C_4^2 = C_2$, $C_6^2 = C_3$, etc. Likewise, any $C_n^{n/2}$ operation is C_2; for example, $C_6^3 = C_2$, $C_8^4 = C_2$, etc.

2. Mirror plane.

The mirror plane σ is the element associated with the operation of reflection. If a molecule has a mirror plane, it will be divided into two halves about that plane, and the two halves will be mirror images of each other. Thus, if there is a point a distance r along a normal to the plane, there will be an equivalent point at the same distance along that normal on the other side of the plane. As with rotations, a molecule may possess several mirror planes. Figure 2 shows the mirror planes of a square plane. The plane of the square is itself a mirror plane, and indeed, all planar molecules must possess at least this type of plane. This plane, which lies perpendicular to the principal axis of rotation (C_4) is designated σ_h, for horizontal. Four vertical planes are shown in Fig. 2, also. The planes marked σ_v (vertical planes) belong to the same class and are customarily defined as the xz and yz planes. The two σ_d planes (dihedral planes) constitute a second class of vertical planes. They bisect the right angles between the two σ_v planes. When both σ_v and

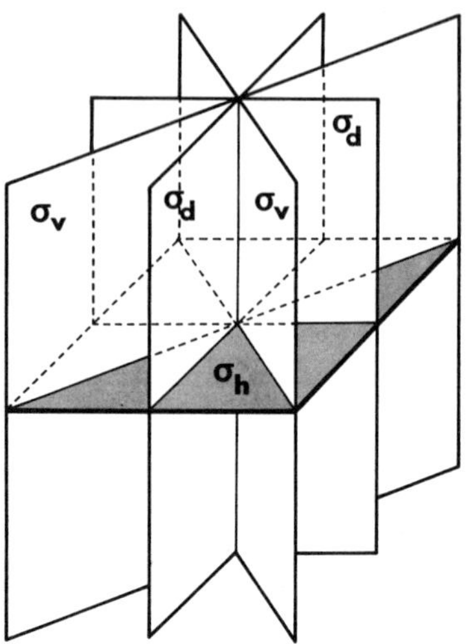

FIG. 2. The mirror planes of a square plane.

σ_d planes exist for the same molecule, the σ_v's are generally defined so as to have bonds lying in them, while σ_d's are defined to
bisect bond angles. However, this notation is not always unambiguous.

3. Improper axis.

The improper axis S_n is associated with an operation which is
actually a composite of operations. It is sometimes called a rotation-
reflection axis, which suggests its nature. An S_n operation is performed by rotating through $2\pi/n$ and then reflecting in a plane perpendicular to the rotational axis; that is, C_n then σ_h. It is apparent that any molecules that possess a C_n axis and a σ_h plane will
also have an S_n axis. However, it is not necessary that either C_n
or σ_h be symmetry operations of the molecule for an S_n axis to exist.

One of the more frequently encountered S_n axes is the S_4 axis
of a tetrahedral molecule, such as methane. One of these (there are
three in methane) is shown in Fig. 3. One hydrogen has been labeled
(*) so that the effect of the operation on it can be correctly visualized. Note that neither the operations C_4 nor σ_h exist for methane, but their combination does result in an indistinguishable configuration (ignoring the labeled hydrogen). Successive performance

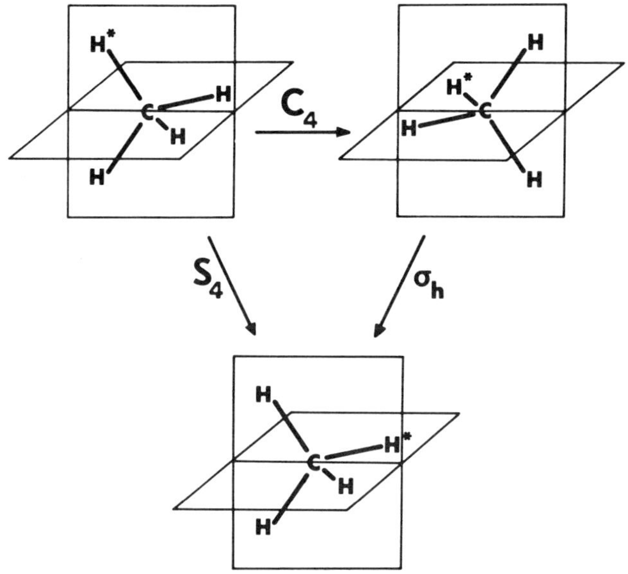

FIG. 3. The S_4 operation of CH_4. One hydrogen is labelled (*) to show the effect of the operation on the individual atoms.

of S_4 is indicated S_4, S_4^2, S_4^3, S_4^4. The performance of S_4^2 carries the molecule into a configuration directly accessible by C_2, and the latter symbol is generally given to such a transformation. In analogy to C_4^3, S_4^3 is equivalent to an S_4 performed with the opposite sense of direction in the rotation step. S_4^4 carries the molecule back into its original configuration. This is designated as identity, rather than an improper rotation.

4. Inversion Center.

The inversion center or center of symmetry is designated i. The corresponding operation is reflection through a point. Thus, in a controsymmetric molecule, a line drawn from any atom through a point at the center will pass through an equilvalent atom at the same distance on the other side of the point. If the center has the Cartesian coordinates (0,0,0) and an atom is located at (x,y,z), there will be an equivalent atom at (-x,-y,-z). Inversion is equivalent to the operation of S_2, as illustrated with trans-1,2-dibromoethene in Fig. 4. The result of i^2 is identity.

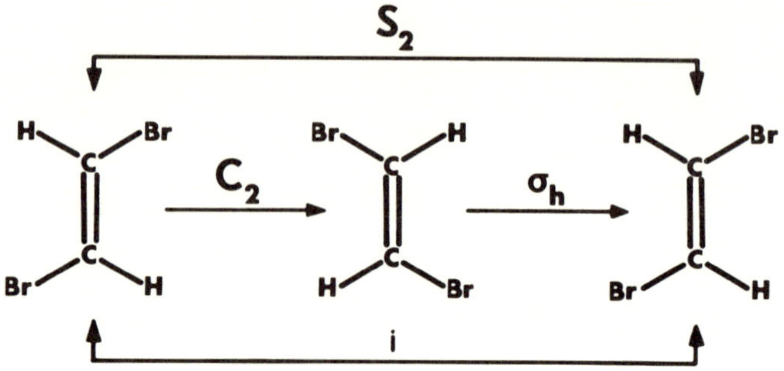

FIG. 4. The equivalence of S_2 and i, demonstrated by trans-1,2-dibromoethene.

5. *Identity.*

Every molecule possesses identity, and if this is the only symmetry, the molecule is regarded as asymmetric. Identity is given
the symbol E, or sometimes I. Although this seems like a trivial
operation, it is required by group theory. We have seen that repeated performance of a particular symmetry operation can carry the
molecule back to its starting configuration, equivalent to identity.
The following are equivalent to identity: C_n^n, σ^2, S_n^n (n even), S_n^{2n}
(n odd), and i^2.

B. Point Groups

The complete set of symmetry operations possessed by any molecule defines a point group. The point group is a mathematical group,
and as such it satisfies the four conditions, or rules, of a group.
These rules are concerned with multiplications between the members
of the group. The term multiplication is used here in the broader
algebraic sense of combination, rather than in the more restrictive
arithmetic sense. In symmetry groups, the multiplication is actually
a succession of symmetry operations, and it is necessary to express
the order in which the operations are to be performed. It is customary to define a right-multiplication sense when writing a product.
Suppose we wish to carry out operation A and then operation B. This
would be written BA. If we were to then do C, the succession of

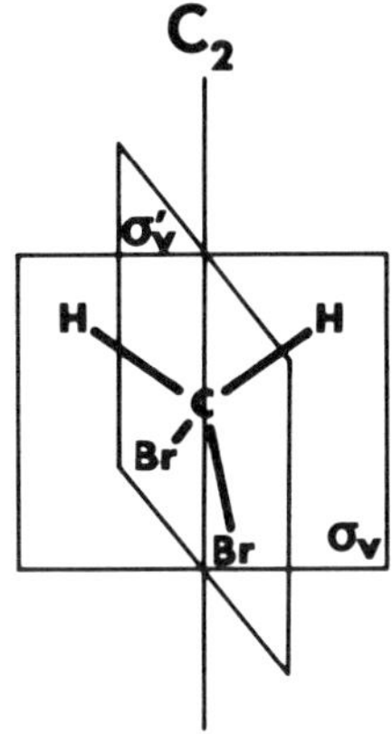

FIG. 5.   Symmetry elements of CH_2Br_2 (point group C_{2v}).

operations would be CBA.   If the single operation X were the same as
the product BA, then we would write BA = X.   In other words, if we
were to follow any particular atom through the course of the operations A and then B, the coordinates of that atom after these operations would be the same as after the operation X.

For some symmetry operations the product depends on the order
of multiplication.   In other words, the commutative law does not hold,
and in general AB ≠ BA.   In spite of this, many operations do commute.
We have already seen that an improper rotation, such as S_4, is defined
as a rotation followed by reflection.   The reverse order leads to the
same transformation; thus, $S_4 = C_4\sigma_h = \sigma_h C_4$.

Consider the symmetry of CH_2Br_2 (Fig. 5).   The symmetry operations of CH_2Br_2 define the group C_{2v}.   We may express the effects of
any two operations on the molecule by construction of a multiplication
table of binary products, such as Table 1.   The order of multiplication in this table is operation at the top followed by operation at
the side, or expressed in a right-multiply sense, (row)(column).*
We will use the symmetry operations of CH_2Br_2 and the multiplication
table for C_{2v} to illustrate the four general rules of a group.

---

*The order is really not important in the present case, since
all binary combinations commute.   In other cases this may not be so.

TABLE 1

Multiplication Table for the Group C_{2v}

C_{2v}	E	C_2	σ_v	σ_v'
E	E	C_2	σ_v	σ_v'
C_2	C_2	E	σ_v'	σ_v
σ_v	σ_v	σ_v'	E	C_2
σ_v'	σ_v'	σ_v	C_2	E

1. Closure.

If any two elements of the group are multiplied together, or if any element is multiplied by itself, the product will also be an element of the group. This is the property of closure. Table 1 shows that this requirement is met by the point group C_{2v}. No matter what series of multiplications are chosen, no operations outside of E, C_2, σ_v, and σ_v' are generated.

Not all collections of symmetry operations result in closure. For example, the collection E, C_2, and C_4 would not comprise a group, since $C_2 \times C_4 = C_4^3$, which is not one of the original elements. All four elements, E, C_2, C_4, and C_4^3, are needed to satisfy this requirement of a group.

2. Identity.

There is an element E which commutes with any element of the group such that, if X is any element of the group, $EX = XE = X$. This is the identity element.

3. Associative Law.

The associative law of combination is valid. Thus if three elements are multiplied together, the product of the first two times the third will be the same as the product of the first times the last two. For the point group C_{2v}, we see $(C_2\sigma_v)\sigma_v' = C_2(\sigma_v\sigma_v') = E$.

4. The Reciprocal.

For every element A there is an element A^{-1}, called the recipro-

cal of A, such that the product of the two is the identity; that is,
$AA^{-1} = E$. Both A and A^{-1} are members of the group, and an element
may be its own reciprocal. Furthermore, every element commutes with
its inverse; that is, $AA^{-1} = A^{-1}A = E$. The symbol A^{-1} does not imply
the inverse in the usual algebraic sense. In the point group C_{2v}, it
is apparent that σ_v and σ_v' are each other's reciprocals, and that C_2
and E are their own reciprocals.

C. Properties and Classification of Point Groups

The number of elements in a group defines its order, denoted h.
The point group C_{2v} has four operations, and h = 4. Within the elements of most groups, it is possible to assemble smaller collections
of elements which also satisfy the mathematical requirements of a
group. These collections of elements are called subgroups, and, obviously, their order is always less than that of the parent group.
Within the elements of C_{2v} we can recognize three unique collections
of elements which satisfy the criteria of a group. These are E; E,C_2;
and E, $\sigma_v \equiv$ E, σ_v'. The order of each subgroup g can be seen to be a
divisor of h, the order of the main group, such that h/g = k, where k
is an integer. This is a general result. As we shall see later, perturbations on a molecule may result in loss of some of its symmetry
through deformations of bond lengths and bond angles. In such a case,
the molecule's symmetry descends to that of a subgroup of its ideal
symmetry, resulting in alterations in the forms and spectral activities of the normal modes.

Point groups have been given labels to identify them. Two systems of nomenclature are commonly used: the Schoenflies and the Hermann-Mauguin notations. The Schoenflies notation is preferred by spec-
tropists and will be used in this chapter. The Hermann-Mauguin notation, preferred by crystallographers, will be introduced only as
needed. The symbol C_{2v}, which was used to designate the point group
of CH_2Br_2, is a Schoenflies notation. The equivalent symbol in the
Hermann-Mauguin notation is *mm*.

Some common point groups (Schoenflies notation) are listed in
Table 2 along with the symmetry operations of which they are composed.

The nonrotational groups (C_1, C_s, and C_i) rank at the bottom of any hierarchy of symmetry groups, as indicated by their low orders (h = 2). The single-axis rotational groups represent a higher order of symmetry and are characterized by the presences of only one axis about which all rotations take place. The notations C_n, C_{nv}, C_{nh}, and S_{2n} indicate families of groups, where n, the order of the principal rotation, may be any integer from 2 to ∞. With the exception of n = ∞, as in $C_{\infty v}$, real molecules are seldom found with one of these point groups having n greater than 6. The point group $C_{\infty v}$ is merely a special case among the family of C_{nv} groups. All linear molecules lacking a center of inversion (or equivalently, σ_h plane) belong to the point group $C_{\infty v}$, for example, HCl, COS, etc. The dihedral rotational groups, including the families of groups D_n, D_{nd}, and D_{nh}, are distinguished by the presence of n twofold axes perpendicular to the principal (n-fold) axis of rotation. Hence, a molecule belonging to one of these groups and in which the principal axis is, say, C_4 will also have four C_2 axes perpendicular to the C_4 axis. The $D_{\infty h}$ group is a special case in the family of D_{nh} groups and is the point group of all centrosymmetric linear molecules; for example, H_2, CO_2, etc. The cubic groups include the point groups which describe tetrahedral (T_d) and octahedral (O_h) molecules. Aside from the distinctive shapes of the molecules with these symmetries, these groups are characterized by the presence of several highest order axes. The last group listed in Table 2, I_h, describes the symmetry of icosahedral molecules, such as $B_{12}H_{12}^{2-}$. The listing in Table 2 is not exhaustive. A more complete listing is given by Herzberg [4].

It is not necessary to enumerate every symmetry element possessed by a molecule in order to determine its point group. The property that certain collections of elements require the presence of other elements reduces the task to searching only for those key elements which characterize the group. The unerring enumeration of these key elements, and hence, the determination of the correct point group for a molecule, is best carried out systematically by looking for particular symmetry elements in a prescribed order. Such an order is summarized in the flow chart shown in Fig. 6. One begins by determining if the molecule has a shape which is immediately identifiable with one of

TABLE 2

Common Point Groups and Their Operations

Nonrotational Groups Operations

C_1	E
C_s	$E,\ \sigma$
C_i	$E,\ i$

Single-axis Rotational Groups ($n = 2,\ 3,\ldots$)

C_n	$E,\ C_n,\ \ldots,\ C_n^{n-1}$
C_{nv}	$E,\ C_n,\ \ldots,\ C_n^{n-1},\ n\sigma_v$ ($n\sigma_v$ and $n\sigma_d$ if C_{2nv})
C_{nh}	$E,\ C_n,\ \ldots,\ C_n^{n-1},\ \sigma_h$ ($C_n \times \sigma_h = S_n,\ S_2 = i$)
S_{2n}	$E,\ S_{2n},\ \ldots,\ S_{2n}^{2n-1}$ ($S_2 = i,\ S_{2n}^n = C_n$, etc.)
$C_{\infty v}$	$E,\ C_\infty,\ \infty\sigma_v$ (noncentrosymmetric linear)

Dihedral Rotational Groups ($n = 2,\ 3,\ldots$)

D_n	$E,\ C_n,\ \ldots,\ C_n^{n-1},\ nC_2 \perp C_n$
D_{nd}	$E,\ C_n,\ \ldots,\ C_n^{n-1},\ S_{2n},\ \ldots,\ S_{2n}^{2n-1},\ nC_2 \perp C_n,\ n\sigma_d$ ($S_2 = i$, $S_{2n}^n = C_n$, etc.)
D_{nh}	$E,\ C_n,\ \ldots,\ C_n^{n-1},\ nC_2 \perp C_n,\ \sigma_h$ ($C_n \times \sigma_h = S_n,\ S_2 = i$)
$D_{\infty h}$	$E,\ C_\infty,\ S_\infty,\ \infty C_2 \perp C_\infty,\ \infty\sigma_v,\ i$ (centrosymmetric linear)

Cubic Groups

T_d	$E,\ 4C_3,\ 4C_3^2,\ 3C_2,\ 3S_4,\ 3S_4^3,\ 6\sigma_d$ (tetrahedron)
O_h	$E,\ 4C_3,\ 4C_3^2,\ 6C_2,\ 3C_4,\ 3C_4^3,\ 3C_2\ (= C_4^2),\ i,\ 3S_4,\ 3S_4^3,\ 4S_6,\ 4S_6^5,\ 3\sigma_h,\ 6\sigma_d$ (octahedron)
I_h	$E,\ 6C_5,\ 6C_5^2,\ 6C_5^3,\ 6C_5^4,\ 10C_3,\ 10C_3^2,\ 15C_2,\ i,\ 6S_{10},\ 6S_{10}^3,\ 6S_{10}^7,\ 6S_{10}^9,\ 10S_6,\ 10S_6^5,\ 15\sigma$ (icosahedron)

the so called "special groups": $C_{\infty v}$ - noncentrosymmetric linear; $D_{\infty h}$ - centrosymmetric linear; T_d - perfect tetrahedron; O_h - perfect octahedron; I_h - perfect icosahedron. The molecule must display the complete symmetry of the group. For example, an octahedrally coordinated MX_4Y_2 species does not belong to O_h. If the molecule does not belong to one of these special groups, elements must be sought in the order

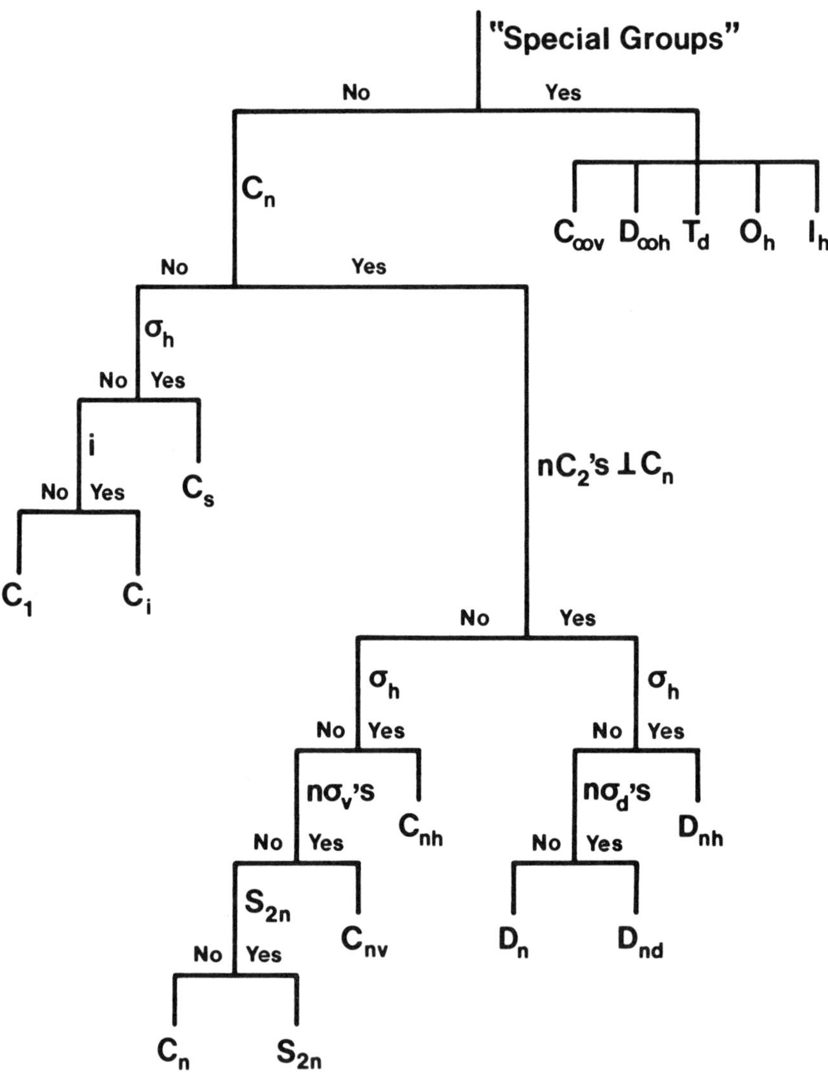

FIG. 6. Flow chart for determining the point group of a molecule.

of the flow chart. At any stage in the process the presence (Yes) or
absence (No) of the indicated element determines which element should
be sought next, until the point group is determined. Figure 6 may be
used to verify the following point group assignments: NH_3 (C_{3v});
BFClBr (C_s); $(C_5H_5)_2Fe$, ferrocene (D_{5d}); *trans*-$C_2H_2Cl_2$ (C_{2h});
$PtCl_4^{2-}$ (D_{4h}).

D. Irreducible Representations and Character Tables

The application of group theory to molecular problems, such as
vibrational selection rules, depends upon the use of a particular
physical or mathematical property as a basis for a representation of
the symmetry group of the molecule. For our purposes, a representation of a group (strictly, an irreducible representation) is any set
of symbols that will satisfy the multiplication table of the group.
The symbols are matrices or the characters (traces) of matrices.*
As such they may be positive, negative, or zero values of any numbers
or functions of which a matrix might be composed. These include integers, square roots of integers, sine and cosine functions, combina-
tions of i and $-i$, $\varepsilon = \exp(2\pi i/n)$, ε^*, ε^2, ε^{*2}, and $(1/2)(1 \pm \sqrt{5})$.

As an example, consider the point group C_{2v} and its multiplication table (Table 1). Suppose we substitute the integer 1 for each
element in C_{2v}. The multiplication table now becomes

C_{2v}	1	1	1	1
1	1	1	1	1
1	1	1	1	1
1	1	1	1	1
1	1	1	1	1

which gives the same results as Table 1. A less trivial sutstitution,
which also satisfies the multiplication table of C_{2v}, is $E = 1$, $C_2 = 1$,
$\sigma_v = -1$, and $\sigma_v' = -1$. The multiplication table now takes on the form

---

*In general, only the complete matrices obey the multiplication
table for the group.

C_{2v}	1	1	-1	-1
1	1	1	-1	-1
1	1	1	-1	-1
-1	-1	-1	1	1
-1	-1	-1	1	1

Other substitutions which may satisfy the C_{2v} multiplication table are $E = 1$, $C_2 = -1$, $\sigma_v = 1$, $\sigma_v' = -1$ and $E = 1$, $C_2 = -1$, $\sigma_v = -1$, $\sigma_v' = 1$. These four sets of 1 and -1 are the irreducible representations of C_{2v}, and a listing of them forms a character table of the group C_{2v}; namely,

C_{2v}	E	C_2	σ_v	σ_v'
A_1	1	1	1	1
A_2	1	1	-1	-1
B_1	1	-1	1	-1
B_2	1	-1	-1	1

The symbols A_1, A_2, B_1, and B_2 are the Mulliken symbols for the irreducible representations. The numbers comprising each representation are the characters of the representation. These four representations are the only representations with ±1 characters for C_{2v}. For example, the set $E = -1$, $C_2 = 1$, $\sigma_v = 1$, and $\sigma_v' = -1$ does not obey the multiplication table of C_{2v}, since this substitution leads to incorrect results such as $EC_2 = E$, $\sigma_v\sigma_v' = \sigma_v'$, etc.

The complete character table for C_{2v}, as usually presented, is shown in Table 3. It contains information beyond the simple tabulation of the characters of the irreducible representations. The second to the last column lists the transformation properties of the three unit vectors of a Cartesian coordinate system. The letter z in this column on the row of A_1 indicates that a unit vector z trans-

TABLE 3

Complete Character Table for C_{2v}

C_{2v}	E	C_2	$\sigma_v(xz)$	$\sigma_v'(yz)$		
A_1	1	1	1	1	z	x^2, y^2, z^2
A_2	1	1	-1	-1	R_z	xy
B_1	1	-1	1	-1	x, R_y	xz
B_2	1	-1	-1	1	y, R_x	yz

forms as the species (representation) A_1 in C_{2v}. Likewise, x transforms as B_1, and y transforms as B_2. The meaning of "transform" in
this connotation is the effects of the symmetry operations on the
named unit vector. Suppose we consider the effects of the operations
of C_{2v} on a unit vector x in the Cartesian coordinate system of Fig.
7, where the z axis is the C_2 axis. The effect of each operation on
x can be represented mathematically by multiplying x, the coordinate
of x on the x axis, by ±1. Thus we have

$$E : \quad (1)(x) = x$$
$$C_2: \quad (-1)(x) = -x$$
$$\sigma_v: \quad (1)(x) = x$$

and

$$\sigma_v': \quad (-1)(x) = -x$$

The value ±1 for each operation is nothing more than the character
for each operation in the representation B_1. Thus, the representation
B_1 expresses the transformations of x by the operations of the point
group C_{2v}. Similar reasoning may be applied to the other unit vectors.

The second to the last column of Table 3 also displays the notations R_x, R_y, R_z. These indicate the transformation properties of
rotations about the x, y, and z axes, respectively. These transformations are determined in a manner similar to that used for the unit
vectors. For example, the transformation of R_z in C_{2v} may be determined by using a curved vector rotating about the z axis (Fig. 8).

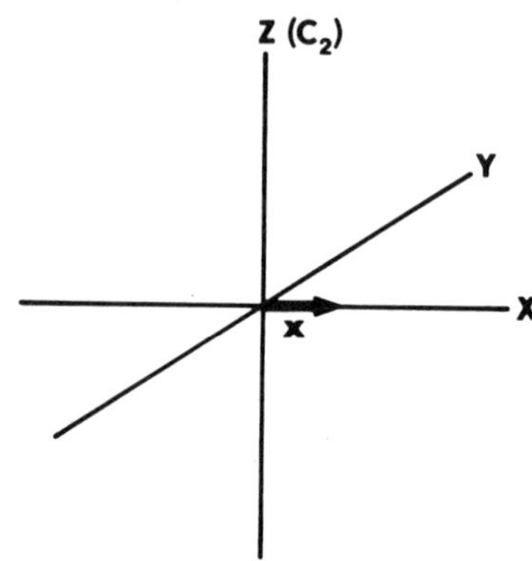

FIG. 7. A unit vector, x, in a system of C_{2v} symmetry.

For each operation, if the direction of the arrow is unchanged, the character is +1, and if the arrow is reversed, the character is -1. For R_z, then, we have

C_{2v}	E	C_2	$\sigma_v(xz)$	$\sigma_v'(yz)$
Γ_{R_z}	1	1	-1	-1

and it follows that $\Gamma_{R_z} = A_2$. In other words, a rotation about z transforms as A_2 in C_{2v}.

The last column of the character table indicates the transformation properties of the squares and direct products of the unit vectors. These may be determined in a manner analogous to that used for the transformations of the unit vectors. Alternately, the direct product transformations may be determined by multiplying together the characters of the irreducible representations by which the appropriate unit vectors transform. Thus, for yz in C_{2v}, the characters for B_2 are multiplied by the characters for A_1, yielding

C_{2v}	E	C_2	$\sigma_v(xz)$	$\sigma_v'(yz)$
Γ_{yz}	1	-1	-1	1

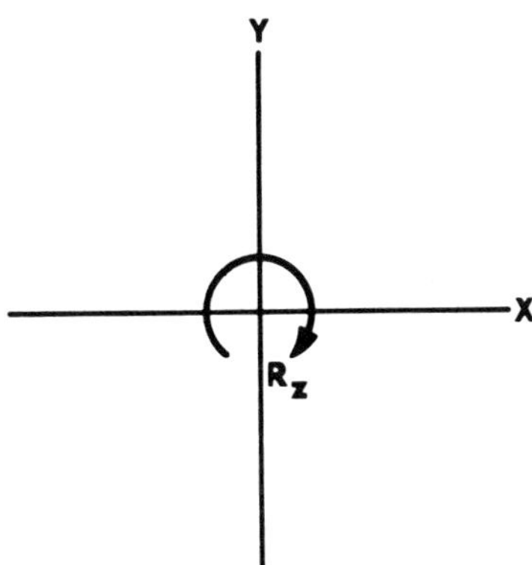

FIG. 8. A rotating vector, R_z, in a system of C_{2v} symmetry.
The C_2 axis (z axis) is perpendicular to the plane of the page.

where it is evident that $\Gamma_{yz} = B_2$.

The character table for the point group T_d (Table 4) exhibits
some features not found in the character table for C_{2v}. The group
T_d consists of 24 elements grouped into five classes. For example,
the notation $6S_4$ indicates that there are six S_4 operations which
belong to the same class. Specifically, these are three orthogonal
S_4 axes and three orthogonal S_4^3 axes, colinear with the S_4 axes.
Likewise, there is a class of three C_2 axes, a class of eight C_3
axes ($4C_3$ and $4C_3^2$), a class of six σ_d planes, and the single-membered
class of the identity, E. A theorem of group theory states that the
number of classes in a group is equal to the number of irreducible
representations. Consequently, there are five irreducible representations for T_d. The representation E is doubly degenerate, and the
representations T_1 and T_2 are triply degenerate.* This means that
if some component of a property transforms as E in T_d, there will be
another component which is indistinguishable and which will transform
simultaneously with the first. We see that in T_d the vectors x, y,
and z transform simultaneously as T_2; i.e., they are triply degenerate. Likewise the three rotational vectors are equivalent by symmetry, transforming as the triply degenerate species T_1.

---

*In some earlier works, triply degenerate irreducible represen-
tations are designated by F.

TABLE 4

Character Table for the Group T_d

T_d	E	$6S_4$	$3C_2$	$8C_3$	$6\sigma_d$		
A_1	1	1	1	1	1		$x^2 + y^2 + z^2$
A_2	1	-1	1	1	-1		
E	2	0	2	-1	0		$(x^2 + y^2 - 2z^2, x^2 - y^2)$
T_1	3	1	-1	0	-1	(R_x, R_y, R_z)	
T_2	3	-1	-1	0	1	(x, y, z)	(xy, xz, yz)

The Mulliken symbols which are used to designate the irreducible
representations convey information about the symmetry of the representations. The symbols A and B indicate symmetry and antisymmetry,
respectively, to the principal axis of rotation. The character for
the principal axis will be +1 for an A representation and -1 for a
B representation. The subscripts g (gerade) and u (ungerade) indicate symmetry and antisymmetry to the operation of inversion. Subscripts (1) and (2) refer to symmetry and antisymmetry to either a
lower order rotational axis (frequently C_2) or a vertical mirror
plane. The superscripts, prime (') and double prime ("), refer to
symmetry and antisymmetry to a horizontal mirror plane. In every
point group there is a representation, usually listed first in the
character table, which has all +1 characters. This is the totally
symmetric representation. Depending on the point group, it is given
the notation A, A_g, A_1, A_{1g}, A', Σ^+, or Σ_g^+. The latter two symbols
are found for the linear groups, $C_{\infty v}$ and $D_{\infty h}$, respectively.

Character tables for commonly used point groups are listed in
Appendix A.

Some point groups contain representations with imaginary characters. In such cases the representation is listed as two rows of complex conjugate pairs of characters. Although these representations
are given the symbol E of a doubly degenerate representation, each
imaginary half is formally regarded as a separate representation in
order to satisfy the requirement that the number of classes must

equal the number of representations. In applying such representations to physical problems, the two imaginary halves are added together to obtain one real representation. For example, the E_g representation of C_{4h} is handled in the added form shown below:

C_{4h}	E	C_4	C_4^3	C_2	i	S_4	S_4^3	σ_h
E_g $\Big\{$	1	i	$-i$	-1	1	$-i$	i	-1
	1	$-i$	i	-1	1	i	$-i$	-1
	2	0	0	-2	2	0	0	-2

E. Reducible Representations

Returning to the point group C_{2v}, consider the transformation of a general vector, v, whose base is at (0,0,0) and whose tip is at (x,y,z), as shown in Fig. 9. We may describe the transformations of this vector in C_{2v} by writing the coordinates of the tip before and after each operation in the form of 3 x 1 column matrix. The transformation of coordinates by each symmetry operation may be expressed mathematically by multiplying the original coordinates by an appropriate 3 x 3 matrix for the operation. In the present case, the following four equations result:

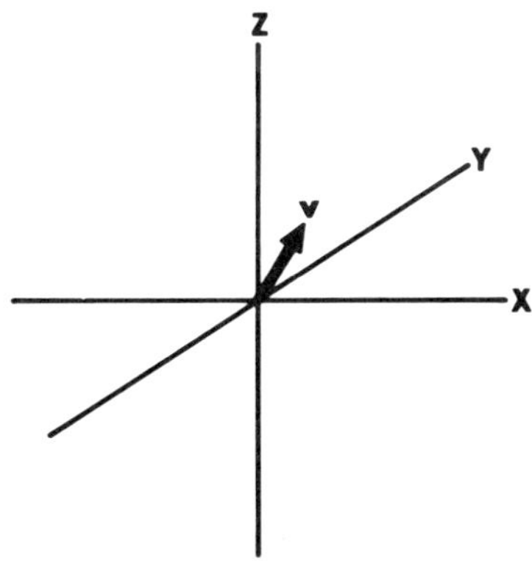

FIG. 9. A general vector, v.

$$
E: \quad \begin{bmatrix} 1 & 0 & 0 \\ 0 & 1 & 0 \\ 0 & 0 & 1 \end{bmatrix} \begin{bmatrix} x \\ y \\ z \end{bmatrix} = \begin{bmatrix} x \\ y \\ z \end{bmatrix}
$$

$$
C_2: \quad \begin{bmatrix} -1 & 0 & 0 \\ 0 & -1 & 0 \\ 0 & 0 & 1 \end{bmatrix} \begin{bmatrix} x \\ y \\ z \end{bmatrix} = \begin{bmatrix} -x \\ -y \\ z \end{bmatrix}
$$

$$
\sigma_v(xz): \quad \begin{bmatrix} 1 & 0 & 0 \\ 0 & -1 & 0 \\ 0 & 0 & 1 \end{bmatrix} \begin{bmatrix} x \\ y \\ z \end{bmatrix} = \begin{bmatrix} x \\ -y \\ z \end{bmatrix}
$$

$$
\sigma_v'(yz): \quad \begin{bmatrix} -1 & 0 & 0 \\ 0 & 1 & 0 \\ 0 & 0 & 1 \end{bmatrix} \begin{bmatrix} x \\ y \\ z \end{bmatrix} = \begin{bmatrix} -x \\ y \\ z \end{bmatrix}
$$

The four 3 x 3 matrices express the effects of C_{2v} on v, and they form a representation of C_{2v}; namely,

C_{2v}	E	C_2	$\sigma_v(xz)$	$\sigma_v'(yz)$
Γ_v	$\begin{bmatrix} 1 & 0 & 0 \\ 0 & 1 & 0 \\ 0 & 0 & 1 \end{bmatrix}$	$\begin{bmatrix} -1 & 0 & 0 \\ 0 & -1 & 0 \\ 0 & 0 & 1 \end{bmatrix}$	$\begin{bmatrix} 1 & 0 & 0 \\ 0 & -1 & 0 \\ 0 & 0 & 1 \end{bmatrix}$	$\begin{bmatrix} -1 & 0 & 0 \\ 0 & 1 & 0 \\ 0 & 0 & 1 \end{bmatrix}$

This is a legitimate representation of C_{2v}, because multiplications of the matrices obey the multiplication table for C_{2v}; for example,

$$
\begin{bmatrix} -1 & 0 & 0 \\ 0 & -1 & 0 \\ 0 & 0 & 1 \end{bmatrix} \begin{bmatrix} 1 & 0 & 0 \\ 0 & -1 & 0 \\ 0 & 0 & 1 \end{bmatrix} = \begin{bmatrix} -1 & 0 & 0 \\ 0 & 1 & 0 \\ 0 & 0 & 1 \end{bmatrix}
$$
$$
C_2 \quad \times \quad \sigma_v(xz) \quad = \quad \sigma_v'(yz)
$$

Taking the characters of the four matrices (the sums of their diagonals) permits us to express Γ_v in a more compact and customary form; namely,

C_{2v}	E	C_2	$\sigma_v(xz)$	$\sigma_v'(yz)$
Γ_v	3	-1	1	1

Γ_v does not correspond to one of the irreducible representations of C_{2v}, but it is a sum of three irreducible representations. Consequently, Γ_v is called a reducible representation of C_{2v}. It may be reduced or decomposed into the sum of $A_1 + B_1 + B_2$, as shown below.

A_1	1	1	1	1
B_1	1	-1	1	-1
B_2	1	-1	-1	1
Γ_v	3	-1	1	1

Since the vector v can be expressed as the resultant of x, y, and z, it is to be expected that the representation for the transformation of v should be the sum of the irreducible representations for the transformations of the unit vectors.

The application of group theory to the vibrational selection rule problem -- and to many other physical problems, as well -- consists of taking vectors on each atom as the basis for a reducible representation in the point group of the molecule. As we have just seen, it is a fairly simple matter to decompose the reducible representation for a single vector located at the center of a Cartesian coordinate system. However, for a polyatomic atom with few, if any, atoms located at the center of the coordinate system, the reducible representation generally is too complex to decompose by inspection. This task may be accomplished by use of the equation,

$$n^I = \frac{1}{h} \sum_i g_i \chi_i^I \chi_i^R,\tag{1}$$

where n^I is the number of times a particular irreducible representation I appears in the reducible representation R, h is the order of group, g_i is the number of elements in the class of the ith operation,

χ_i^I is the character of the ith operation in the irreducible representation, and χ_i^R is the character of the ith operation in the reducible representation.

The task of carrying out the decomposition of a reducible representation by Eq. (1) can be facilitated by using a tabular format
to organize the multiplications and additions involved. To illustrate the procedure, suppose we have a reducible representation in
the point group T_d with the following characters

T_d	E	$6S_4$	$3C_2$	$8C_3$	$6\sigma_d$
Γ_R	8	2	0	2	2

The T_d character table, from which we obtain the χ_i^I characters, is
shown in Table 4. Table 5 shows the tabular organization of the work
of Eq. (1) to yield the decomposition of Γ_R. The rows of numbers for
each species in Table 5 are the products, $g_i\chi_i^I\chi_i^R$, for each operation
in T_d. The number appearing under a particular operation and opposite a particular species is obtained by multiplying the character
of the operation in the reducible representation χ_i^R by the character

TABLE 5

Reduction of a Reducible Representation, Γ_R,

in the Point Group T_d

	E	$6S_4$	$3C_2$	$8C_3$	$6\sigma_d$	T	T/24
Γ_R	8	2	0	2	2	T	T/24
A_1	8	12	0	16	12	48	2
A_2	8	-12	0	16	-12	0	0
E	16	0	0	-16	0	0	0
T_1	24	12	0	0	-12	24	1
T_2	24	-12	0	0	12	24	1

of that operation χ_i^I given for the particular species in the T_d character table, and then multiplying that product by the number of elements in the class of the operation g_i. The numbers appearing in the column under T are the sums of rows, and represent $\Sigma g_i \chi_i^I \chi_i^R$. The column labelled T/24 represents the division of the sum by the order of the group (h = 24 for T_d) to give n^I. Thus Γ_R decomposes as $2A_1 + T_1 + T_2$.

This procedure is, of course, general to applications of point group theory. It is particularly helpful to organize the work in this way when dealing with point groups having large numbers of classes. Errors can be avoided by realizing some general results which can serve as internal checks of the arithmetic. The most obvious is that $T/h = n^I$ must be an integer. If it is not, an error has been made in $g_i \chi_i^I \chi_i^R$ or in construction of the reducible representation itself. Furthermore, the sum of the degeneracies of the irreducible representations must equal the character of the reducible representation for the identity. This is simply a quick check that the irreducible representations add up to the reducible representation. For the representation $\Gamma_R = 2A_1 + T_1 + T_2$, we have 2(1) + 3 + 3 = 8.

III. SELECTION RULES AND STRUCTURE

A. Vibrational Activities of Free Molecules

1. *The Application of Group Theory*

The 3n degrees of freedom possessed by a molecule can be classified according to the irreducible representations by which they transform under the point group of the molecule. For a nonlinear molecule, three degrees of freedom will correspond to pure translations of the entire molecule and three additional degrees of freedom will correspond to pure rotations. The symmetry species of these six nongenuine modes can be recognized among the 3n degrees of freedom, since they will be the same as those of the translational and rotational unit vectors, as listed in the next to last column of the appropriate

character table. The classification by symmetry species of the remaining 3n - 6 degrees of freedom allows the application of symmetry-
based selection rules to determine the infrared and Raman activities.

The classification into symmetry species is accomplished by
locating a set of Cartesian vectors (x_i, y_i, z_i) at each atom, and
taking the complete set of these 3n vectors as a basis for the representation in the point group of the molecule.

Figure 10 shows a water molecule with such a set of 3n vectors.
A 9 x 9 matrix is required to express the transformations of these
vectors under the operations of C_{2v}. For the identity element we
have

$$
\begin{bmatrix}
1 & 0 & 0 & 0 & 0 & 0 & 0 & 0 & 0 \\
0 & 1 & 0 & 0 & 0 & 0 & 0 & 0 & 0 \\
0 & 0 & 1 & 0 & 0 & 0 & 0 & 0 & 0 \\
0 & 0 & 0 & 1 & 0 & 0 & 0 & 0 & 0 \\
0 & 0 & 0 & 0 & 1 & 0 & 0 & 0 & 0 \\
0 & 0 & 0 & 0 & 0 & 1 & 0 & 0 & 0 \\
0 & 0 & 0 & 0 & 0 & 0 & 1 & 0 & 0 \\
0 & 0 & 0 & 0 & 0 & 0 & 0 & 1 & 0 \\
0 & 0 & 0 & 0 & 0 & 0 & 0 & 0 & 1
\end{bmatrix}
\begin{bmatrix}
x_1 \\ y_1 \\ z_1 \\ x_2 \\ y_2 \\ z_2 \\ x_3 \\ y_3 \\ z_3
\end{bmatrix}
=
\begin{bmatrix}
x_1 \\ y_1 \\ z_1 \\ x_2 \\ y_2 \\ z_2 \\ x_3 \\ y_3 \\ z_3
\end{bmatrix}
$$

the character of which is $\chi_E^R = 9$. It can be seen that this 9 x 9
matrix consists of smaller 3 x 3 block matrices and many zero elements. Only those blocks which lie along the diagonal contribute to
the character; furthermore, only elements along the diagonals of such
blocks will contribute. The significance of this can be seen with
the matrix for C_2:

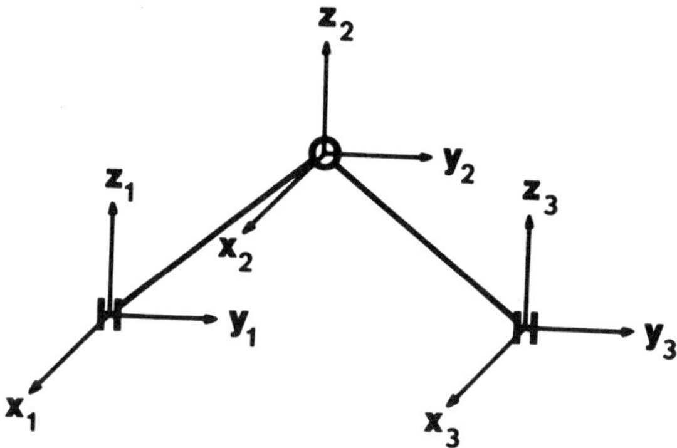

FIG. 10. Cartesian coordinates for a water molecule.

$$
\begin{vmatrix}
\begin{matrix} -1 & 0 & 0 \\ 0 & -1 & 0 \\ 0 & 0 & 1 \end{matrix} & & \\
& \begin{matrix} -1 & 0 & 0 \\ 0 & -1 & 0 \\ 0 & 0 & 1 \end{matrix} & \\
& & \begin{matrix} -1 & 0 & 0 \\ 0 & -1 & 0 \\ 0 & 0 & 1 \end{matrix}
\end{vmatrix}
\begin{vmatrix} x_1 \\ y_1 \\ z_1 \\ x_2 \\ y_2 \\ z_2 \\ x_3 \\ y_3 \\ z_3 \end{vmatrix}
=
\begin{vmatrix} -x_3 \\ -y_3 \\ z_3 \\ -x_2 \\ -y_2 \\ z_2 \\ -x_1 \\ -y_1 \\ z_1 \end{vmatrix}
$$

Inspection of the 3 x 3 blocks shows that they are all the same, that is, C_2 results in 3 x 3 matrices with characters of -1. However, only the block along the diagonal contributes to the character of the 9 x 9 matrix, for which $\chi_{C_2}^{R} = -1$. This occurs because the two hydrogens are exchanged by the operation of C_2, while the oxygen remains unshifted. Only the vectors on oxygen contribute to the character of the 9 x 9 matrix. This illustrates the general result that the character of the 3n x 3n matrix, χ_i^{R}, for the ith operation of the group is the product of the number of atoms, N_i, unshifted by the operation times the contribution per unshifted atom, χ_i, that is,

$$\chi_i^R = N_i \chi_i \tag{2}$$

Furthermore, atoms which do not lie on the symmetry element will be shifted and will contribute nothing to the character.

From the two matrices shown, it is evident that $\chi_E = 3$ and $\chi_{C_2} = 1$. This result is independent of the point group, since the effect of a particular operation on a set of Cartesian vectors will be the same, regardless of the point group in which it is found. Values of the contribution per unshifted atom for all operations have been tabulated [5]. However, these values are easily obtained from any character table containing the operation of interest: simply add up the characters for the operation in the irreducible representations by which the three unit vectors transform. For example, to find the contribution per unshifted atom for the operation of reflection σ, we consider the characters for this operation in any convenient character table, such as C_{2v} (cf. Table 3). In C_{2v}, x transforms as B_1, y as B_2, and z as A_1. Adding up the characters for these species under either $\sigma_v(xz)$ or $\sigma_v'(yz)$ gives +1, which is χ_σ in any point group, namely,

C_{2v}	E	C_2	$\sigma_v(xz)$	$\sigma_v'(yz)$	
A_1			1	1	z
B_1			1	-1	x
B_2			-1	1	y
			1	1	

Equation (2) permits the generation of the reducible representation for the 3n vectors, Γ_{3n}, without the need to construct the n x n matrices. Table 6 shows the generation of Γ_{3n} for the H_2O molecule and its subsequent reduction by Eq. (1) to give $\Gamma_{3n} = 3A_1 + A_2 + 2B_1 + 3B_2$. Six degrees of freedom arise from pure translations and pure

TABLE 6

Classification of the Nine Degrees of Freedom for H_2O

C_{2v}	E	C_2	$\sigma_v(xz)$	$\sigma'_v(yz)$			
N_i	3	1	1	3			
χ_i	3	-1	1	1			
$\Gamma_{3n}(\chi_i^R)$	9	-1	1	3	T	T/4	
A_1	9	-1	1	3	12	3	z
A_2	9	-1	-1	-3	4	1	R_z
B_1	9	+1	1	-3	8	2	x, R_y
B_2	9	+1	-1	3	12	3	y, R_z

$$\Gamma_{3n} = 3A_1 + A_2 + 2B_1 + 3B_2$$

rotations of the entire molecule; i.e., the 3n - 6 genuine modes are given by

$$\Gamma_{vib} = \Gamma_{3n} - \Gamma_{trans} - \Gamma_{rot} \tag{3}$$

Both Γ_{trans} and Γ_{rot} are given in the second to the last column of the character table. In C_{2v}, the translations transform as $\Gamma_{trans} = A_1 + B_1 + B_2$, and the rotations ($R_x$, R_y, R_z) transform as $\Gamma_{rot} = A_2 + B_1 + B_2$. Thus the three genuine modes for water have the symmetry $\Gamma_{vib} = 2A_1 + B_2$. The motions of the atoms in these three normal modes are illustrated in Fig. 11.

2. Selection Rules

Once the 3n - 6 normal modes have been classified by symmetry species, their infrared or Raman activities can be determined by applying two selection rules expressed in terms of group therapy. These rules result from considerations of the symmetry of the quantum mechanical expressions for the transition moments for the infrared and Raman processes. The basic selection rule is that a vibrational

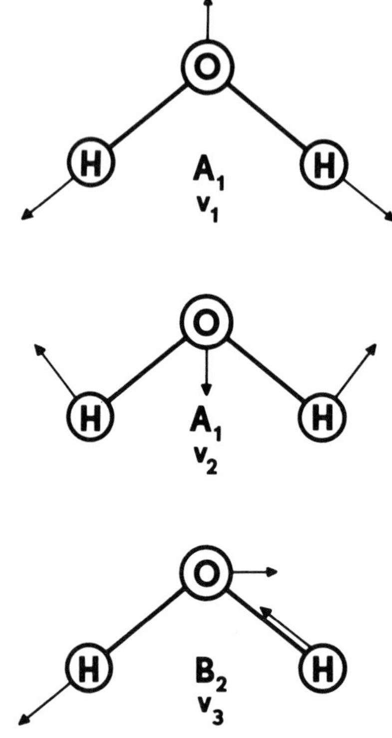

FIG. 11.   The normal modes of H_2O.

transition will have nonzero intensity in either the infrared or
Raman spectrum if the appropriate transition moment is nonzero.   The
expressions for these transition moments involve the electric dipole
moment and polarizability tensor for the infrared and Raman spectra,
respectively.   These properties can be expressed as the sums of several components, and an expression for the transition moment can be
written for each component.   In terms of symmetry, the transition
moment will be nonvanishing over all space, if any of its components
transforms as the totally symmetric representation of the point group
of the molecule.

For infrared absorption, the moment for a fundamental transition
(quantum number change from 0 to 1) is given by

$$M_{01} = \int \psi_0 \mu \psi_1 \, d\tau \tag{4}$$

where ψ_0 and ψ_1 are the wave functions for the states and μ is the electric dipole moment vector as a function of the normal coordinate. The vector μ can be viewed as the resultant of three components μ_x, μ_y, and μ_z, and an integral of the form of Eq. (4) applies to each component. Thus, if any of the corresponding moments M_x, M_y, or M_z is nonvanishing, M will not vanish. The ground state wave function ψ_0 is totally symmetric for all molecules (except free radicals) and the excited state wave function ψ_1 has the same symmetry as the normal mode. Thus the symmetry of the integral depends upon $\mu_i \psi_1$ (i = x, y, or z). Now the product of two species of the same symmetry always contains the totally symmetric representation, so the integral will be nonvanishing only if μ_i and ψ_1 transform as the same species under the point group of the molecule. Thus we may state: *a normal mode will be active in the infrared if it belongs to the same species as one of the components of the electric dipole moment.* These components transform by the same irreducible representations as the unit vectors listed in the second to the last column of the character tables. In the case of water, all three modes will be infrared active, since they belong to the same representation as either the z component (A_1) or the y component (B_2).

A similar reasoning can be applied to the case of Raman activity. Here the transition moment P is given by

$$P = \int \psi_0 \alpha E \psi_1 \, d\tau = E \int \psi_0 \alpha \psi_1 \, d\tau \tag{5}$$

where E is the electric vector of the exciting radiation and α is the polarizability tensor. Since α is a tensor, it may be represented by the array

$$\alpha = \begin{bmatrix} \alpha_{xx} & \alpha_{xy} & \alpha_{xz} \\ \alpha_{yx} & \alpha_{yy} & \alpha_{yz} \\ \alpha_{xx} & \alpha_{zy} & \alpha_{zz} \end{bmatrix}$$

where $\alpha_{ij} = \alpha_{ji}$. If any component of the induced moment P_{ij} is nonvanishing for a particular normal mode, the transition will be Raman
active. Considering the symmetries of ψ_0 and ψ_1 as before, *a normal*
mode will be active in the Raman effect if it belongs to the same
species as one of the components of the polarizability tensor. These
components have the same symmetry as the direct products of unit vectors listed in the last column of the character table for the point
group. In the case of water, ν_1 and ν_2 (A_1) belong to the same spe-
cies as the components α_{xx}, α_{yy}, and α_{zz}; and ν_3 (B_2) belongs to the
same species as α_{yz}. Thus all three modes of water are Raman active.

One important consequence of the selection rules is the so-called
rule of mutual exclusion: if the molecule is centrosymmetric, normal
modes which are infrared active will not be Raman active and vice
versa. This result is apparent upon insepcting the character table
for any point group containing the element i (for example, D_{4h}, O_h).
It can be seen in such cases that the unit vectors always belong to
species different from the direct products of the unit vectors.

3. The Depolarization Ratio

Measurement of the depolarization ratio ρ of an observed Raman
band is a useful aid in assigning the spectrum. The depolarization
ratio is customarily defined in terms of the right-angle scattering
arrangement of most Raman spectrometers. If the incident exciting
radiaiton is propagated in the z direction and the scattered radiation
is observed in the xy plane, then the depolarization ratio is defined
as

$$\rho = \frac{I_\perp}{I_{||}} \tag{6}$$

where $I_\perp$ is the intensity of the scattered light polarized perpendi-
cular to the xy plane and $I_{||}$ is the intensity of the scattered light
parallel to this plane. A rotatable polarizer is used to make the
measurements.

The symmetry of the polarizability tensor as it changes during
the execution of a particular mode results in the phenomenon of the
depolarization ratio. In a fluid, α will be randomly oriented

throughout the population, and so it is necessary to express ρ in terms of invariants, which are independent of orientation. If the incident radiation is plane polarized (as with lasers), then the depolarization ratio is given by

$$\rho_\ell = \frac{3(\gamma')^2}{45(\bar{\alpha}')^2 + 4(\gamma')^2} \tag{7}$$

where the primes indicate partial derivatives with respect to the normal coordinate, and $\bar{\alpha}'$ and γ' are defined as

$$\bar{\alpha}' = \frac{1}{3}(\alpha'_{xx} + \alpha'_{yy} + \alpha'_{zz})$$

and

$$(\gamma')^2 = \frac{1}{2}[(\alpha'_{xx} - \alpha'_{yy})^2 + (\alpha'_{yy} - \alpha'_{zz})^2 + (\alpha'_{zz} - \alpha'_{xx})^2$$
$$+ 6(\alpha'^2_{xy} + \alpha'^2_{yz} + \alpha'^2_{zx})]$$

Inspection of Eq. (7) shows that $\rho_\ell^{max} = 3/4$ when $\alpha' = 0$, and $0 < \rho_\ell < 3/4$ when $\bar{\alpha}' \neq 0$. These two conditions correspond to a depolarized band and a partially or completely polarized band, respectively. Only in the case of a totally symmetric normal mode will $\bar{\alpha}'$ be non-zero; therefore, *only normal modes belonging to the totally symmetric species will result in polarized Raman bands.* Normal modes belonging to other species will have $\rho_\ell = 3/4$. In the case of water, the two A_1 modes are polarized.

Molecules belonging to one of the cubic point groups (for example, T_d, O_h, etc.) represent a special case, since they are isotropic ($\gamma' = 0$). For such molecules, Eq. (7) leads to the conclusion that $\rho_\ell = 0$ for totally symmetric normal modes and $\rho_\ell = 3/4$ for all others. In actual practice, departures from total polarization are often observed due to departures from perfect cubic-group symmetry.

For natural (unpolarized) light (as with arc sources), Eq. (7) is replaced by

$$\rho_n = \frac{6(\gamma')^2}{45(\alpha')^2 + 7(\gamma')^2} \tag{8}$$

Depolarization ratios are then in the range $0 < \rho_n < 6/7$. Otherwise the conclusions are the same.

Experimental factors influencing the precision with which depolarization ratios may be measured have been fully discussed by Dawson[6].

4. *Determining Activities for Linear Molecules*

Linear molecules belong to one of the infinite point groups, $C_{\infty v}$ or $D_{\infty h}$. In these cases it is no longer possible to use Eq. (1) to decompose the reducible representation, and an alternate route must be taken to determine the symmetry species of the 3n - 5 normal modes.

Strommen and Lippincott [7] have suggested a convenient way around this problem. The reducible representation Γ_{3n} is generated in a subgroup (G) of the infinite point group (G°) with the assumption that the infinite axis of G° becomes z in G. Although any subgroup can be used, convenient choices are C_{2v} for $C_{\infty v}$ and D_{2h} for $D_{\infty h}$. The representation is decomposed by Eq. (1), yielding the species in the point group G. The species of the three translational vectors (x, y, z) and the two rotational vectors (R_x, R_y) are subtracted, leaving the species of the 3n - 5 genuine normal modes. The species in G are converted into the species in G° by matching (correlating) irreducible representations in the two groups which have the same basis vectors. Finally, spectral activities are determined in the usual manner.

As an example, consider the linear molecule O=C=C=C=O, carbon suboxide. Since $G° = D_{\infty h}$, we may assume $G = D_{2h}$. The reducible representation in D_{2h} is

D_{2h}	E	$C_2(z)$	$C_2(y)$	$C_2(x)$	i	$\sigma(xh)$	$\sigma(xz)$	$\sigma(yz)$
Γ_{3n}	15	-5	-1	-1	-3	1	5	5

This reduces as

$$\Gamma_{3n} = 2A_g + 3B_{1u} + 2B_{2g} + 3B_{2u} + 2B_{3g} + 3B_{3u}$$

The translations and rotations are

$$\Gamma_{trans} = B_{1u} + B_{2u} + B_{3u}$$

$$\Gamma_{rot} = B_{2g} + B_{3g}$$

This leaves

$$\Gamma_{vib} = 2A_g + 2B_{1u} + B_{2g} + 2B_{2u} + B_{3g} + 2B_{3u}$$

Consulting the transformation properties of vectors in D_{2h} and $D_{\infty h}$ permits construction of the following correlation:

D_{2h}	Basis	$D_{\infty h}$
A_g	z^2	Σ_g^+
B_{2g}	xz	Π_g
B_{3g}	yz	
B_{1u}	z	Σ_u^+
B_{2u}	y	Π_u
B_{3u}	x	

Therefore, in $D_{\infty h}$ we have

$$\Gamma_{vib} = 2\Sigma_g^+ + \Pi_g + 2\Sigma_u^+ + 2\Pi_u$$

Note that B_{2g} and B_{3g} in D_{2h} become degenerate as Π_g in $D_{\infty h}$, and only one frequency will be observed. Likewise, B_{2u} and B_{3u} modes in D_{2h} become doubly degenerate Π_u modes in $D_{\infty h}$. The selection rules indicate that Σ_g^+ and Π_g modes are Raman active, and Σ_u^+ and Π_u modes are infrared active. In summary, we expect three Raman bands (two of which are polarized), three infrared bands, and no coincidences (rule of mutual exclusion).

5. Complications and Limitations

The use of group theory for determining the spectroscopic activities of vibrational fundamentals is a very powerful structural tool,

but not one that can be applied näively. There are a variety of
situations which can occur and which lead to either a smaller or
larger number of observed frequencies than might be expected. Familiarity with the more commonly encountered causes of these disparities
between expectation and observation is essential to correct assignment and interpretation of the spectra.

Weak Intensities. Perhaps the most obvious cause of observing
fewer bands than expected is that an allowed transition may yield a
spectroscopic band of too weak intensity to be detected above the
instrument background noise. Vanishing intensity for allowed transitions is usually not predictable, a priori. However, oscillations
causing only small changes in the electric dipole will give rise to
weak bands in the infrared, while those causing small changes in the
polarizability will result in weak bands in the Raman.

On this basis a few rather crude generalizations can be made
concerning the expected intensities of commonly encountered types of
active modes. A caveat is in order, since many exceptions occur.
Nonetheless, totally symmetric stretching modes tend to have high intensity in the Raman and low intensity in the infrared (when allowed),
while the reverse tends to be the case for asymmetric stretching
modes. Bending modes tend to be weak in the Raman and strong in the
infrared; and deformations such as group rocking and twisting tend
to be weak in both spectra. Perhaps the most widely recognized of
these generalizations is that totally symmetric modes are strong in
the Raman. Notable exceptions are the C-H out-of-plane mode in ferrocene, and the A_1 (CO) mode of $Mn(CO)_5Br$ in which the axial CO group
 stretches out of phase with the four CO groups in the plane [8]. Both
have extremely weak intensities.

Modes of low spectral intensity can sometimes be detected by
taking spectra on another phase of the sample. This technique is
not without hazards. Frequently bands may result from splitting of
degenerate modes and intermolecular associations. For example, the
Cr-O-Cr asymmetric bridge stretch in $Cr_2O_7^{2-}$ is not detectable in the
Raman spectra of aqueous solutions [9]. This mode is detectable as

very weak bands in the Raman spectra of the $M_2Cr_2O_7$ salts (M = K, Rb, Cs); but the observed frequencies range from 730 cm^{-1} to 774 cm^{-1} depending on the compound and crystal form, and two bands are observed in some cases [10]. Furthermore, gross structural changes may occur on change of phase. For example, the gas-phase Raman spectra of the trihalides of aluminum, gallium, and indium show marked differences from those of the melt, due to the presence of increased amounts of monomeric MX_3 in equilibrium with the dimeric M_2X_6 species at higher temperatures [11].

Accidental Degeneracy. An accidental degeneracy results when two different modes fortuitously have the same frequency. The degeneracy is not predictable from symmetry arguments alone and arises from purely dynamic considerations. Accidentally degenerate modes usually belong to different species. As might be expected, accidental degeneracies are most common among modes of similar motional type (stretching, bending, etc.), and the probability of two or more modes having the same frequency increases with the number of allowed vibrations. For example, the 21 normal modes expected for the $Cr_2O_7^{2-}$ ion give rise to only five bands in the Raman spectrum of aqueous solutions [9]. In this case, as with many others, several modes differ only by the phase with which identical functional groups in the molecule execute the same motion. In effect, "group frequencies" are observed.

The problem can be compounded by inadequate spectrometer resolution and band breadth which can result in nearly degenerate modes appearing to be truly degenerate. In those cases where narrower slits do not suffice to resolve a multicomponent band, graphical resolution may be successful, aided by an analog computer curve resolver. Computer programs [12] have also been developed for this purpose. The danger in this approach lies in not knowing how many bands comprise the envelope, since it is regrettably common that different choices for the number of component bands can result in what appears to be successful resolution of the band contour.

Band narrowing and, subsequently, improved resolvability can sometimes be achieved by cooling the sample. This approach is most often successful for lower frequency modes, where band broadening is at least in part due to transitions from thermally excited states, as allowed by the Boltzmann distribution. Of course, cooling may result in structural changes and consequently a different spectrum. While such changes may be interesting in themselves, they are not helpful to the problem at hand.

Combinations and Overtones. Although the fundamental selection rule for the harmonic oscillator allows only transitions for which $\Delta v = \pm 1$, anharmonicity in the oscillations of real molecules gives rise to weak bands in the spectra due to combinations and overtones. Combination bands result from the sum or difference of two different fundamentals, while overtones result from multiples of a single fundamental (that is, $\Delta v > 1$). A combination or overtone band may be ovserved if it contains the irreducible representation of a component of the electric dipole (infrared active) or a component of the polarizability (Raman active). Although many combination and overtone bands may be allowed by summetry, generally only a few have sufficient intensity to be observed. Furthermore, such bands tend to be more frequently observed in the infrared than the Raman.

The symmetry of a combination band is determined by taking the direct product of the irreducible representations of the two fundamentals. If both fundamentals are nondegenerate, the direct product will be nondegenerate. However, if one or both are degenerate, the direct product may not be an irreducible representation of the group. In such cases, the representation of the direct product can be decomposed into irreducible representations by inspection or application of Eq. (1).

The procedure can be illustrated by using methane as an example. The fundamentals of the tetrahedral CH_4 molecule are listed in Table 7. Consider a combination $\nu_1 + \nu_4$. The direct product is obtained by multiplying the characters of A_1 and T_2 (see Table 4):

A_1	1	1	1	1	1
T_2	3	-1	-1	0	1
T_2	3	-1	-1	0	1

TABLE 7

Fundamental Frequencies of CH_4

Assignment	Frequency, cm^{-1}	Activity
$\nu_1(A_1)$	2914	R (pol.)
$\nu_2(E)$	1520	R
$\nu_3(T_2)$	3020	R, IR
$\nu_4(T_2)$	1306	R, IR

The combination $\nu_1 + \nu_4$ contains T_2, and it is therefore formally active in both the Raman and infrared. An infrared band at 4216 cm^{-1} is assigned to this combination. It is evident from this example that any combination with a totally symmetric mode will have the same symmetry and spectral activity as the asymmetric mode of the combination.

The combination of $\nu_2 + \nu_4$ does not belong to a single representation, as shown by the direct product of E and T_2.

$$
\begin{array}{c|ccccc}
E & 2 & 0 & 2 & -1 & 0 \\
T_2 & 3 & -1 & -1 & 0 & 1 \\
\hline
 & 6 & 0 & -2 & 0 & 0
\end{array}
$$

This reduces by Eq. (1) to $T_1 + T_2$. T_2 is an infrared and Raman active species and the combination $\nu_2 + \nu_4$ is formally active in both spectra. An infrared band at 2823 cm^{-1} has been assigned to this combination. The treatment is the same for combinations of more than two fundamentals.

The symmetry of overtones of nondegenerate fundamentals can be determined from the direct product of the irreducible representation of the mode with itself. If Γ_n is the irreducible representation of the nth overtone of a nondegenerate fundamental, then $\Gamma_n = (\Gamma_f)^{n+1}$, where Γ_f is the irreducible representation of the fundamental. The characters of nondegenerate representations are either +1 or -1, and their binary products will be +1. Thus all first overtones of nondegenerate

fundamentals belong to the totally symmetric representation and will
be observed as polarized Raman bands (if, indeed, they are observed
at all). Furthermore, inspection of the character tables (Appendix
A) reveals that these first overtones will be infrared inactive, except for molecules belonging to the point groups C_1, C_s, C_n, and C_{nv}.
These are the only point groups for which totally symmetric modes are
active in both spectra.

In some cases the presence of overtones is useful for determining the frequencies of "silent" modes, which are inactive in both the
infrared and Raman spectra. For example, the plane deformation mode
(B_{1u}) for a square planar molecule is inactive (see Fig. 12) but its
first overtone is allowed in the Raman spectrum. A weak feature at
442 cm^{-1} in the Raman spectrum of solid XeF_4 has been tentatively
assigned as this overtone, implying a frequency of about 221 cm^{-1}
for the fundamental [14].

Determining the symmetry species of overtones of degenerate
fundamentals is less direct. The procedure is described fully by
Wilson et al [13].

Fermi Resonance. Fermi resonance can occur when an overtone or
combination band has the same symmetry and nearly the same frequency
as a fundamental. The Raman spectrum of carbon dioxide offers the

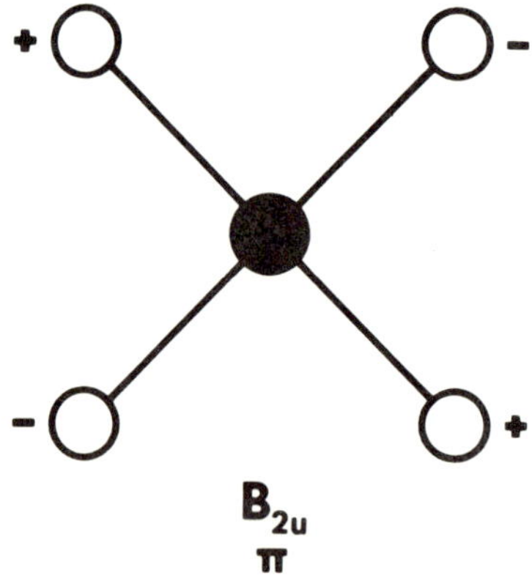

FIG. 12. The inactive deformation mode π of a square planar
molecule.

classic example of this phenomenon. Linear CO_2 belongs to point group $D_{\infty h}$, and one polarized band (ν_1, Σ_g^+) is expected in the Raman spectrum. Two bands are allowed in the infrared. The infrared-active bending mode (ν_2, Π_u) occurs at 667 cm^{-1}, and its first overtone, $2\nu_2$, is in Fermi resonance with ν_1. As a result, the overtone "borrows intensity" from the fundamental, and the two levels are repulsed. Thus, instead of the one band expected in the Raman spectrum, two strong bands are observed at 1388 cm^{-1} and 1285 cm^{-1}. The assignment of one band to ν_1 and the other to $2\nu_2$ has been the subject of some debate [15-17].

Another well known example of Fermi resonance occurs in both the infrared and Raman spectra of CCl_4. In this case the asymmetric stretching mode $\nu_3(T_2)$ interacts with the combination $\nu_1 + \nu_4(T_2)$. Two depolarized bands of medium intensity appear in the Raman spectrum (Fig. 13), while two very strong bands appear in the infrared spectrum (Fig. 14). The frequencies of the pair are the same within experimental error in both spectra, as expected for coincident modes of a noncentrosymmetric molecule.

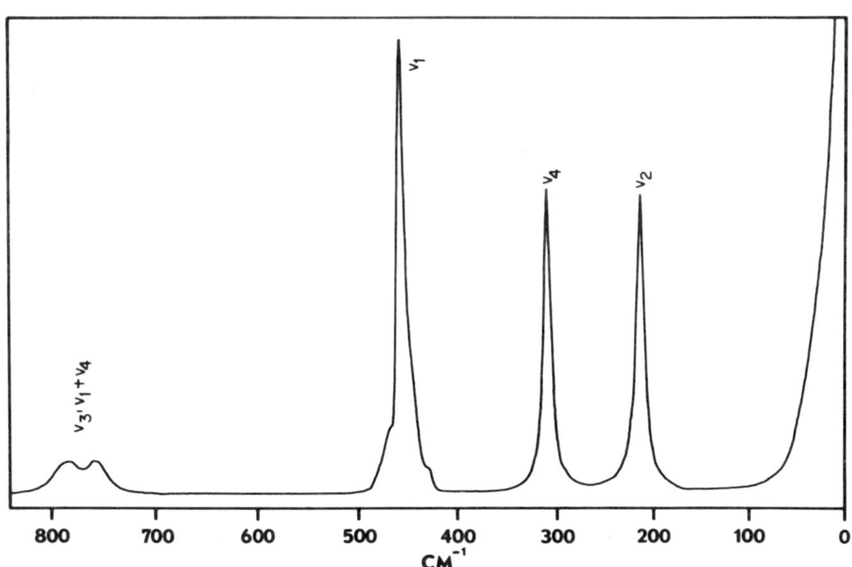

FIG. 13. Medium resolution Raman spectrum of liquid CCl_4.

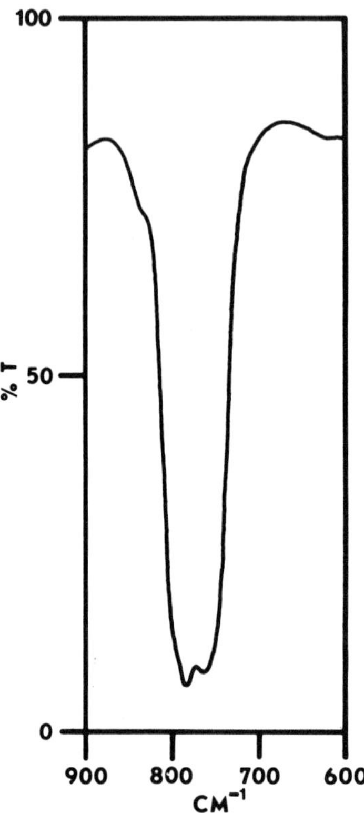

FIG. 14. Infrared spectrum of CCl_4 in the region of asymmetric
stretching vibrations, showing Fermi resonance of ν_3 and $\nu_1 + \nu_4$.
(Thin film between CsI plates.)

Isotopic Splitting. The spectra of compounds containing elements with two or more isotopes of significant natural abundance
may show an apparent splitting of some bands. Again, CCl_4 offers
one of the better known examples of this phenomenon. Naturally occurring chlorine consists of ^{35}Cl (75.53%) and ^{37}Cl (24.47%); thus
carbon tetrachloride is a mixture of the species $C^{35}Cl_4$, $C^{35}Cl_3{}^{37}Cl$,
$C^{35}Cl_2{}^{37}Cl_2$, $C^{35}Cl^{37}Cl_3$, and $C^{37}Cl_4$. Under high resolution the ν_1
band in the Raman spectrum (Fig. 15) is seen to consist of five bands
whose intensities reflect the abundance of each species.

Strictly speaking, this fine structure is not true splitting,
since no degeneracies between modes have been lifted. It is simply
the superimposition of the spectra of five different molecular spe-

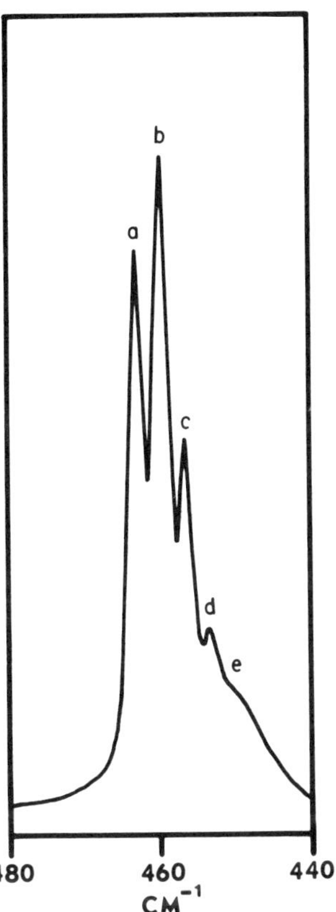

FIG. 15. High resolution Raman spectrum of CCl_4 showing fine structure of ν_1 due to ^{35}Cl and ^{37}Cl isotopes: (a) $^{35}CCl_4$; (b) $^{35}CCl_3$; (c) $^{35}CCl_2{}^{37}Cl_2$; (d) $^{35}CCl{}^{37}Cl_3$; (e) $^{37}CCl_4$.

cies. However, CCl_4 does have degenerate modes, and the possibility of splitting these by lowering the symmetry should be considered. While $C^{35}Cl_4$ and $C^{37}Cl_4$ have true T_d symmetry, $C^{35}Cl_3{}^{37}Cl$ and $C^{35}Cl{}^{37}Cl_3$ are C_{3v}, and $C^{35}Cl_2{}^{37}Cl_2$ is C_{2v}. Table 8 shows the Correlation of the three active species of T_d with the species of C_{3v} and C_{2v}. It can be seen that all degeneracies are lifted under C_{2v}, and the triply degenerate T_2 of T_d is split under C_{3v}. However, splitting of the bands for degenerate modes of CCl_4 (ν_2, ν_3, ν_4) from isotopic substitution is not observed. The mass differences between ^{35}Cl and ^{37}Cl are not sufficient to cause observable frequency differences. Indeed, even the degenerate modes of $C^{35}Cl_4$ and $C^{37}Cl_4$ are not sufficiently different from one another to be resolved in the spectra of the liquid.

TABLE 8

Correlation of Active Species of T_d

with Those of C_{3v} and C_{2v}

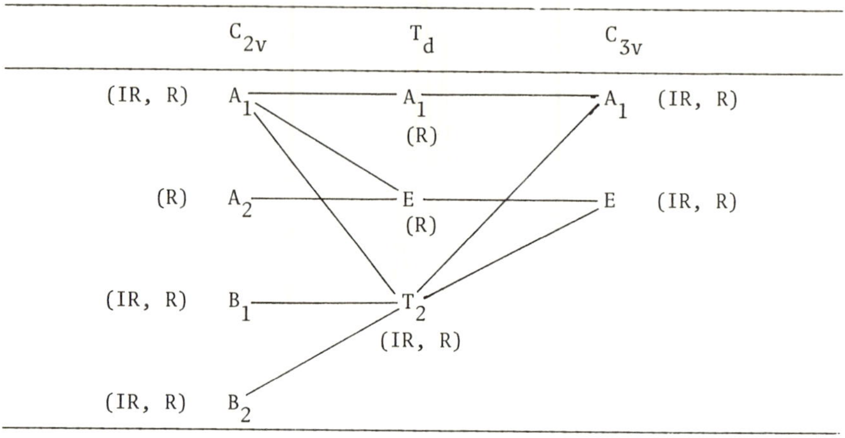

	C_{2v}	T_d	C_{3v}	
(IR, R)	A_1	A_1 (R)	A_1	(IR, R)
(R)	A_2	E (R)	E	(IR, R)
(IR, R)	B_1	T_2 (IR, R)		
(IR, R)	B_2			

The compounds of boron are also susceptible to isotopic fine
structure resulting from a natural abundance of 19.78% ^{10}B and
80.22% ^{11}B. The frequencies of the planar boron trihalides (D_{3h})
and descriptions of the normal modes are given in Table 9. The modes
ν_3 and ν_4 are active in both spectra, while ν_1 is only Raman active
and ν_2 is only infrared active. No isotope effect is noted for ν_1
frequencies, since the boron atom does not move in this oscillation.
A small effect is noted for ν_4 with BF_3 and BCl_3, but the greatest
effects are evident for ν_2 and ν_3, where the boron atom has appreciable amplitude.

Nonideal Fluids. The derivation of spectroscopic acitvities for
a molecule implicitly presumes that the physical environment has no
effect on the vibrating system. This ideal state exists only for
very dilute gases. In most real samples neighboring molecules interact to some extent, causing perturbations to the vibrational modes.
Where the molecules have little affinity for one another, these perturbations are usually minor, resulting in small frequency shifts or
band broadening. In such cases, even changes of state may have no

TABLE 9

Vibrational Frequencies[a] of Boron Trihalides [18-22]

$^{10}BF_3$	$^{11}BF_3$	$^{10}BCl_3$	$^{11}BCl_3$	$^{10}BBr_3$	$^{11}BBr_3$	$^{10}BI_3$	$^{11}BI_3$	Assignment
888	888	471	471	278	278	190	190	$\nu_1(A_1')$, symmetric stretch (ν_s)
718	691	480	460	395	375	352	336	$\nu_2(A_2'')$, plane deformation (π)
1505	1454	995	956	856	820	737	704	$\nu_3(E')$, asymmetric stretch (ν_d)
482	480	244	243	150	150	100	100	$\nu_4(E')$, bend (δ_d)

[a]Units in cm^{-1}.

gross effect on the number of bands observed. Indeed, the five-band
fine structure of ν_1 for CCl_4 remains intact at about the same frequencies when the sample is cooled to 77°K, and at that temperature
the band widths and resolution are much sharper.

Frequently molecular interactions cause changes in the symmetry
of the vibrating species. These changes can occur as a result of
interactions with molecules of the same or different type. If no
bonds are made or broken, the perturbed molecule will generally have
a lower symmetry. The descent in symmetry is to a subgroup of the
point group of the unperturbed molecule. Degenerate modes may split
in the descent and inactive modes may become active. The correlation
between the irreducible representations of the point group of the unperturbed molecule and those of the subgroup can be found in correlation tables, such as those in Appendix B. The spectroscopic activities can be deduced for the species of the subgroup in the usual
manner.

Spectroscopic evidence of descent in symmetry can be a useful
tool for studying molecular or ionic interaction. This approach has
been taken with infrared and Raman studies of aqueous solutions of
metal nitrates to determine the nature of interaction between the
nitrate ion and a number of cations. The frequencies and activities
for the free nitrate ion (planar, D_{3h}) are listed in Table 10. The
observed spectra in aqueous solution are seldom this simple [23].
The $\nu_1(A_1')$ mode, slightly shifted to lower frequency, appears as a
moderately strong band in the infrared of concentrated aqueous solutions. The $\nu_3(E')$ band is split into two other bands [24,25], and in
solutions with polyvalent cations the $\nu_4(E')$ band also may be split
[23,25]. These effects are consistent with cation-anion association
resulting in reduction of the nitrate symmetry to C_{2v}. The descent
to C_{2v} from D_{3h} removes all degeneracies and all six modes become
active in both the infrared and Raman spectra. Table 11 shows the
splitting of modes and observed frequencies for 7.2M $Ca(NO_3)_2$ solution [23].

TABLE 10

Vibrational Modes for Free NO_3^- (D_{3h})

Frequency, cm^{-1}	Activity	Assignment	Description
1384	IR, R	$\nu_3(E')$	NO_2 symmetric and asymmetric stretch
1048	R	$\nu_1(A_1')$	NO stretch
825	IR	$\nu_2(A_2'')$	plane deformation
719	IR, R	$\nu_4(E')$	NO_2 symmetric and asymmetric bend

TABLE 11

Splitting of the Degenerate Modes of the

Nitrate Ion in 7.2M $Ca(NO_3)_2$ Solution

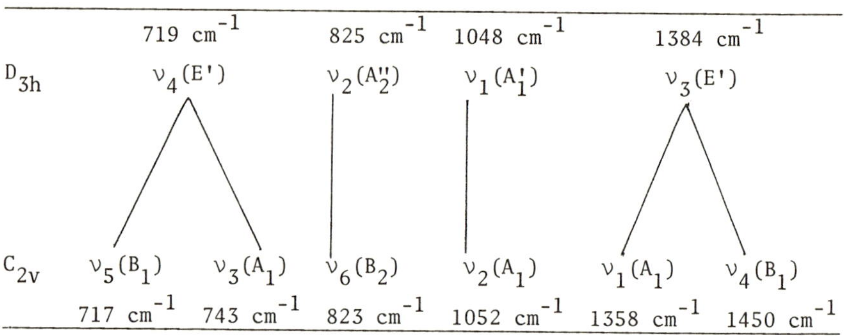

The separation between $\nu_1(A_1)$ and $\nu_4(B_1)$ (C_{2v} symmetry) has been
shown to increase with increasing charge/radius ratio of the cation
in solid nitrates [26], and it has been taken as a measure of the
dissymmetry of the nitrate group in solution [27,28]. Accordingly,
the following order of ability to cause nitrate ion distortion has
been given [29]: $Th^{4+}>In^{3+}>Cu^{2+}>Hg^{2+}>Ce^{3+}>Ca^{2+}>Zn^{2+}$, Al^{3+}, $Ag^+>Na^+$,
K^+, NH_4^+. Association with solvated water appears to be sufficient
to remove the degeneracy of $\nu_3(E')(D_{3h})$, while cation association in
concentrated solution (>7.0M) is necessary to cause splitting of
$\nu_4(E')(D_{3h})$ [30]. The cation-anion associations in solutions of Bi^{3+},
Ce^{4+}, and Sc^{3+} also result in the appearance of cation-oxygen stretching frequencies [31-33], and covalent bonding by bidentate nitrate
has been argued. Similar studies of reduction in nitrate ion symmetry have been made on ionic melts [34].

Band Contours in the Vapor Phase. Although symmetry-derived
selection rules are exact for dilute gases, the observed bands in the
infrared and Raman spectra of vapors frequently display complicated
shapes which are characteristic of the molecular symmetry and vibration type. For simple molecules observed at high resolution, the
bands may be seen to be structured with many closely spaced peaks.
Figure 16, which shows the stretching band of HBr in the infrared,

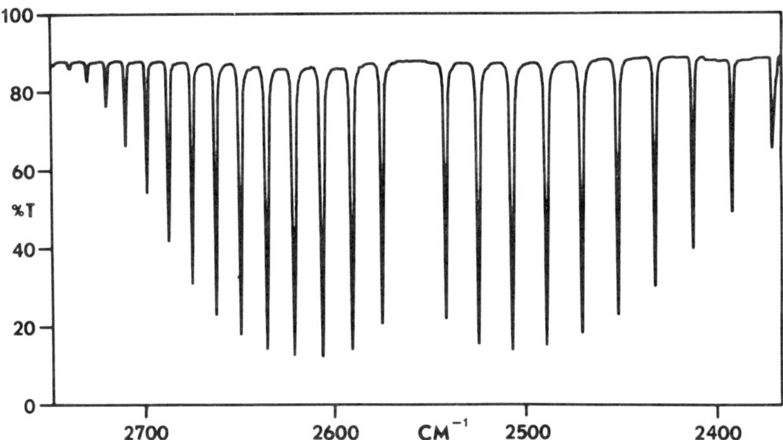

FIG. 16.   Infrared spectrum of HBr gas (1 atm) showing rotational
fine structure and absent Q branch.

is fairly typical. This fine structure is, of course, the result of
simultaneous rotational excitations accompanying the fundamental
vibrational transitions. The characteristic band structures result
from the symmetry along the three moments of inertia, which may be
identified for the rotating molecule (I_x, I_y, I_z or I_A, I_B, I_C), and
the restrictions on the rotational quantum number(s) for the rotor
type. A discussion of the theoretical basis of rotational fine structure is beyond the scope of this chapter and may be found elsewhere
[3,4,35]. Only a very brief outline of the results is presented here.

The types of rigid rotors and the defining restrictions on their
moments of inertia are given in Table 12. For all cases, the type of
observed band structure depends upon the selection rules for the rotational quantum number J which specifies the square of the total angular momentum of the rotor. If a vibrational fundamental ($\Delta v = 1$)
occurs with no change in rotational state ($\Delta J = 0$), a sharp band at
the fundamental frequency will occur (if allowed), which is called a
Q band. A series of lines of lower frequency than Q for which $\Delta J =$
-1 define a P branch, while those of higher frequency for which $\Delta J =$
+1 define a R branch. Only P, Q, and R branches are observed in the
infrared. However, in the Raman spectrum it is also possible to have

TABLE 12

Types of Rotors

Type	Moments	Point Groups	Examples
Linear	$I_x = I_y$, $I_z = 0$	$C_{\infty v}$, $D_{\infty h}$	HCl, CO_2
Spherical	$I_x = I_y = I_z$	Cubic Groups	CH_4
Symmetric (oblate)	$I_x = I_y < I_z$	$n > 3$: C_n, C_{nv},	BF_3
Symmetric (prolate)	$I_x < I_y = I_z$	C_{nh}, S_n, D_n, D_{nd}, D_{nh}, D_{2d}	$ClCH_3$
Asymmetric	$I_A < I_B < I_C$	All others	H_2CO

branches from $\Delta J = -2$ (O branch) and $\Delta J = +2$ (S branch). The band
structures of symmetric and asymmetric rotors also depend upon the
quantum number K, which specifies the component of angular momentum
along the z axis.

The rotational selection rules for various rotors and the expected band contours are summarized in Tables 13 and 14. The quantum numbers J' and J" in Table 14 refer to the initial and final rotational levels of the transition. It can be seen from Table 13 that
certain vibrational modes can be classified as $\parallel$ or $\perp$, depending on
whether their net dipole moment change is parallel or perpendicular
to the principal symmetry axis of the molecule. In the case of linear or symmetric molecules, a distinction between $\parallel$ and $\perp$ can be
made from the band contours, at least in principle. Caution should
be exercized, however, since complicating circumstances such as
Coriolis coupling, inversion doubling, and multiple Q branches can
be encountered. These topics are fully discussed by Herzberg [4].

Except for the simplest molecules, the rotational fine structure
is usually incompletely resolved, and in many cases only the band
envelopes of allowed branches are apparent. This is particularly
true in the Raman spectra where OPQRS structures may occur. Fig. 17

TABLE 13

Rotational Infrared Selection Rules

Rotor Type	Band Type	Components	Selection Rules	Band Contour[a]
Linear	nondegenerate ($\|\|$)	z	$\Delta J = \pm1$	PR
	doubly degenerate ($\perp$)	(x,y)	$\Delta J = 0, \pm1$	PQR
Spherical	triply degenerate	(x,y,z)	$\Delta J = 0, \pm1$	PQR
Symmetric	nondegenerate ($\|\|$)	z	$\Delta J = 0, \pm1; \Delta K = 0 \ (K \neq 0)$	$\Big\}$PQR
			$\Delta J = \pm1; \Delta K = 0 \ (K = 0)$	
	doubly degenerate ($\perp$)	(x,y)	$\Delta J = 0, \pm1; \Delta K = \pm1$	PQR
Asymmetric	nondegenerate $(\|\|_x, \ \|\|_y, \ \|\|_z)$	x;y;z	$\Delta J = 0, \pm1$	b

[a] Q = strong Q spike.

[b] See Ref. 4, pp. 468-491.

TABLE 14

Rotational Raman Selection Rules

Rotor Type	Band Type[a]	Components	Selection Rules	Band Contour[b]
Linear	totally symmetric (p)	xx + yy, zz	$\Delta J = 0, \pm2$	OQS
	doubly degenerate	(yz, xz)	$\Delta J = 0, \pm1, \pm2$	OPQRS
Spherical	totally symmetric (p)	xx + yy + zz	$\Delta J = 0$	Q
	doubly degenerate	(xx + yy − 2zz, xx − yy)	$\Delta J = 0, \pm1, \pm2$ $J' + J'' \geq 2$	OPQRS
	triply degenerate	(xy, yz, xz)	$\Delta J = 0, \pm1, \pm2$ $J' + J'' \geq 2$	OPQRS
Symmetric	totally symmetric (p)	xx + yy, zz	$\Delta J = 0, \pm1, \pm2$ $J' + J'' \geq 2; \Delta K = 0$	OPQRS
	nondegenerate	xx − yy, xy	$\Delta J = 0, \pm1, \pm2$ $J' + J'' \geq 2; \Delta K = \pm2$	OPQRS
	doubly degenerate	(xx−yy, xy)	$\Delta J = 0, \pm1, \pm2$ $J' + J'' \geq 2; \Delta K = \pm2$	OPQRS
		(yz, xz)	$\Delta J = 0, \pm1, \pm2$ $J' + J'' \geq 2; \Delta K = \pm1$	OPQRS
Asymmetric			$\Delta J = 0, \pm1, \pm2$ $J' + J'' \geq 2$	c

[a]p = polarized.

[b]Q = strong Q spike.

[c]See Ref. 4, pp. 468-491.

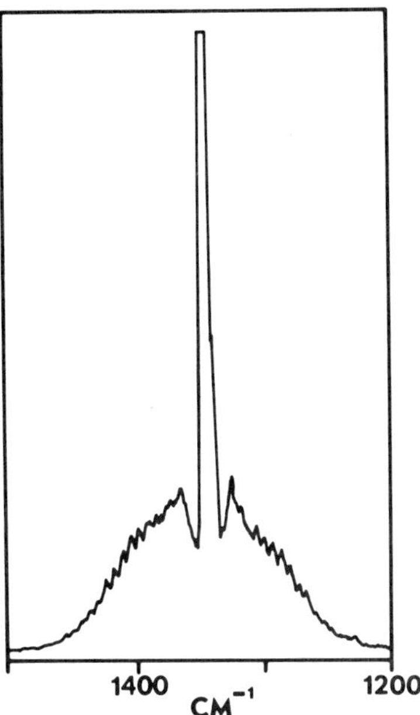

FIG. 17. Raman spectrum of $\nu_3(A_g)$ of ethylene gas (1 atm). (Courtesy of Spex Industries, Inc.)

shows the $\nu_3(A_g)$ band structure in the Raman spectrum of ethylene, which displays a typical OPQRS contour.

In the past, intensity difficulties with Raman spectra limited most vapor phase studies to the infrared. The advent of high intensity laser sources and multipass gas cells has resulted in increasing use of vapor phase Raman studies. Renewed activity in this area has also produced advances in the theory of vapor-phase Raman band contours [36].

6. Examples.

A few simple examples, chosen more or less at random, are presented here to illustrate the use of vibrational spectroscopy for deducing the structures of inorganic species. Some of the problems encountered in such applications are also illustrated.

XeF$_2$ and KrF$_2$. Two structures are possible for these triatomic molecules: linear ($D_{\infty h}$) or bent (C_{2v}). The selection rules for both models are shown in Table 15 and the observed frequencies are

TABLE 15

Spectroscopic Activities for Linear

and Bent Models of XeF_2 and KrF_2

Model	Mode	Activity
Linear $(D_{\infty h})$	$\nu_1 \ (\Sigma_g^+)$	Raman (pol.)
	$\nu_2 \ (\Sigma_u^+)$	IR $(\perp)$
	$\nu_3 \ (\Pi_u)$	IR $(\parallel)$
Bent (C_{2v})	$\nu_1 \ (A_1)$	Raman (pol.), IR
	$\nu_2 \ (A_1)$	Raman (pol.), IR
	$\nu_3 \ (B_2)$	Raman, IR

listed in Table 16. The data [37,38] for both indicate the linear
geometry. In keeping with the selection rules, only one band is observed in the Raman spectrum. The assignment of the two infrared-
active modes and the linear structure itself are confirmed by the
band contours. Only P and R branches are observed for ν_3 modes, as
expected for a $\parallel$ band of a linear molecule. Since no Q branch is

TABLE 16

Observed Vapor-phase Frequencies

of XeF_2 and KrF_2 [37,38]

Frequencies, cm^{-1}		Assignment
XeF_2	KrF_2	
515 (R)	449 (R)	ν_1
213.2 (IR)	232.6 (IR)	$\nu_2 \ (\perp)$
$\left.\begin{matrix} 550 \\ 564 \end{matrix}\right\}$ 557 (IR)	$\left.\begin{matrix} 580 \\ 596 \end{matrix}\right\}$ 588 (IR)	$\nu_3 \ (\parallel)$
1070 (IR)	1032 (IR)	$\nu_1 + \nu_3$

observed for this type of band, the ν_3 fundamental is taken as the midpoint between the P and R maxima. The appearance of the combination $\nu_1 + \nu_3$ in the infrared spectrum is allowed, because $\Sigma_g^+ \times \Pi_u = \Pi_u$. The linear geometry is unambiguously proved, and it has also been confirmed by x-ray diffraction on the solids [39,40].

XeO_2F_2. XeF_2 and KrF_2 present "textbook" illustrations of structure determinations by vibrational spectroscopy. Everything about them is ideal: the spectra are conveniently obtained in the gas phase; the band contours confirm the assignment; only two models are reasonable; and the selection rules for the two models are radically different. Most compounds are not this accommodating. A more typical example is provided by another noble gas compound, XeO_2F_2.

The preparation follows the reaction [41]

$$XeO_3 + XeOF_4 = 2XeO_2F_2$$

The product is a solid at room temperature with a melting point of 30.8°C. The vapor pressure of XeO_2F_2 is too low to permit study of the vapor phase spectra, and spectroscopic investigation was confined to the liquid and solid [42]. The problem was further complicated by the possibility that the compound might be a complex, such as $XeO_3 \cdot XeOF_4$, or that it might be polymerized by elimination of xenon-oxygen double bonds with formation of xenon-oxygen-xenon linkages. The two molecular models which were considered are the trans-planar structure (D_{2h}) and an out-of-plane distortion of this structure (C_{2v}). Both structures are shown in Fig. 18, and the spectroscopic selection rules are summarized in Table 17.

The reported spectra [42] are given in Table 18. The experimental arrangement for the Raman spectra prohibited accurate measurement of depolarization ratios, which accounts for the absence of such data in the table. The infrared spectrum was obtained from a sample of XeO_2F_2 deposited in a solid argon matrix. This technique dilutes the concentration of XeO_2F_2 and reduces the intermolecular interactions that are expected for the neat solid. Evidence for such interactions can be seen by comparing the Raman frequencies for

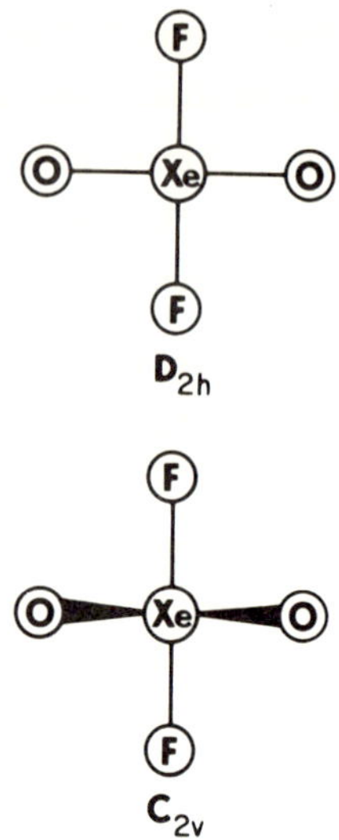

FIG. 18.   Two possible structures for XeO_2F_2.

the solid and liquid. The magnitude of frequency shifts in the
Raman spectrum of the solid suggests intermolecular perturbations,
and the deduction of the structure was based primarily on a comparison of the liquid-state Raman with the matrix isolated solid-state
infrared. The assignment of the spectrum was made by comparison
with the spectra of xenon compounds in which the Xe-O and Xe-F vibrations have been reliably assigned [43], and also by comparison with
the assignments for SO_2F_2 [44], CrO_2F_2 [45], and SeO_2F_2 [46].

In spite of the different physical states of the samples, there
are a sufficient number of frequencies in the infrared and Raman
spectra which agree within experimental error to rule out a centrosymmetric model, such as the planar D_{2h} structure. However, it is
not immediately obvious that the data demonstrate the C_{2v} structure.
The nonplanar structure does become more likely from valence-shell
electron-pair repulsion (VSEPR) considerations [47]. Accordingly,
the molecule is expected to have a roughly linear F-Xe-F moiety with
three electron pairs in the equator about xenon, where two of these
pairs are engaged in bonding with oxygens. To the degree to which

TABLE 17

Spectroscopic Activities for D_{2h}

and C_{2v} Models of XeO_2F_2

Model	Species	Activity
Planar (D_{2h})	$2A_g$	Raman
	B_{1g}	Raman
	$2B_{1u}$	IR
	$2B_{2u}$	IR
	$2B_{3u}$	IR
Nonplanar (C_{2v})	$4A_1$	Raman, IR
	A_2	Raman
	$2B_1$	Raman, IR
	$2B_2$	Raman, IR

it is possible to view Xe-F and Xe-O motions independently, this
model suggests that a symmetric Xe-F stretching mode should be intense in the Raman and weak (by vitrue of its small dipole moment
change) in the infrared. This agrees with the observed intense band
in the Raman at 490 cm^{-1} (an appropriate frequency for Xe-F stretching), which is not observed in the infrared. The infrared counterpart may be obscured by matrix background bands. Furthermore, the
asymmetric Xe-F stretch $\nu_8(B_2)$ is expected to be strong in the infrared and weak in the Raman, in agreement with the assignment given
in Table 18. Although it is tempting to see further evidence for
the C_{2v} structure in the expected failure to observe $\nu_5(A_2)$ in the
infrared, the absence of a band is always tenuous evidence. In the
present case, the expected frequency (assuming this mode is correctly assigned in the Raman) would occur near the limits of the spectrophotometer, reducing its probability of being detected if present.
Nonetheless, the combination of chemical intuition and observed

TABLE 18

Observed Spectra of XeO_2F_2 (frequencies in cm^{-1})[a]

Raman		Infrared[b]	Assignment
Solid	Liquid		
205 ms	198 w		$\nu_4 (A_1)$
224 w	223 vw		$\nu_5 (A_2)$
315 vs	313 ms	317 ms	$\nu_9 (B_2)$
		224 s	$\nu_7 (B_1)$
350 ms	333 ms		$\nu_3 (A_1)$
537 vs	490 s		$\nu_2 (A_1)$
		537 vw	$\nu_5 + \nu_9 \ (B_1)$
		550 w	$\nu_5 + \nu_7 \ (B_2)$
		574 w	impurity (?)
	578 w	585 vs	$\nu_8 (B_2)$
769 w	788 vw		$\nu_8 + \nu_4 \ (B_2)$ or impurity
814 w			?
850 vs	845 vs	848 ms	$\nu_1 (A_1)$
882 s	902 w	905 s	$\nu_6 \ (B_1)$
		1023 w	$\nu_1 + \nu_4 \ (A_1)$ or impurity
		1444 w	$\nu_1 + \nu_8 \ (B_2)$
		1496 vw	impurity

[a]Abbreviations: v-very, m-moderately, w-weak, s-strong.
[b]From argon matrix at 4°K.

spectroscopic data is sufficient to strongly suggest the proposed C_{2v} structure, if not "prove" it.

Squarate Ion, $C_4O_4^{2-}$. The squarate ion $C_4O_4^{2-}$ is the smallest member of a limited series of cyclical aromatic ions with the general formula $C_nO_n^{2-}$. The possibility of resonance stabilization over the entire ion suggested a planar (D_{4h}) structure [48], which was subsequently demonstrated by infrared and Raman spectroscopy [49]. The possible structures which were considered are shown in Fig. 19, and the consequent selection rules are summarized in Table 19.

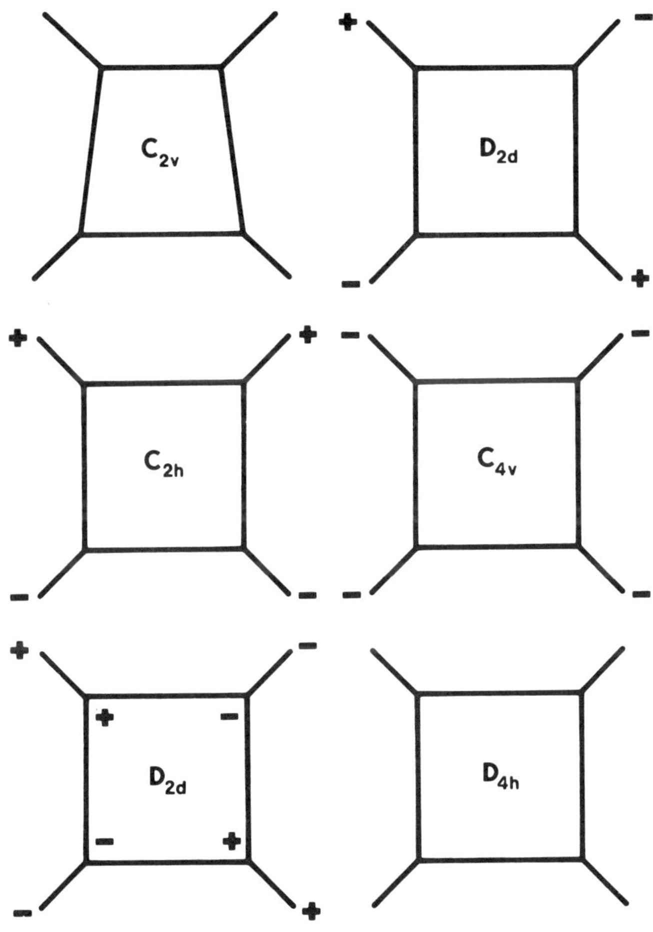

FIG. 19. Possible structures for the squarate ion, $C_4O_4^{2-}$.

TABLE 19

Possible Structures for $C_4O_4^{2-}$

and the Numbers of Expected Bands

Model	Raman		Infrared	Coincidences
	Polarized	Depolarized		
C_{2v}	7	11	15	15
C_{4v}	3	10	7	7
D_{2d}	4	9	7	7
C_{2h}	5	4	9	0
D_{4h}	2	7	4	0

The Raman spectrum was obtained from a 10% by weight aqueous solution of $K_2C_4O_4$, and the infrared spectrum was obtained from Nujol mulls of the salt [49]. The observed frequencies are given in Table 20. It is immediately apparent that no coincidences between infrared and Raman frequencies occur, and the magnitudes of the disparities argue against frequency shifts from difference in physical states of the samples. Therefore, the molecule must be centrosymmetric. The observation of only two polarized Raman bands indicates the D_{4h} structure, rather than the C_{2h} structure which should have five such bands. The planar structure and vibrational assignments indicated in Table 20 lead to a successful normal coordinate treatment [49], adding further confidence to the vibrational analysis. Later x-ray investigation of $K_2C_4O_4 \cdot H_2O$ confirmed that the squarate ion is, indeed, completely planar [50].

B. Vibrational Selection Rules in Solids*

1. *Introduction to the Spectra of Solids*

The vibrational spectra of solids exhibit certain features which are not predicted by free-molecule selection rules. For example, consider the tetrahedral chromate ion, CrO_4^{2-}. As a consequence

---

*Adapted in part from Ref. 51 with permission of the copyright holder.

TABLE 20

Observed Frequencies (cm^{-1}) for $C_4O_4^{2-}$.[a]

Raman	Infrared	Assignment (D_{2h})
	259 s	$\nu_4(A_{2u})$ out-of-plane CO bending
294 w		$\nu_6(B_{1g})$ in-plane CO bending
	350 m	$\nu_{14}(E_u)$ in-plane CO bending
647 s		$\nu_{10}(B_{2g})$ ring bending
(662) vw[b]		$\nu_{11}(E_g)$ out-of-plane CO bending
723 s (pol.)		$\nu_2(A_{1g})$ ring breathing
	1090 s	$\nu_{13}(E_u)$ CC stretching
1123 vs		$\nu_5(B_{1g})$ CC stretching
1329 vw		$\nu_2 + \nu_{11}$
	1530 vs	$\nu_{12}(E_u)$ CO stretching
1593 s		$\nu_9(B_{2g})$ CO stretching
	1700 vw	?
1794 w (pol.)		$\nu_1(A_{1g})$ CO stretching
	2200 w	$\nu_5 + \nu_{13}$

[a]Abbreviations: v-very, w-weak, m-moderate, s-strong.
[b]Only observed in the solid.

of its T_d symmetry, the chromate ion has four normal modes of vibration, whose spectral activities and observed frequencies are given in Table 21. The infrared and Raman spectra of solid potassium chromate, by contrast, are not as simple, as can be seen from Figs. 20 and 21, respectively. Single bands in the spectrum of the free chromate ion generally are replaced by multiplets in the infrared and Raman spectra of solid potassium chromate. Bands corresponding to ν_1 and ν_2 appear in the infrared spectrum of potassium chromate even

TABLE 21

Selection Rules and Observed Vibrational

Frequencies for the Chromate Ion

Mode	Activity	Frequency[a], cm^{-1}
ν_1	Raman	847
ν_2	Raman	348
ν_3	IR & Raman	884
ν_4	IR & Raman	368

[a]Raman data of Ref. 52.

though these modes are expected to be infrared inactive on the basis of the free-ion selection rules (cf. Table 21). Furthermore, a number of bands of very low frequency are evident in the Raman spectrum. Similar low frequency bands occur in the far infrared (usually below 250 cm^{-1}), although these are beyond the limits of the spectrophotometer used for Fig. 20. All these unique features of the solid state

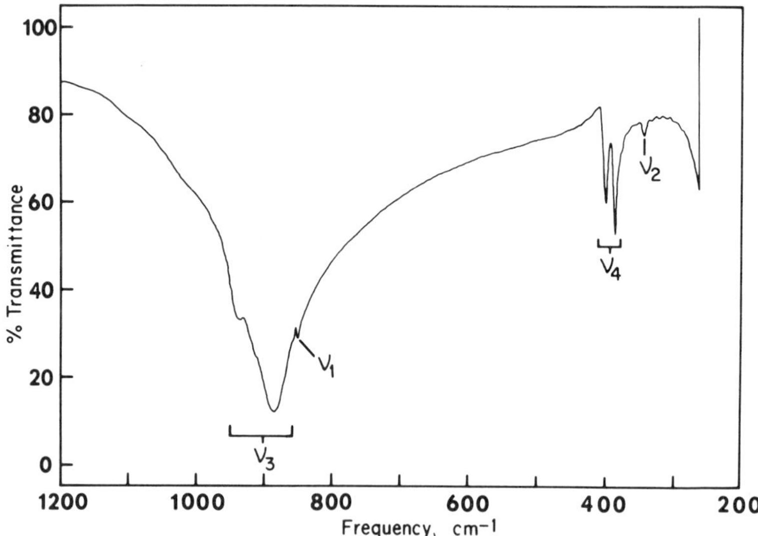

FIG. 20. Infrared spectrum of K_2CrO_4 (KBr-pellet sample).

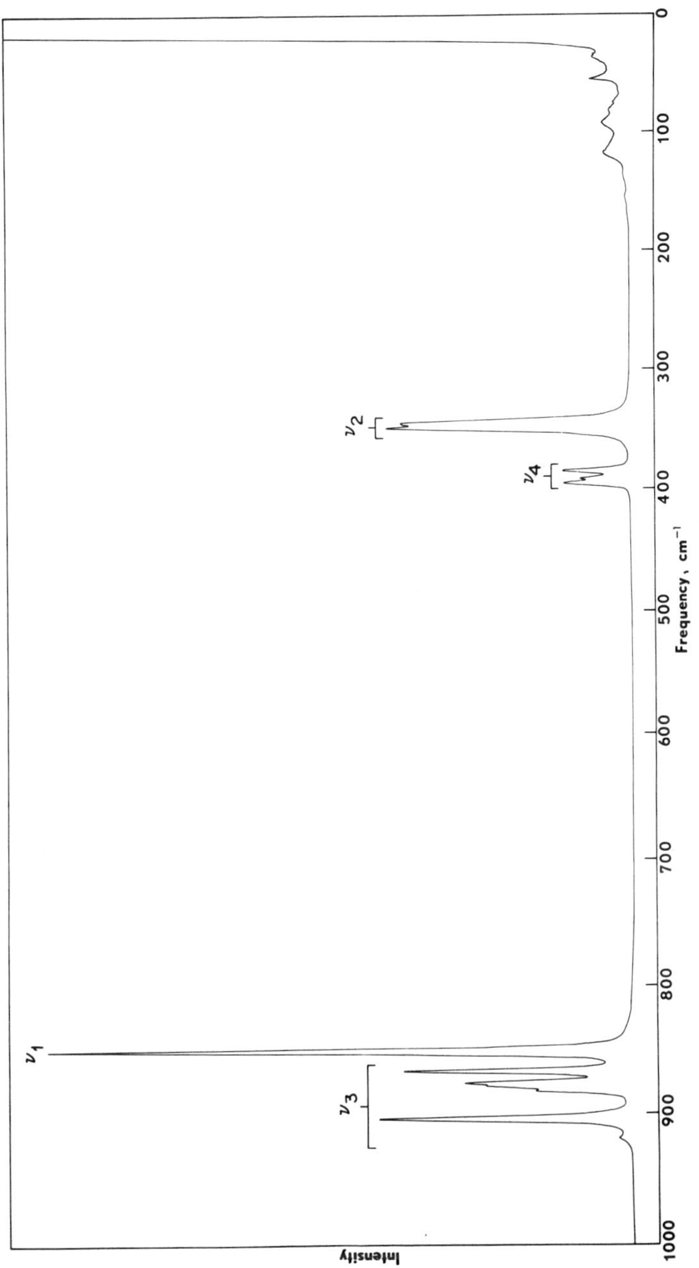

FIG. 21. Raman spectrum of solid K_2CrO_4 (unoriented fine crystals).

spectra result from the existence of a different symmetry environment within the crystal and the presence of significant intermolecular forces not present in the gas.

The frequencies observed in the vibrational spectra of solids are conveniently divided into two types, internal and external modes. Internal modes or molecular modes arise from the motions of the atoms relative to one another within a molecule or polyatomic ion. These modes usually have frequencies which are very close to those of the free molecule or ion. The low frequency vibrations, which are unique to solids, are called external modes or lattice modes. These modes arise from the motions of entire molecules or ions relative to one another. Lattice modes are subdivided into translatory modes and rotatory modes. Translatory modes arise from the translational motions of molecules or ions relative to one another, while rotatory modes arise from the librational motions of molecules of polyatomic ions (not single-atom ions) relative to one another.

Two approaches in group theory have been advanced to account for the features peculiar to the vibrational spectra of solids. These are site group analysis and factor group analysis. Although each of these approaches is based upon a unique set of assumptions, both are founded in the recognition that the symmetry governing the vibrations in solids must be the symmetry of the crystal. Consequently, it is necessary to have an appreciation for the types of symmetry present in crystals before preceeding with any discussion of these methods of deriving selection rules.

2. Symmetry in Crystals

Both site group analysis and factor group analysis depend upon prior knowledge of the symmetry present in the crystal of interest. This information is usually obtained from the analysis of the x-ray structure of the crystal. The complete symmetry of the crystal is described by one of the 230 crystallographic space groups. The space groups are composed of the usual point group symmetry elements plus the unique elements of translation, screw axis, and glide plane.*

---

*An explanation of these elements may be found in most books on x-ray analysis and crystallography. An interesting presentation using the graphic works of M. C. Escher has been provided by Glasser [53].

Every atom in the unit cell occupies a specific position or site.
Since one or more symmetry elements (including identity) pass through
each site, the symmetry of the site is described by a site group.
The site group is always one of the 32 crystallographic point groups
and must be a subgroup of the space group. Since any point in a unit
cell is related to one or more other points in the unit cell by at
least one element of symmetry (except in space group P1 - C_1^1)*, sites
of a particular site group symmetry will occur in sets. If n sites
of the same symmetry are related to one another by some element or
elements of symmetry in the cell, the set of sites is said to be n-
fold, where n = 2, 3, 4, 6, 8, 12, 16, 24, 32, 48, 64, 96. There
may be more than one n-fold set of sites of the same symmetry in the
same unit cell. These sets of the same symmetry may differ both in
the coordinates of their positions and the particular symmetry elements which relate the members of each set to one another. The 230
space groups and their sites have been tabulated [54-56].

It is not necessary to know the site symmetry of each and every
atom for the purposes of deriving vibrational selectional rules.
Since the atoms may be grouped logically into molecules and ions, it
is necessary only to know the symmetry of the sites occupied by the
centers of gravity (or affixes) of these chemical species in order
to describe the symmetry about them. This information is usually,
but not always, obtained from the x-ray structure study. In those
cases where the site symmetry of the molecular affixes is not available, it sometimes can be deduced. This is possible because the site
group must be a subgroup of both the symmetry group of the free molecule and the space group of the crystal. Furthermore, the molecular
affixes must occupy a complete set of sites of the proper site group
symmetry. These two conditions often, but not always, suffice to fix
the site symmetry of the molecules or ions in a crystal for which the
space group and number of formula units per unit cell are known.

---

*The symbol P1 is the Hermann-Mauguin notation and C_1^1 is the
Schoenflies notation for the most trivial space group.

3. *Site Group Analysis*

Halford [55] developed site group analysis in order to provide
a simple model for interpreting the features of vibrational spectra
of solids. The analysis may be carried out if the site symmetry of
the molecular or ionic affixes in the crystal are known or can be
deduced. Halford [55] has tabulated the available site symmetries
for the 230 space groups and Couture [56] has made corrections to the
list.

The primary assumption of the site group approach is that intermolecular coupling of vibrational motions is negligible in the
crystal. The internal (molecular) modes are found by applying the
usual point group procedure for free molecules, but now using the
site symmetry of the particular molecule of polyatomic ion in the
crystal. Since the site symmetry is usually lower than the free-
molecule symmetry, the effect of the site group analysis for internal modes is to split degeneracies of the normal modes of the free
molecule. However, if the free molecule has no higher than a twofold rotational axis of symmetry, no splitting of degenerate modes
is possible, because degenerate normal modes do not occur for molecules of such low symmetry. The group D_{2d} is an exception to this
statement.

The site symmetry approach also predicts the number of external
(lattice) modes. The lattice modes are considered to be the result
of the oscillations of the individual molecules or ions. Since no
appreciable interactions are assumed, molecules or polyatomic ions
give rise to three rotatory and three translatory oscillations, while
monatomic ions produce only three translatory oscillations. Degeneracies are possible. The number of frequencies and their spectral
activities are determined by considering the molecule or ion under
the site symmetry. The species of the three translatory modes arising from each type of molecule or ion are those of the transformations
of the three translational unit vectors under the site group symmetry,
indicated by x, y, and z in the next to the last column of the char-
actor table for the point group. Likewise, the species of the three
rotational modes arising from each type of molecule or polyatomic

ion are those of the transformations of the rotational unit vectors, indicated by R_x, R_y, and R_z in the next to the last column of the character table.

Infrared and Raman activities are determined in the usual manner by inspecting the vector transformation properties listed in the next to the last and last columns of the character table for the site group. A normal mode will be active in the infrared spectrum if it transforms as one or more of the translational unit vectors. It should be pointed out that the implication of this for site symmetry selection rules is that all translatory lattice modes will be infrared active, since they obviously have the same species as the translational unit vectors. A normal mode will be active in the Raman spectrum if it transforms as one or more of the direct products of the translational unit vectors, x^2, xy, yz, etc.

The site symmetry approach is sometimes adequate to interpret the spectra observed for molecular crystals, but it is usually incapable of describing the spectra of ionic crystals. The requirement of site group analysis that intermolecular interactions be weak usually breaks down in ionic crystals. The result is that many more lines may appear in the spectrum than predicted. The failure of this approach is particularly pronounced in crystals containing molecules or ions having no higher than a twofold rotational axis of symmetry, since no new internal-mode frequencies are expected.

4. *Factor Group Analysis*

The logical alternative to the site group analysis assumption of weak intermolecular interaction is the assumption of complete vibrational coupling. This is the basic assumption of factor group analysis, sometimes called the unit-cell approach. The term "unit-cell approach" arises from the further assumption that all meaningful frequencies of the crystal may be determined by considering only the atoms contained in a single primitive unit cell.*

---

*Bertie and co-workers [57] have pointed out that the term unit-cell approach is more appropriate. However, the term factor group analysis, although inaccurate, has precedent in the literature, and will be used here.

The justification for this is not difficult to understand, at least on a qualitative level. In the bulk crystal, the oscillations of equivalent atoms will add destructively, except in those cases where they move identically in phase. Two atoms are deemed equivalent, in terms of crystal symmetry, if one can be obtained from the other by the operation of translation. If the unit cell is defined as the smallest collection of atoms which are not related to one another by translation, all the oscillations of the unit cell represent the smallest set of vibrations which are not subject to destructive addition due to phase problems. The oscillations of the bulk crystal, then, are the oscillations of a unit cell carried through the crystal, in the same phase, by the operations of translation. Consequently, considering the normal modes of vibration of the unit cell suffices to enumerate the genuine modes of the crystal as a whole.

It is important to note that the unit cell must be primitive, which is not always the case with the customary crystallographic unit cell. In the case of a nonprimitive crystallographic cell, the number of primitive cells comprising it must be recognized and the contents of the crystallographic cell must be divided by the appropriate number (2 or 4) to give the number of formula units per primitive unit cell. Here the Hermann-Mauguin space group notation is helpful. Singly primitive unit cells are indicated by a notation beginning with P. Doubly primitive cells are indicated by A, B, C, or I, depending upon the Bravais lattice of the crystal. Face centered cells, indicated by F in the Hermann-Mauguin notation of their space groups, are quadruply primitive.

The assumption that complete vibrational coupling occurs in the unit cell implies an inherently different definition of the nature of vibrations in crystals than that of the site group approach. In terms of factor group analysis, the observed frequencies in the vibrational spectra of solids are presumed to result from the motions of all atoms in the unit cell (and hence the whole crystal) acting in concert, and not from the isolated motions of individual molecules. In a sense, the atoms of the unit cell may be thought to comprise a large molecule. Hence, all that is needed in group theory is a means of extending the usual method for determining selection rules to include the pecularities of crystal symmetry. The factor group approach represents such an extension.

The atomic positions in the unit cell are taken as the basis set for the formation of reducible representations of a group composed of the symmetry elements of the cell. The group which describes the symmetry of a single unit cell is a finite factor group of the crystal's space group. In terms of the mathematics of group theory, the factor group is formed from the cosets of the invariant subgroup comprised of the translational elements of the space group.* The effect is that all the translational elements of the space group (which are infinite in number) form the identity element of the factor group. However, in actual practice one never needs to derive the factor group from the space group, and knowledge of the mathematics involved is not required. The reason for this is that the factor group is always isomorphous with one of the 32 crystallographic point groups, for which the character tables are well known. Furthermore, the isomorphous point group may be identified simply by deleting the superscript from the Schoenflies notation of the space group. For example, if the space group is C_{2h}^{5}, then the factor group is isomorphous with the point group C_{2h}, and the C_{2h} character table would be used as the character table for the factor group. Although the character tables for the factor groups and point groups are identical, the operations of rotation and reflection in the point group may correspond to screw and glide, respectively, in the factor group. The exact nature of the factor group operations may be determined, when needed, by consulting "International Tables for X-ray Crystallography" [54] for the symmetry elements of the corresponding space group. For example, for a crystal belonging to space group D_{3d}^{6}, the point group character table D_{3d} would be used as the factor group character table, with the understanding that the reflection planes of the point group correspond to glide planes in the factor group.

----

*A concise discussion of invariant subgroups, cosets, and factor groups may be found in Ref. 58.

The factor group analysis may be accomplished by two different
procedures, both of which give the same results. As originally set
down by Bhagavantam and Venkatarayudu [59], a factor group analysis
is operationally quite similar to the point group analysis for free
molecules. Since this method has been adequately described elsewhere [59,60], only a brief summary will be given here. The method
of Bhagavantam and Venkatarayudu has largely been replaced by a correlation chart method, which has its theoretical justification in
the work of Hornig [61], and Winston and Halford [62]. A more complete account of the correlation chart approach will be given.

In the methodology of Bhagavantam and Venkatarayudu [59] the
atoms of the unit cell are considered successively in three groups.
The unit cell is considered to consist of p atoms, which form s molecules or ions, of which v are monoatomic. The three sets, which are
taken individually as bases for representations are: (1) the set of
p atoms which forms a representation for the totality of modes, numbering $\sum n_i$; (2) the set of s molecules or ions which forms a representation for the translational lattice modes, numbering $\sum T'$; and
(3) the set of (s - v) molecules or ions which forms a representation
for the rotational lattice modes, numbering $\sum R'$. The summations are
over all symmetry species of the factor group. The number of pure
translations $\sum T$ is three or less, depending on degeneracy. The symmetry species to which these translations belong may be determined
by inspection of the transformation properties of the translational
unit vectors in the character table. The characters of the representations, which reduce to give n_i, T', and R' for each symmetry species, are generated in each case by determining the number of invariant particles which remain after each symmetry operation. In the
cases of the representations for $\sum T'$ and $\sum R'$, each molecule or ion
as a whole constitutes a particle. A particle is deemed invariant
if it is transformed into its equivalent in the same or adjacent
unit cell. The number of particles unshifted after each operation
is multiplied by the matrix contribution per unshifted particle for
the particular operation. Formulas for these matrix contributions

are given by Bhagavantam and Venkatarayudu [59]. After the three
representations have been reduced in the usual manner, the number of
internal modes n'_i belonging to each symmetry species may be determined by $n_i = n'_i - (T + T' + R')$. The Raman and infrared activities
are determined, as usual, by inspection of the appropriate transformation properties listed in the character table. This method has
been extended by Mitra [63] to apply to crystals containing linear
groups.

This method of applying factor group analysis has several serious disadvantages. It is virtually impossible to proceed with the
analysis unless a map of the unit cell is available or can be con-
structed. Frequently the published x-ray data are not this detailed,
particularly if the data are obtained from powder patterns. Furthermore, even when a map of the unit cell is available, it is extremely
difficult to visualize the effects of the three-dimensional space
group operations on a two-dimensional drawing. Use of a three-di-
mentional model may help, but carrying out the procedure in even a
simple case is very tedious.

Fortunately, the correlation chart method of factor group analysis does not suffer from these disadvantages. The only requirements
made of the x-ray data are knowledge of the space group and site
group symmetries, and the number of molecules per unit cell. A map
of the unit cell is completely unnecessary, since no manipulation of
symmetry operations is required. Tables of the correlations between
the irreducible representations of the group and their subgroups,
such as given in Appendix B, are convenient but not always necessary,
since frequently the correlations can be easily seen by inspecting
the character tables.

The number of normal modes and their symmetries are determined
first for any polyatomic molecules or ions, using the appropriate
point group for the free species. Employing the notation used previously, the results for the free molecule or ion are multiplied by
s - v, the number of molecules or polyatomic ions per unit cell, to
give (s - v)(3n - 6) internal modes, where n is the number of atoms in

the molecule or ion.* These structured groups also give rise to
3(s - v) rotational external modes and 3(s - v) translational external
modes (including pure translations). The symmetry species of these
external modes are the same as those of the transformations of the
translational and rotational unit vectors listed in the character
table of the point group. The symmetry species for the internal and
external modes under the point group are then correlated to the symmetry species of the site group of the polyatomic molecules or ions.
The symmetry species of the site group are subsequently correlated
to the symmetry species of the factor group, which are the same species as those of the factor group's isomorphorous point group. The
correlation of a species under one symmetry to a species under another symmetry may be determined by using the correlation tables.
The normal modes, both internal and external, will either split or
become degenerate during the correlation process according to the
changes in the symmetry species (i.e., irreducible representations)
to which they belong under the various groups.

The monoatomic atoms or ions in the crystal, which number v,
will give rise to 3v translational modes, including pure translations.
Correlation in this case is made only from the site group of the monoatomic ions to the factor group. The symmetry species of these trans-
laboty modes under the site symmetry are the same as those of the
translational unit vector transformations in the character table of
the site group. Correlation between the symmetry species of the site
group and the symmetry species of the factor group is made in the
usual manner.

After the correlation between the various symmetries has been
made, the resultant external modes under the factor group symmetry
will include three degrees of translational freedom which are not
spectroscopically active vibrations. These three degrees of freedom
represent translations of the entire unit cell in the three cardinal
directions of the Cartesian coordinate system. These translations,
which are not active in the infrared or Raman spectra, make up the

----

*The number will be (s - v)(3n - 5) in the case of linear mole-
cules or ions.

acoustic modes. Their symmetry species are the same as the transformations of the translational unit vectors in the factor group
character table. Once the acoustic modes have been identified, they
are subtracted from the translational modes of the same species,
leaving only genuine translatory lattice modes.

The spectral activites of the internal and external modes under the factor group symmetry are determined by inspecting the transformation properties of the translational unit vectors and direct
products in the character table of the point group isomorphous with
the factor group. Infrared-active modes have the same species as
the transformations of the translational unit vectors, and Raman-active modes have the same species as the transformations of the direct
products. In contrast with the site symmetry approach (vide ante),
all translational modes are not infrared active, ipso facto.

The following example will illustrate the procedure. Potassium
chromate, K_2CrO_4, has space group symmetry D_{2h}^{16}-Pnma with four formula
units per unit cell [64]. The point group which is isomorphous with
the factor group of D_{2h}^{16} is D_{2h}. The x-ray data reveal that the potassium ions occupy two sets of fourfold sites of C_s symmetry, and
the chromium centers of the chromate ions occupy another fourfold
set of sites of C_s symmetry. The symmetry of the free chromate ions
is T_d. Consequently, the task of determining factor group selection
rules by the correlation chart approach involves correlating the T_d
free-ion symmetry to the C_s site symmetry and, subsequently, to the
factor group symmetry isomorphous with D_{2h}.

We begin with the artifice of considering the internal modes
of the crystal under the point group of the free ion. A free chromate ion of T_d symmetry has nine normal modes of vibration constituting four frequencies of three different symmetry species, namely,
$\nu_1(A_1)$, $\nu_2(E)$, $\nu_3(T_2)$, $\nu_4(T_2)$. Since there are four chromate ions
per unit cell, 36 internal modes constituting 16 frequencies of the
same three symmetry species as the free ion will result. The symmetry species of the external modes arising from the chromate ions
(12 translatory and 12 rotatory) are found by inspecting the character table for T_d (Table 4). It can be seen from the next to the last

column of the character table that the 12 translatory modes belong to the species T_2. Likewise, the 12 rotatory modes belong to species T_1. It should be recalled that T_1 and T_2 are triply degenerate, and any frequency of these types actually represents three degenerate modes. Consequently, the 12 translatory modes and 12 rotatory modes each give rise to four frequencies under the T_d symmetry. The correlation of the free chromate ion symmetry species under T_d to the species under site group symmetry C_s and factor group symmetry D_{2h} is shown in Table 22. The correlations have been determined by consulting tables, such as those in Appendix B. The rotatory and translatory external frequencies, regardless of degeneracies, are indicated by R and T, respectively. The notation $2\nu_3$, etc., indicates two internal modes arising from the same free-ion mode, ν_3, and not the first overtone of ν_3. Note in Table 22 that the number of frequencies of one species under a particular symmetry may be divided between two or more species under another symmetry, and degeneracies may be eliminated in the correlation process. The split-

TABLE 22

Correlation Diagram for the Chromate Modes

of Potassium Chromate

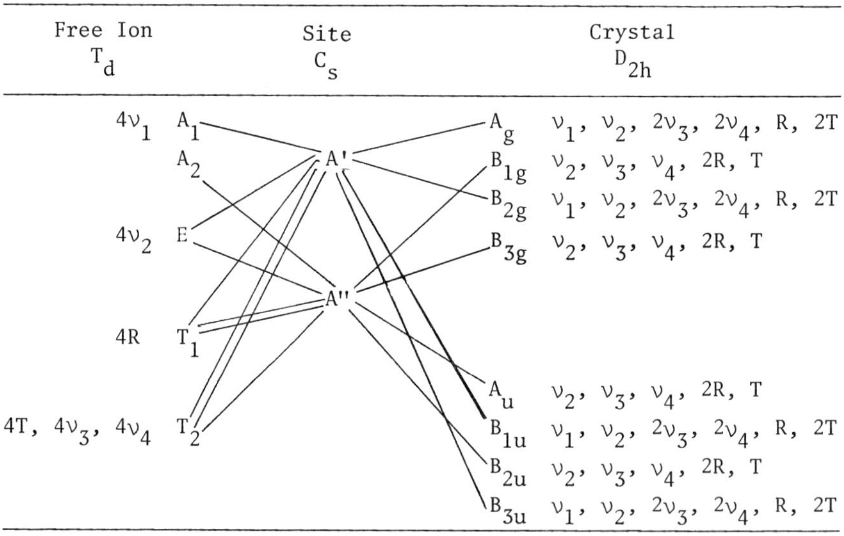

Free Ion T_d	Site C_s	Crystal D_{2h}	
$4\nu_1$	A_1	A_g	ν_1, ν_2, $2\nu_3$, $2\nu_4$, R, 2T
	A_2	B_{1g}	ν_2, ν_3, ν_4, 2R, T
		B_{2g}	ν_1, ν_2, $2\nu_3$, $2\nu_4$, R, 2T
$4\nu_2$	E	B_{3g}	ν_2, ν_3, ν_4, 2R, T
$4R$	T_1		
$4T$, $4\nu_3$, $4\nu_4$	T_2	A_u	ν_2, ν_3, ν_4, 2R, T
		B_{1u}	ν_1, ν_2, $2\nu_3$, $2\nu_4$, R, 2T
		B_{2u}	ν_2, ν_3, ν_4, 2R, T
		B_{3u}	ν_1, ν_2, $2\nu_3$, $2\nu_4$, R, 2T

ting of frequencies follows the splitting of the species. It is important to realize that the number of modes is unaffected by the correlation; only the number of frequencies and their degeneracies are affected.

The correlation diagram shown in Table 22 is incomplete, since the contributions of the potassium ions have been ignored up to this point. The eight unstructured potassium ions will contribute 24 translatory modes. Customarily, the species of the translatory modes arising from monoatomic ions are found by consulting the character table for their site symmetry in the crystal, and then correlating those species with the factor group species. However, when both cations and anions have the same site symmetry, it is unnecessary to construct a separate column for the site symmetry of the monoatomic ions. In such cases the monoatomic ions may be treated as if they had the same free-ion symmetry as the structured ions, thereby avoiding needless complication of the correlation diagram. Consequently, the 24 external modes arising from the potassium ions may be considered as giving rise to eight frequencies of species T_2 under T_d symmetry. The correlations for these translatory modes follow the paths already established for the chromate ions.

The complete correlation diagram for potassium chromate is shown in Table 23. The notation (-T) indicates that one pure translation (acoustic mode) has been subtracted from the number of translatory modes of the species B_{1u}, B_{2u}, and B_{3u}. That the three pure translations of the lattice belong to these species can be seen by inspecting the translational unit vector transformation properties in the character table of D_{2h} (cf. Appendix A). The infrared and Raman activities listed in Table 23 are determined by consulting the transformational properties of the translational unit vectors and their direct products, respectively, as listed in the character table of D_{2h}.

The selection rules for potassium chromate illustrate some features commonly encountered in factor group analyses. It should be noted from Table 23 that all Raman active modes are infrared inactive, and vice versa. This is the result of the rule of mutual

TABLE 23

Correlation Diagram and Selection Rules

for Potassium Chromate

Free Ion T_d		Site C_s	Crystal D_{2h}		Activity
$4\nu_1$	A_1		A_g	ν_1, ν_2, $2\nu_3$, $2\nu_4$, R, 6T	Raman
	A_2		B_{1g}	ν_2, ν_3, ν_4, 2R, 3T	Raman
		A'	B_{2g}	ν_1, ν_2, $2\nu_3$, $2\nu_4$, R, 6T	Raman
$4\nu_2$	E		B_{3g}	ν_2, ν_3, ν_4, 2R, 3T	Raman
$4R$	T_1	A''	A_u	ν_2, ν_3, ν_4, 2R, 3T	
			B_{1u}	ν_1, ν_2, $2\nu_3$, $2\nu_4$, R, 5T, (-T)	IR
$12T$, $4\nu_3$, $4\nu_4$	T_2		B_{2u}	ν_2, ν_3, ν_4, 2R, 2T, (-T)	IR
			B_{3u}	ν_1, ν_2, $2\nu_3$, $2\nu_4$, R, 5T, (-T)	IR

exclusion in crystals: if the crystal has inversion symmetry, frequencies active in the Raman spectrum will not be active in the infrared spectrum, and vice versa. Note that in the case of potassium
chromate mutual exclusion is expected even though the chromate ions
themselves do not possess inversion symmetry. Since mutual exclusion
is used frequently as a test for inversion symmetry in molecules, it
is important to realize that mutual exclusion does not necessarily
indicate the presence of centrosymmetric molecules when the spectra
are obtained from solid samples. In these cases mutual exclusion
only indicates a centrosymmetric crystal structure, and indicates
nothing with regard to inversion symmetry in the molecules of which
the crystal is composed.

The use of the free-ion vibrational numbering system in refer-
ring to the internal modes of the crystal deserves comment. The use
of this system is a matter of convenience. It is not meant to imply

that the observed frequencies in the crystal arise from the isolated motions of individual molecules or polyatomic ions. To the contrary, the observed spectrum results from all the atoms in the unit cell executing unique crystal vibrations. The internal modes, however, are based upon the normal modes of the free molecules or ions and will appear in the spectrum very near to the frequencies of the free species. Internal modes of different symmetry species which are built upon the same free-molecule modes often differ chiefly in the phase relations between the individual motions of the molecules or ions. These phase differences may result in significant differences in energies among the several modes. The selection rules for potassium chromate indicate several instances where internal modes arising from the same free-ion mode also belong to the same symmetry species. In such cases the distinction between the modes is difficult to visualize, but the difference usually involves the magnitude of the motions in different crystallographic directions. The frequency difference between two such modes is not necessarily insignificant and depends upon the physical conditions in the crystal.

It may be useful to compare the actual spectroscopic results for K_2CrO_4 with the factor group selection rules of Table 23. The Raman and infrared frequencies for K_2CrO_4 and their assignments are given in Table 24. The Raman data were obtained from oriented single crystals of K_2CrO_4 [64], which enabled determination of the symmetry species of the vibrational modes. The theory of oriented crystal work is discussed in the next section. The infrared data were obtained from KBr-pellet samples in the region of KBr transmittance (above 250 cm^{-1}). All Raman-active internal modes predicted by the factor group analysis can be identified. However, it is important to note that in several cases two modes of different symmetry have the same frequency, i.e., accidental degeneracies occur. Similarly, the number of infrared frequencies shown in Table 24 is fewer than the number of internal modes allowed by the selection rules. This is the result of accidental degeneracies and near degeneracies combined with the lower resolution obtainable from pellet samples in the infrared. The results of Table 24 are fairly typical. The vibra-

TABLE 24

Vibrational Frequencies of Crystalline K_2CrO_4

Raman Frequencies[a] (cm^{-1})	Infrared Frequencies[b] (cm^{-1})	Assignment	Number of Modes Expected
918 (B_{2g})	936	ν_3	6 Raman, 5 IR
903 (A_g)	910		
881 (B_{2g})	883		
878 (B_{3g})			
876 (B_{1g})			
867 (A_g)	859		
851 (A_g, B_{2g})	850	ν_1	2 Raman, 2 IR
396 (A_g, B_{2g})	398	ν_4	6 Raman, 5 IR
392 (B_{1g})			
387 (B_{3g})			
386 (A_g, B_{2g})	382		
350 (B_{1g}, B_{2g})		ν_2	4 Raman, 3 IR
346 (B_{3g})	342		
345 (A_g)			
157 (A_g)	[c]	External modes	24 Raman, 16 IR
138 (B_{3g})			
119 (B_{1g}, B_{3g})			
116 (A_g)			
114 (B_{2g})			
109 (A_g)			
99 (B_{2g})			

TABLE 24 (continued)

Raman Frequencies[a] (cm^{-1})	Infrared Frequencies[b] (cm^{-1})	Assignment	Number of Modes Expected
93 (A_g)			
91 (B_{3g})			
85 (B_{3g})			
83 (A_g)			
67 (B_{1g})			
54 (B_{1g}, B_{3g})			
37 (A_g)			

[a]Data of Ref. 64.

[b]Unpublished results, this laboratory. KBr-pellet sample run on a Perkin-Elmer Model 225 spectrophotometer.

[c]No data available.

tional spectra of most crystalline compounds show fewer frequencies than the number of allowed modes due to accidental degeneracies.

Only Raman frequencies for the external modes are shown in Table 24, due to the experimental difficulties of obtaining high resolution infrared spectra below 250 cm^{-1}. Assignment as rotatory and translatory modes usually is not possible from the results of the infrared and Raman spectra, and it is customary to assign these frequencies simply as "external modes" or "lattice modes." Table 24 shows 14 frequencies which can be identified with 16 external modes. Evidently eight modes are "missing." Accidental degeneracies and weak intensities may account for these eight modes. Accidental degeneracies are particularly troublesome in the region of external modes, because in most compounds many modes of this type are allowed and occur within a relatively limited range of frequencies. With regard to weak intensities, it is good to keep in mind that a spectroscopically active mode does not necessarily give rise to a strong band

in the spectrum. Consequently, factor group selection rules must
be taken as giving what is possible, and not what is experimentally
detectable. This, of course, is true for spectroscopic selection
rules in general.

5. *Raman Spectra of Oriented Single Crystals*

The well-known general selection rule of Raman spectroscopy is
that a normal mode of vibration will be active in the Raman effect
if it involves a change in the polarizability of the molecule, defined by the relationship

$$P = \alpha E \tag{9}$$

where P is the induced moment of the scattered radiation, E is the
electric vector of the incident radiation, and α is the polarizability. Both P and E are vectors, and consequently α is, in general, a
tensor. The components of P are given by the following:

$$
\begin{aligned}
P_x &= \alpha_{xx}E_x + \alpha_{xy}E_y + \alpha_{xz}E_z \\
P_y &= \alpha_{yx}E_x + \alpha_{yy}E_y + \alpha_{yz}E_z \\
P_z &= \alpha_{zx}E_x + \alpha_{zy}E_y + \alpha_{zz}E_z
\end{aligned}
\tag{10}
$$

Only six of the nine components of the polarizability tensor are
unique, since $\alpha_{ij} = \alpha_{ji}$. The tensor may be visualized in terms of
these six components by means of a polarizability ellipsoid of the
form

$$\alpha_{xx}x^2 + \alpha_{yy}y^2 + \alpha_{zz}z^2 + 2\alpha_{xy}xy + 2\alpha_{yz}yz + 2\alpha_{zx}zx = 1 \tag{11}$$

Equation (11) represents a triaxial ellipsoid in the most general case of anisotropic media. In general, P and E will have different directions, except where their components coincide along the
axes of the ellipsoid, as can be seen from Eq. (10). Examination of
Eq. (10) reveals that any orientation of E_j (where E_j is parallel to
any one of the axes of the ellipsoid) will induce a moment P, each
of whose components along the axes of the ellipsoid (P_x, P_y, and P_z)
will be associated with a particular component of the polarizability
tensor. This fact suggests a powerful tool for interpreting Raman
spectra of anisotropic substances, if the polarizability tensor can
be oriented.

Special conditions obtain for the special case of isotropic media. In this case, $\alpha_{xx} = \alpha_{yy} = \alpha_{zz} \neq 0$ and $\alpha_{ij} = 0$. Therefore Eq. (11) reduces to the formula for a sphere, and Eq. (10) becomes

$$P_x = \alpha_{xx} E_x$$
$$P_y = \alpha_{yy} E_y \qquad (12)$$
$$P_z = \alpha_{zz} E_z$$

Equation (12) shows that for isotropic media the directions of P and E are always parallel and α is a scalar quantity. The spherical symmetry of the polarizability figure implies that there are no unique orientations.

Factor group selection rules for vibrations in solids assert that the normal modes of a crystal result from the motions of all the atoms of the crystal moving in concert. Consequently, the polarizability tensor is a bulk property of the crystal. Furthermore, since the positions of all atoms are fixed with respect to the crystallographic axes, the polarizability ellipsoid has a specific orientation in the crystal. The three principle axes of the polarizability figure coincide with the symmetry axes of the crystal, which in turn coincide with the axes of the optical indicatrix of optical crystallography [65]. The orientations of the ellipsoid axes with respect to the crystallographic axes are subject to restrictions which vary with crystal class, and the exact orientation of the axes and their identity can be determined by the procedures of optical crystallography for any crystal of interest. Table 25 lists some properties of the polarizability ellipsoid in crystals of various classes.

The ability to orient the axes of the polarizability tensor with respect to the directions of propagation and polarity of the incident radiation in the Raman experiment permits identification of the polarizability tensor components associated with each normal mode of vibration. Since the form of these components involves one of the quadratic functions of the Cartesian coordinates (that is, x^2, y^2, z^2, xy, yz, zx), all will belong to the same symmetry representations under the factor group symmetry of the crystal as do the correspond-

TABLE 25

Properties of the Polarizability Figure in Each Crystal Class

Crystal Class	Polarizability Figure	Axis Orientations of the Polarizability Figure
Cubic	Sphere	Not orientable
Tetragonal	Biaxial ellipsoid	Optic axis parallel to c; c = z
Trigonal	Biaxial ellipsoid	Optic axis parallel to c; c = z
Hexagonal	Biaxial ellipsoid	Optic axis parallel to c; c = z
Triclinic	Triaxial ellipsoid	Orientation unrestricted
Monoclinic	Triaxial ellipsoid	One axis parallel to b
Orthorhombic	Triaxial ellipsoid	All axes parallel to crystallographic axes

ing direct product transformations listed in the character table for
the factor group of the crystal of interest. Consequently, identifying the observed normal modes with their tensor components makes
known the symmetry species of the modes. This can be used for assigning the observed Raman spectrum.

Useful information regarding polarizability tensor components
can be obtained from the Raman spectrum of an oriented crystal only
when the propagation direction of the incident light, the electric
vector of the incident radiation, the faces of the crystal, and the
direction of observation of the scattered radiation are parallel or
perpendicular to the axes of the polarizability figure. Any deviations from these relations will cause birefringence and refraction
within the crystal, resulting in excitations of all components of
the tensor and loss of integrity in the polarity of the components
of the scattered radiation. Within these restrictions two configurations of incident to scattered radiation are allowed: (1) perpendicular to one another or (2) parallel to one another. Only the first
of these will be discussed here, since it is the most common experimental arrangement for Raman spectroscopy. The case of parallel incident and scattered radiation is treated by the same principles and
has been discussed elsewhere [65,66].

Consider a crystal cut and oriented as in Fig. 22, where we will
assume the general case of a triaxial polarizability figure. This is
only one of the useful orientations. As indicated in Fig. 22, the
polarity of the incident radiation may be E_x or E_z. With either vector orientation, the scattered radiation at right angles in the xy-
plane will have intensity components corresponding to P_y and P_z. The
components of the scattered radiation may be observed separately by
using an analyzer of Polaroid film. The notation of Damen et al.
[67] is useful to indicate the various orientations of crystal axes
and light vectors. For example, in the present case (Fig. 22) the
notation y(xz)x means that the spectrum is produced with incident
radiation propagated in the y direction and polarized in the x direction, and that the scattered light is observed from the x direction with the polaroid analyzer placed so as to pass only the z

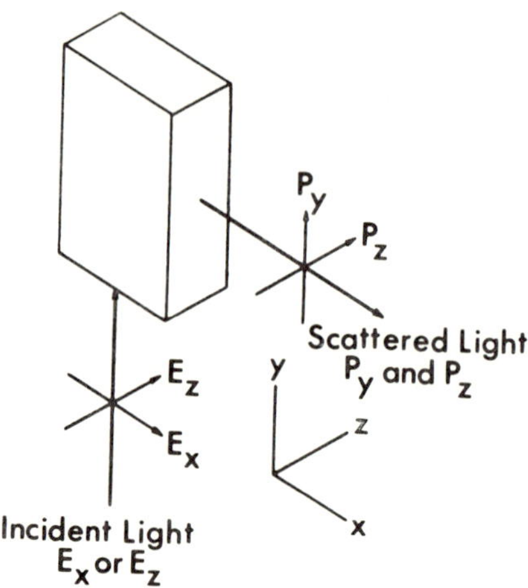

FIG. 22.   A single crystal oriented for Raman spectroscopy.

component.   If the analyzer were changed to another position at 90°
to the first, the notation would be y(xy)x.   Symbols outside the
parentheses refer to propagation directions; symbols inside the parentheses refer to light polarization directions.

The orientation shown in Fig. 22 requires that $P_x = 0$, since
there can be no electric vector component in the direction of propagation of the light.   Furthermore, since this orientation forbids
excitation by E_y, it follows that $\alpha_{yy}E_y = 0$ and $\alpha_{zy}E_y = 0$.   Consequently, Eq. (10) reduces to

$$P_y = \alpha_{xy}E_x + \alpha_{yz}E_z \neq 0$$
$$P_z = \alpha_{zx}E_x + \alpha_{zz}E_z \neq 0 \tag{13}$$

Now if the incident radiation is polarized in the x direction, E_x,
then Eq. (13) reduces to

$$P = P_y = \alpha_{yx}E_x \quad \text{when } y(xy)x \tag{14}$$

and

$$P = P_z = \alpha_{zx}E_x \quad \text{when } y(xz)x \tag{15}$$

Likewise, if the incident radiation is polarized in the z direction,
E_z, then

$$P = P_y = \alpha_{yz}E_z \quad \text{when } y(zy)x \tag{16}$$

and

$$P = P_z = \alpha_{zz}E_z \quad \text{when } y(zz)x \tag{17}$$

Equations (14) through (17) show that it is possible with the arrangement of Fig. 22 to observe separately the Raman spectra produced by each one of four components of the polarizability tensor. An additional feature of the Damen-Porto-Tell notation now should be evident: the symbols in the parentheses indicate which tensor component is being measured.

The crystal shown in Fig. 22 could be oriented at right angles to the incident radiation in two other ways which are useful for Raman observation. These are $y(mn)z = z(mn)y$ and $x(mn)z = z(mn)x$, where either $m = n$ or $m \neq n$. Considerations of the type made for $y(mn)x$ (vide supra) indicate that any orientation $i(mn)j$ of a crystal with a triaxial polarizability ellipsoid permits isolation of the spectra due to each of three α_{mn} components and one α_{mm} component ($i \neq m \neq j$). The specific identity of the four components will depend upon the orientation. If each and every one of the six unique components of the polarizability tensor are to be measured on such a crystal, three separate orientations of the crystal are required.

Selective observation of only those normal modes involving a change in a particular tensor component enables complete assignment of the spectrum to be made. Suppose the crystal of Fig. 22 is orthorhombic and has factor group symmetry D_{2h}, such as K_2CrO_4. The direct product transformation properties listed in the character table (cf. Appendix A) indicate the symmetry species of the corresponding tensor components. Each of the four measurements made with the orientation of Fig. 22 yields only the spectrum of modes of one symmetry species. The results for the present case are:

$y(xy)x$ gives B_{1g} modes

$y(xz)x$ gives B_{2g} modes

$y(zy)x$ gives B_{3g} modes

$y(zz)x$ gives A_g modes

One orientation of the crystal is sufficient to make the complete assignment of the spectrum in this case, even though α_{xx} and α_{yy} are not measured. Figure 23 shows the spectra of K_2CrO_4 in the region of ν_1 and ν_3 (stretching) vibrations for the crystal orientation $y(mn)z$. Measurements of this type [64] lead to the assignments given in Table 23. Similar classification of normal modes into four symmetry species is possible for any crystal having one of the other orthorhombic factor group symmetries (namely, D_2 and C_{2v}).

The polarizability figure for an orthorhombic or other optically biaxial crystal is a triaxial ellipsoid and $\alpha_{xx} \neq \alpha_{yy} \neq \alpha_{zz}$. Since changes in any α_{jj} component will be along the particular j axis, some information concerning the magnitudes of the changes along these axes for the A_g modes can be gained by comparing the resultant intensities from all α_{jj}. In the present case, $y(xx)z$ and $x(yy)z$ would add this information to that already obtained by $y(zz)x$.

The other optically biaxial crystal classes have fewer symmetry distinctions between normal modes. In triclinic crystals all Raman active modes belong to the same symmetry species, and consequently oriented crystal spectroscopy does not aid assignment of the spectrum. The Raman active normal modes of monoclinic crystals (factor groups C_s, C_2, and C_{2h}) are divided between two symmetry species. Symmetric modes involve a change in α_{xx}, α_{yy}, α_{zz}, and α_{xy}. Asymmetric modes involve a change in α_{yz} and α_{xz}. Thus, the symmetry distinction in monoclinic crystals can be determined by only two measurements in a single orientation, for example, $y(zy)x$ and $y(zz)x$. If it is desired, information concerning the relative magnitudes of the changes in the polarizability tensor components may be obtained for either triclinic or monoclinic crystals by using the three necessary orientations of the crystal, that is, $x(mn)y$, $y(mn)z$, $z(mn)x$.

Similar considerations apply to optically uniaxial crystals, e.g., hexagonal, trigonal, and tetragonal. However, the polarizability figure for uniaxial crystals is a biaxial ellipsoid. The x and y axes are equal and indistinguishable. Consequently, α_{xx}, α_{yy}, and α_{xy} cannot be separately isolated, and $z(xx)y$, $z(yy)x$, $z(xy)x$, and $z(yx)y$ are indistinguishable. Here these notations simply mean that

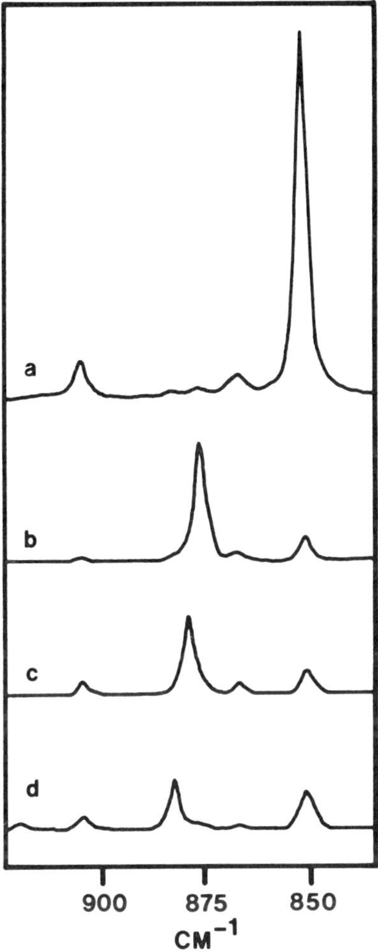

FIG. 23. CrO_4-stretching Raman bands of K_2CrO_4 for the crystal orientation y(mn)z: (a) y(xx)z(A_g);   (b) y(xy)z̄(B_{1g});   (c) y(zy)z(B_{3g}); (d) y(zx)z(B_{2g}).

the incident radiation is parallel to the optic axis (crystallographic c) and that the Raman scattering is observed at right angles.   It
is presumed that the crystal faces are cut normal to these directions.
Observation of z(jj)i on uniaxial crystals will always yield a spectrum displaying modes of two or three different symmetry species,
one of which will be the totally symmetric species (A, A_1, A_g, etc.).

The totally symmetric modes may be distinguished by a measurement of the type y(zz)x, since α_{zz} belongs to the totally symmetrical representation in every factor group. If z(jj)i excites three species of normal modes, as occurs in tetragonal crystals of certain factor group symmetries, only the totally symmetric modes can be distinguished from the rest. A further consequence of the symmetry of the uniaxial polarizability figure is that α_{xz} and α_{yz} are degenerate, and any measure of one is a measure of the other, also. These generalizations for uniaxial crystals may be verified by inspection of the direct product transformation properties listed in the character tables. (The factor group of a uniaxial crystal has either a three-, four-, or sixfold principal axis of symmetry, with the exceptions of the tetragonal factor groups D_{2d} and S_4.)

From an experimental standpoint, it should be kept in mind that both the incident and scattered radiation will be subject to the rules of refraction and birefringence. These two phenomena determine whether or not the electric vector polarities and wave normal directions retain the desired orientations inside the crystal. If it occurs that the propagation direction is parallel to one of the axes of the optical indicatrix, but the electric vector deviates from parallelism with one of the two orthogonal axes, then the electric vector will have components along both axes. In this case, more than the intended components of the polarizability tensor will be involved in producing Raman scattering. Hence, "forbidden" lines will appear in the spectrum. The intensity of these undesired lines will vary with the square root of the magnitude of the undesired electric vector component. This means that small deviations in the orientation of the electric vector from parallelism with a tensor axis may be tolerated, since the "forbidden" intensities will be weak.

If the wave normal of the incident beam upon entering the crystal is not parallel to one of the axes of the optical indicatrix, then, in general, it will be split into two rays polarized at right angles to each other. If the incident light is polarized in one of the privileged directions, which are given by the Biot-Fresnel rule

[68], then only one ray is produced, but its direction of propagation will not coincide with the wave normal direction. In either case, unwanted components of the polarizability tensor will be excited, and "forbidden" lines will appear. The intensity of these lines will depend upon the extent of the misalignment.

There are several experimental causes for deviations of the propagation direction and electric vector direction of the incident radiation inside the crystal. The most obvious of these is misalignment of the crystal. However, the quality of the crystal sample is another important source of problems. Not only must the axes of the polarizability tensor be aligned with respect to the directions of incidence and observation, but the crystal faces must be cut normal to these directions. If the faces deviate from normality to these directions, refraction will occur at the air-crystal interface, and the propagation directions and electric vector orientations inside the crystal will not be the ones desired. Irregularities in the crystal faces may be sufficient to excite "forbidden" components. Furthermore, real crystals are seldom perfect and may contain striations, dislocations, and a variety of other defects which would cause results to deviate from theoretical expectations. Convergence of the exciting radiation is another source of unexpected components.

The crystal must be of sufficient size to allow lapidary preparation. However, many substances do not form sufficiently large or perfect crystals, due to low solubility or tendency to grow with defects. Sometimes this problem can be surmounted if geological samples of the substance of interest are known. Otherwise, powdered crystalline samples, from which no polarizability data may be obtained, must be used for the Raman spectra.

Noncentrosymmetric crystals can present additional difficulties. Here some modes will be active in both the Raman and infrared, and perturbation of the polarizability tensor by the oscillating electric dipole can result. The theory of these interactions has been elegantly set down by Loudon [69] and by Mitra [70] in a wider ranging review. An example of such perturbations is provided by α-quartz

(D_3 crystal symmetry) where each of the eight E modes (infrared and Raman active) are split into a longitudinal (LO) and transverse(TO) component in the Raman spectra of appropriately oriented samples [71]. Furthermore, noncentrosymmetric crystals belonging to one of the enantiomeric space groups will be optically active, thereby rotating the polarity of the incident and scattered radiation. In such cases, single crystal Raman studies are usually impractical.

6. *Infrared Spectra of Oriented Single Crystals*

An infrared-active vibration in isotropic medium involves an oscillating dipole, μ, which may be resolved into three components,

$$\mu = \mu_x + \mu_y + \mu_z \tag{18}$$

where the subscripts x, y, and z refer to Cartesian coordinates. In anisotropic medium, μ occurs along a specific direction, resolvable into only one or two of the components of Eq. (18). In such a medium, plane polarized infrared radiation with its electric vector parallel to μ will result in maximum absorption at the frequency of the normal mode, while radiation polarized in a plane perpendicular to μ will result in no absorption. For nondegenerate normal modes, μ will be aligned along x, y, or z as defined by the symmetry of the crystal system. For doubly degenerate modes, μ will be perpendicular to z (the principal symmetry axis), lying in the xy plane. This suggests that polarized infrared radiation can be used to identify the orientation of μ and, consequently, the symmetry species of each active normal mode. Thus, for example, for a tetragonal crystal of D_{4h} symmetry (cf. Appendix A), radiation polarized parallel to z should allow only A_{2u} modes to absorb, while radiation polarized perpendicular to z should allow only E_u modes to absorb. Of course, it must be experimentally feasible to have radiation of the desired polarizations passing through the anisotropic medium without suffering refraction or change of character (elliptically polarized, circularly polarized, etc.).

Newman and Halford [72] have summarized the restrictions of crystal class on polarization of the incident radiation within the medium. In uniaxial crystals (hexagonal, trigonal, and tetragonal)

and biaxial crystals of the orthorhombic system, all distinguishable
polarization axes are fixed by symmetry along the crystallographic
axes. If the directions of incidence and polarization of the infrared radiation are along these axes, then in principle only those
modes associated with the polarization direction will absorb.

For monoclinic crystals, one axis is fixed by symmetry, and
the other two are unrestricted. There are no restrictions on x, y,
and z in triclinic crystals. In these cases it is necessary to
align the incident radiation with respect to the axes of the optical
indicatrix, which requires a crystal with faces cut normal to those
directions. The orientation of the optical indicatrix may be determined by standard methods of optical crystallography [73].

If the orientation of μ is not parallel to one of the crystallographic axes, it can sometimes be determined by measuring the dichroic ratio with respect to an axis in the plane of the crystal.
The dichroic ratio is defined as $I_\perp/I_\parallel$, where $I_\perp$ and $I_\parallel$ are the
intensities perpendicular and parallel to the reference axis, respec-
tively. These electric vector orientations, again, must be allowed
within the crystal without suffering refractive loss of polarization.
When the assignment of the spectrum is reliably known, measurement
of the dichroic ratio can sometimes be used to deduce the orientation
of the molecules within the crystal in the absence of x-ray data.
The analysis of the data frequently assumes the so-called "oriented-
gas model" [74], where the molecules are presumed to be oriented in
the lattice but with negligible interactions. This approach is most
fruitful with molecular crystals [74]. Analysis based on site group
or factor group selection rules tends to be more meaningful for more
ionic crystals, where the presumption of significant intermolecular
interactions is made.

Experimental difficulties encountered with polarized infrared
spectroscopy are similar to those encountered with single-crystal
Raman spectroscopy. The problems of refraction have already been
noted. The preparation of the sample is another problem. Growth of
a sufficiently large crystal specimen of known orientation, capable
of withstanding sectioning to a thickness allowing transmission in

"nonabsorbing" regions, is often an experimental difficulty that
cannot be overcome. Incomplete polarization by the AgCl or wire
polarizer and monochromator polarization of the transmitted beam can
result in the appearance of forbidden components in the spectra [75].
Convergence of the sample beam within the crystal is another source
of unexpected components. This problem becomes increasingly significant with micro sampling, where beam convergence is appreciable [76].

7. *Applications*

Aside from the proper assignment of solid-state vibrational
spectra, selection rules for crystals offer an additional tool for
studying structural aspects of solids. One such application is the
use of the rule of mutual exclusion to determine crystal centrosymmetry. Demonstrating the presence of inversion symmetry in crystals
has been a traditionally difficult problem of x-ray analysis, associated with the so-called "phase-problem" of the structure factors.
The use of solid-state infrared and Raman spectra offers a simple
test. For example, significant differences between infrared and
Raman frequencies has been used to confirm centrosymmetry in the
crystal structures of the dichromates of potassium, rubidium, and
cesium [10]. This test, however, is not always unambiguous. Frequency disparities are usually small between the infrared and Raman
spectra of noncentrosymmetric molecules in a centrosymmetric lattice,
even where appreciable intermolecular interaction may occur. Furthermore, centrosymmetric molecules in a noncentrosymmetric lattice may
exhibit mutual exclusion when intermolecular interactions are negligible. With these limitations in mind, however, differences signifi-
cantly greater than the experimental errors of the spectrometers can
indicate centrosymmetry, at least for ionic crystals.

Solid-state spectra also have been used to provide evidence of
random orientation of molecular or ionic species in crystals. In
the single-crystal infrared investigations of NH_4NO_3 (IV) and
NH_4NO_3 (III), Newman and Halford [72] found that absorptions by molecular modes of nitrate ion were strongly polarized in the directions
expected from the site group selection rules. However, absorptions
by molecular modes of ammonium ion were not polarized, suggesting
that they are disoriented in these crystals.

Factor group considerations have been used to suggest the structural nature of compounds with x-ray determined pseudosymmetry [77]. Pseudosymmetry arises when isoelectronic ions of different mass occurring in the same compound become indistinguishable in the x-ray determination. Since a false equivalence of nonequivalent nuclei results, the space group assignment reflects a higher symmetry than possible. Two examples are $KCrO_3F$ [78] and $KOsO_3N$ [79]. In a Raman study [77] of these compounds and the related compound $KCrO_3Cl$ (which does not have pseudosymmetry), the $KCrO_3Cl$ spectrum showed sharp bands and excellent agreement with the operative selection rules for the C_{2h} structure of this compound. The spectra of $KCrO_3F$ and $KOsO_3N$ showed broader bands and only fair agreement with the selection rules for a C_{2h} structure, which would be present if the MO_3X^- ions had specific orientation in these crystals. These results are expected if the X and O atoms occupy the available sites about each central M atom in a random fashion. The conclusion is that CrO_3F^- and OsO_3N^- ions do not have preferred orientations in $KCrO_3F$ and $KOsO_3N$, respectively. Thus any attempt to distinguish between the isoelectronic nuclei by x-ray diffraction should be unsuccessful, yielding an "averaged structure."

Beattie and co-workers [80] have used single crystal Raman measurements of $NbOCl_3$ to suggest that the space group $P4_2/mnm$ (D_{4h}^{14}), assumed in the x-ray studies [81], is incorrect. The solid-state structure of $NbOCl_3$ is an infinite chain polymer with two adjacent strands of *trans* Nb-O-Nb linkages cross linked by the two *cis* chlorine bridges connecting each Nb atom with its counterpart in the adjacent strand. The site group of the discreet polymer is D_{2h} if the space group is D_{4h}^{14}, and only one Nb-O-Nb stretching mode retaining inversion symmetry (i.e., Raman active) is permitted. This mode involves oxygens in one strand moving up the chain (z axis) while oxygens in the adjacent strand are moving down the chain. The zz component for this mode should be zero. However, a strong band at 770 cm^{-1} with a large zz component is reliably assigned to Nb-O-Nb stretching. Although this is inconsistent with the assumed D_{4h}^{14} structure, it is compatible with the expectations for one of the other space groups considered in the x-ray study. Thus, the Raman data suggest a space group $P4_2nm$ (C_{4v}^4) with C_{2v} site symmetry for the discrete chain. This is consistent with alternately long and short Nb-O links in the chain.

IV. SOME EXPERIMENTAL CONSIDERATIONS
AND TECHNIQUES

A. Characteristic Frequencies

A number of techniques for making spectroscopic assignments have been discussed in the preceeding section. These include depolarization ratios, band contours, relative intensities, and single crystal techniques. Except in the simplest cases, these aids are not sufficient by themselves to assign the complete vibrational spectrum. Some knowledge of approximately what frequency a particular mode is likely to have is usually required, and often additional means of verifying the assignment are needed.

The use of group frequencies for spectroscopic assignment in organic chemistry is a well-established tradition. Relatively few elements make up the bulk of organic compounds, and so the tabulation of group frequencies reaches a relatively high degree of specificity without becoming excessively complex. Indeed, most organic chemists who have even infrequent use of vibrational spectroscopy are familiar with the more commonly encountered group frequencies. By contrast, the elemental composition of inorganic compounds literally extends across the entire periodic table. Thus any attempt at complete tabulation is impractical. Nonetheless, tables of inorganic group frequencies for various arrays of inorganic species can be found widely in the literature [82-85], and these can be useful for "fingerprinting" applications.

The task of complete assignment of the vibration spectra generally requires more specific information than can be provided by group frequency tables, particularly when structural elucidation is the goal. In such cases it is generally more useful to compare the observed frequencies with the published data for compounds of similar

structural and compositional type. Compendia of complete vibrational
frequencies for many inorganic compounds are available. The older
literature is well covered by Herzberg's classic text [4], and more
recent data can be found in the highly recommended books by Nakamoto
[85] and Siebert [86]. Continuing reviews of the current literature
appear biannually in the journal *Analytical Chemistry,* and more exhaustive coverage is now provided annually by the Chemical Society's
Specialists Periodical Report series [87].

B. Importance and Problems of the Low Frequency Region

1. Far Infrared

Successful use of vibrational spectroscopy for deducing the
shapes of molecules depends upon the ability to observe all or nearly all of the structurally important frequencies. Routine investigation of organic compounds can often be accomplished by looking at
the region above 600 cm^{-1}, since common organic compounds are composed of relatively light atoms with strong bonds. Thus many organic
chemists are quite content with a small infrared spectrophotometer
covering the range 600 to 4000 cm^{-1}. However, inorganic compounds
tend to consist of more massive elements, often with weaker bonds,
and the fundamentals lie at lower frequencies. Indeed the complete
spectrum of many simple inorganic compounds occurs in the region below 1000 cm^{-1}. In addition to fundamentals of simple species, the
low frequency region yields information on metal-ligand vibrations,
chain vibrations in polymers, lattice modes, barriers to rotation
and inversion, rotations in gases, and other data of stereochemical
and thermodynamic importance. From this it is clear that the routine
instrumentation which is satisfactory for organic studies is generally too limited for inorganic applications.

Relatively inexpensive routine infrared spectrophotometers with
extended capability to 250 cm^{-1} are now commercially available, and
research grade instruments with this range have been available for a
number of years. However, operation in the truly far infrared region

(<250 cm^{-1}) requires fairly sophisticated instrumentation at considerable cost. Fortunately, the inconvenient far infrared instrumentation available before 1970 has now given way to grating spectrophotometers and interferometers capable of routine operation in this region.

Sampling in the low frequency infrared region requires special window materials. Some commercially available materials and their useful transmission ranges are potassium bromide (>340 cm^{-1}), cesium bromide (>250 cm^{-1}), cesium iodide (>200 cm^{-1}), KRS-5 (>250 cm^{-1}), silver bromide (>300 cm^{-1}), and polyethylene (33 to 625 cm^{-1}). The properties and applications of these materials are more fully discussed in the introductory chapter of this book.

Polyethylene is the most commonly used material in the truly far-infrared region. It is available as sheets, as a powder, and in a variety of prefabricated sampling devices. Its chief virtues are cheapness and inertness. Its chief liability is the ease with which it is scratched, making it difficult to clean thoroughly. One time use is generally advisable. Solid samples can be mulled with Nujol, petroleum jelly, or a number of other organic solvents [88] and held on a polyethylene substrate. Also, solids can be mixed with powdered polyethylene and melted at moderate temperatures ($\sim$100°C) to form pellets of any convenient size. These are particularly easy to handle and store. A simple die apparatus for making polyethylene pellets has been described by Barish and co-workers [89].

Excellent discussions of low frequency infrared spectroscopy and its applications to inorganic chemistry can be found in the monographs by Adams [90] and Ferraro [91]. These texts provide review of the literature, as well as details of instrumentation and technique.

2. *Low-frequency Raman*

The scattered radiation in the Raman effect is visible light, making it relatively easy to detect low frequency vibrations. Furthermore, unlike the infrared, no special window materials or sampling devices are required to gain access to this region. Nonetheless, observing Raman bands in this region is not devoid of problems.

The proximity to the exciting frequency demands that the spectrometer be efficient at stray light rejection. Stray light causes broadening of the apparent width of the band of the exciting frequency, possibly obscuring sample frequencies. This is particularly a problem with highly reflecting samples, such as many solids. Some relief can be obtained by closing down the slits, with subsequent loss of intensity, but there is no substitute for a good spectrometer. Most modern Raman instruments employ at least a double monochromator arrangement, wherein the increased dispersion and additional slits result in high resolution and stray light rejection. Further improvement in stray light rejection, often of dramatic proportions, is achieved by triple monochromator spectrometers, now commercially available. Figure 24 shows low-frequency Raman spectra of ℓ-cystine with double and triple monochromators. The improvement obtained with the triple monochromator is evident. Of course, the reduction of stray light has its price, both in money and lessened light through-put.

High levels of light reflected into the spectrometer can result in the appearance of generally sharp bands in the recorded spectrum from grating ghosts. Ghosts are transmissions of small amounts of light of the exciting frequency at grating angles which should block transmission. They arise from imperfections which are inevitably present in all replicated gratings. Considerable effort is expended by manufacturers to produce "ghost-free" gratings, but "ghost-free" generally means that spurious intensities have been minimized, rather than totally eliminated. The distinction between ghosts and genuine bands in a recorded spectrum is not always unambiguous, since the intensities of the spurious lines depend to a large extent on the reflectance properties of the scattering medium. A standard procedure for determining the characteristic ghosts of a particular spectrometer with a certain laser excitatin is to scatter the beam into the spectrometer off a block of magnesium carbonate or similar material. While ghosts tend to be most prevalent and troublesome "near in" to the exciting line, they can occur with considerable intensity at higher displacements. Constant improvements in grating technology

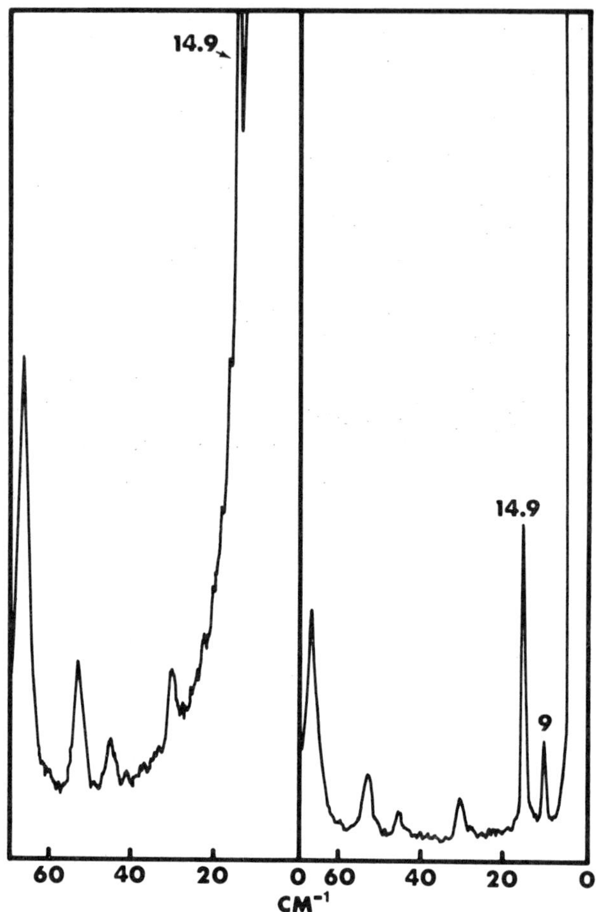

FIG. 24. Low frequency spectrum of ℓ-cystine obtained with
double monochromator (left) and triple monochromator (right) and
4880Å excitation. (Courtesy of Spex Industries, Inc.)

in recent years have minimized ghosts, and the newer triple monochromator spectrometers are virtually free of this problem.

Another problem frequently encountered in the low frequency
region of the Raman spectrum is the appearance of sharp peaks from
the nonlasing emission lines of the laser. Although the output of
a gas laser has its principal intensity at the lasing frequency, the
characteristic emission frequencies of the gas or gases comprising

the laser are also transmitted, usually with very weak intensities.
The gas emission frequencies are tabulated in most chemistry and
physics handbooks. The elimination of these frequencies from the
Raman spectrum can be accomplished by passing the laser beam through
an interference filter designed to pass the lasing frequency. Interference filters have two disadvantages: the intensity of the laser
light is always reduced in passing through the filter, sometimes by
50% or more, and the filters break down over long periods of use.
An alternative provided by some instrument manufacturers is a so-
called fore-prism or Claassen filter. These devices are essentially
small prism monochromators which disperse the unwanted frequencies
of the laser before reaching the sample. Even without the aid of
these devices, some reduction in nonlasing intensity may be achieved
by placing the laser at some distance from the sample and directing
the beam through a slit or hole just wide enough to pass it without
vignetting. The rays from the nonlasing frequencies, unlike the
collimated laser beam, have significant divergences, and at least
that portion which is not colinear with the beam will be blocked by
the slit jaws or mask.

C. Heavy Metal Isotopic Substitution

Isotopic substitution has long been employed as an aid to vibrational assignment. If we consider the simple case of a diatomic
harmonic oscillator, it can be seen that the ratio of the vibrational
frequency of a molecule composed of isotope A to that of such a molecule composed of isotope B is

$$\frac{\nu_A}{\nu_B} = \left(\frac{\mu_B}{\mu_A}\right)^{1/2} \tag{19}$$

where μ is the reduced mass. Ignoring radioactive isotopes, the mass
ratio in Eq. (19) reaches its highest value for hydrogen ($_1^1H:_1^2D = 2:1$),
and pronounced shifts in vibrational frequency are expected with
deuterium substitution. For elements of higher nuclear charge, the
ratio of the masses of any two isotopes ($m_A < m_B$) is less than 2 and,
consequently, smaller isotopic shifts are observed. As the nuclear

charges increase, the mass ratios begin to approach 1 and increasing
demands are placed on spectroscopic resolution in order to use isotopic substitution as an aid to assignment. Thus, the use of this
technique has largely been confined to substitution of elements in
the first three periods.

Recently Nakamoto and co-workers [92] have developed metal isotope substitution as a method for studying metal-ligand vibrations
in a large number of transition-metal complexes. Nonradioactive isotopes exist in significant natural abundance for many of the transition elements, and these can be obtained in high purity and at reasonable cost from Oak Ridge National Laboratory. Thus it is feasible
to prepare a compound with two different isotopes of the same transition element and compare the spectra of both. The magnitudes of the
observed isotopic frequency shifts will depend upon the nature of
the vibrational mode and the relative masses of the metal isotopes.
The shift is expected to be greater for those modes in which the
metal atom has appreciable amplitude, and in general the effect will
be greater for antisymmetric modes than for symmetric modes. For
the compounds studied by Nakamoto and co-workers, M-L frequency shifts
were 0 to 2 cm^{-1} for bending modes and 2 to 10 cm^{-1} for stretching
modes. It is to be expected that these isotopic shifts would be enhanced by increasing the mass difference between the two isotopes.
While this would seem to be an important criterion governing the
choice of isotopes, the selection is limited by the number of existing
nonradioactive isotopes, their availability by virtue of their natural
abundance, and ultimately their cost. Even so, significant and useful
shifts have been observed for isotopes separated by only 2 amu, for
example, ^{63}Cu - ^{65}Cu [92g].

The far-infrared and Raman study of the zinc halide complexes of
pyridine, $Zn(py)_2X_2$, by Saito, Cordes, and Nakamoto [92h] is typical
of the application of this technique. The $Zn(py)_2X_2$ complexes (X =
Cl, Br, I) are tetrahedral (C_{2v}) and give rise to nine normal modes,
if the pyridine ring is assumed to act as a single atom with the mass
of the C_5H_5N group. These are

$A_1 : \nu_s(\text{Zn-X}), \; \nu_s(\text{Zn-N}), \; \delta(\text{NZnN}), \; \delta(\text{XZnX})$

$A_2 : \delta(\text{XZnN})$

$B_1 : \nu_a(\text{Zn-N}), \; \delta(\text{XZnN})$

$B_2 : \nu_a(\text{Zn-X}), \; \delta(\text{XZnN})$

where ν_s is symmetric stretching, ν_a is antisymmetric stretching,
and δ is bending. A_2 is active in the Raman only, while all others
are active in both spectra. The spectroscopic assignment was based
in part on the metal isotope effect between ^{64}Zn and ^{68}Zn compounds.
The relative magnitudes of the shifts can be expected to be $\nu_a > \nu_s \gg \delta$.
In addition, deuterium substitution in the pyridine (d_5) should cause
a shift in all those modes involving it; namely, Zn-N stretching,
NZnN bending, and XZnN bending. Likewise, changing the halogen should
cause a shift in the frequencies for Zn-X stretching. XZnX bending,
and XZnN bending. These simple predictions break down if vibrational
coupling occurs, as between the two A_1 modes $\nu_s(\text{Zn-X})$ and $\nu_s(\text{Zn-N})$.
In this case, coupling would cause some shift in $\nu_s(\text{Zn-X})$ with pyridine (d_5) substitution.

The ^{64}Zn - ^{68}Zn isotope effect on the far-infrared and Raman
spectra of the $\text{Zn(py)}_2\text{X}_2$ complexes is shown in Figs. 25 and 26, respectively. Frequencies, isotopic shifts, and assignments are listed
in Tables 26 and 27. Not all frequencies and isotopic shifts were
observed, due to problems with band shape, overlapping bands, and
weak intensities. Raman investigation of $\text{Zn(py)}_2\text{I}_2$ was hampered by
sample decomposition in the green argon laser beam (5145Å). Nonetheless, the data available in Tables 26 and 27 are consistent with
the expectations for the given assignments. The assignments were
further confirmed by an approximate normal coordinate treatment.

D. Problems with Pellets

Potassium bromide and similar salt pellets (NaCl, CsBr, etc.)
are frequently used for obtaining the infrared spectra of solid samples, because they offer a number of advantages over Nujol mulls and

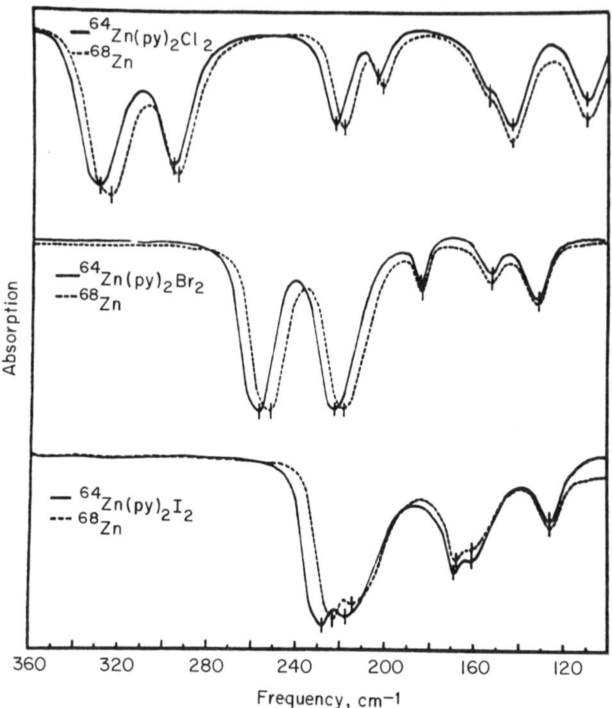

FIG. 25. Far-infrared spectra of $^{64}Zn(py)_2X_2$ (X = Cl, Br, I) and their ^{68}Zn analogs. (Reproduced from Ref. 92h with permission.)

deposited films. Sample preparation is easy, there are no intefering solvent bands in the spectrum, and the sample can be stored indefinitely in a desiccator for future reference. Nonetheless, such pellets are susceptible to a number of failings which tend to crop up more often with inorganic samples than with others.

Chief among the difficulties is that the salt sometimes is not inert with respect to the sample. Bromide is easily oxidized to bromine by many samples. Of course this is to be expected with compounds which are strong oxidizing agents. However, oxidation-reduction is not always obviously predictable, since the customarily tabulated reduction potentials for aqueous solutions can be markedly different from those in the solid. When bromide oxidation occurs

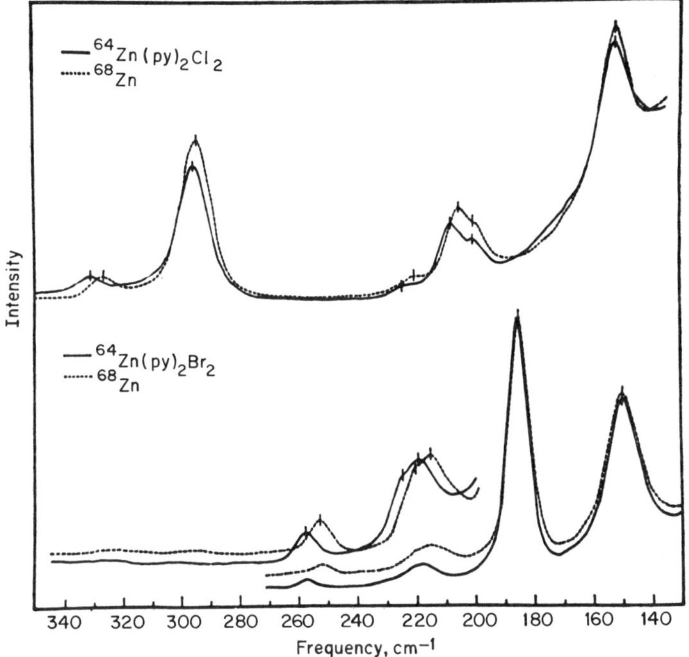

FIG. 26.   Raman spectra of $^{64}Zn(py)_2X_2$ (X = Cl, Br) and their ^{68}Zn analogs.   (Reproduced from Ref. 92h with permission.)

the pellet will appear yellow to brown, often with brown spots. Colored samples may mask these indications, and it is generally advisable to run a comparison Nujol mull spectrum when the sample is a potential oxidizing agent.

Ionic samples pose the problem of possible ion exchange with the diluting salt of the pellet. In some cases this may cause a change in color, but more typically the appearance of the pellet gives no indication that ion exchange has occurred. The spectroscopic result is generally a shifting of band positions, shapes, and intensities, occasionally accompanied by the appearance of new bands. The spectrum is neither that of the sample nor the salt of the sample with an ion of the diluting alkali halide. Ion exchange and its effects are most extensive when the salt-sample mixture has been excessively ground (as often with vibrator grinders), when the pellet has been heated, or when the preparation involved excessively high pressure for long

TABLE 26

Far-infrared Frequencies, Isotopic Shifts,

and Assignments for $Zn(py)_2X_2$ Complexes

(All values in cm^{-1}) [92h]

$Zn(py)_2Cl_2$			$Zn(py)_2Br_2$			$Zn(py)_2I_2$			Assignment
$\nu(^{64}Zn)$	$\Delta\nu_{Zn}$	$\Delta\nu_D$	$\nu(^{64}Zn)$	$\Delta\nu_{Zn}$	$\Delta\nu_D$	$\nu(^{64}Zn)$	$\Delta\nu_{Zn}$	$\Delta\nu_D$	
329.2(vs)	4.8	0.8	257.0(vs)	5.3	1.7	227.4(vs)	4.4	1.9	ν_a(Zn-X), B_2
296.5(s)	2.4	0.2	223.0(vs)	3.7	1.3				ν_s(Zn-X) + ν_s(Zn-N), A_1*
222.4(m)	3.6	3.8				217.0(vs)	2.5	3.1	ν_a(Zn-N), B_1
203.9(w)	2.4	4.1	184.5(w)	0.7	3.2	168.8(m)	0.7	7.2	ν_s(Zn-N) + ν_s(Zn-X), A_1*
154(sh)		∿ 7	152.5(w)	0.0	9.1	161.0(sh)		6.0	δ(NZnN), A_1
143.3(m)	0.3	4.7	131.6(m)	0.0	6.8	126.0(w)	0.2	6.1	δ(XZnN), B_1, B_2
109.9(m)	0.0	0.7	77.0(m)	0.0	0.5	66.5(m)	0.0	0.3	δ(XZnX), A_1

*Not coupled in the chloro complex.

Abbreviations: $\nu(^{64}Zn)$ - frequency of ^{64}Zn complex, $\Delta\nu_{Zn}$ - ^{64}Zn - ^{68}Zn isotopic shift, $\Delta\nu_D$ - py-py(d_5) isotopic shift, v - very, s - strong, m - medium, w - weak, sh - shoulder.

TABLE 27

Raman Frequencies, Isotopic Shifts, and Assignments

for $Zn(py)_2X_2$ Complexes   (All values cm^{-1}) [92h]

$Zn(py)_2Cl_2$			$Zn(py)_2Br_2$			$Zn(py)_2I_2$		Assignment
$\nu(^{64}Zn)$	$\Delta\nu_{Zn}$	$\Delta\nu_D$	$\nu(^{64}Zn)$	$\Delta\nu_{Zn}$	$\Delta\nu_D$	$\nu(^{na}Zn)$	$\Delta\nu_D$	
330.3(vw)	3.9	1.0	257.4(vw)	4.9	1.3			$\nu_a(Zn-X)$, B_2
294.8(w)	1.3	0.4	217.8(w)	2.2				$\nu_s(Zn-X) + \nu_s(Zn-N)$, A_1*
224.1(vw)	4.4	5.0	224(sh)	~ 3				$\nu_a(Zn-N)$, B_1
207.6(w)	3.2	4.2	185.6(vs)	0.5	3.4	168(sh)	~ 6	$\nu_s(Zn-N) + \nu_s(Zn-X)$, A_1*
200(sh)	~ 0	~ 10				160(sh)	~ 8	$\delta(XZnN)$, A_2
151.2(vs)	0.0	6.6	150.0(vs)	0.2	8.9	149.5(vs)	3	$\delta(NZnN)$, A_1

*Not coupled for the chloro complex.

Abbreviations: $\nu(^{64}Zn)$ - frequency of ^{64}Zn complex, $\nu(^{na}Zn)$ - frequency of natural abundance zinc complex, $\Delta\nu_{Zn}$ - ^{64}Zn-^{68}Zn isotopic shift, $\Delta\nu_D$ - py-py(d_5) isotopic shift, v - very, s - strong, w - weak, sh - shoulder.

periods of time. Satisfactory pellets of ionic samples usually can be prepared when these extreme conditions are avoided. However, it is a prudent practice to compare the spectrum from such a pellet with the spectrum obtained from a mull.

Conventionally prepared pellets are not suitable for hydroscopic samples, unless the entire preparation and spectroscopic use take place in a dry atmosphere. However, Maas and Tolk [93] have shown that a single hydroscopic crystal completely surrounded by potassium bromide in a pellet does not begin to absorb water until the relative humidity reaches 57 percent. This suggests that encapsulation in pure potassium bromide can be used to prepare pellets of hydroscopic powdered samples. Nortia and Kontas [94] have described a technique for adapting a standard KBr die for the preparation of such pellets. A cup-shaped and a lid-shaped KBr pellet are formed separately in a standard 13-mm KBr die using two appropriately shaped brass inserts. The cup is made upside down by centering a slightly conical 10-mm brass disk on the bottom of the pellet die and adding 600 to 800 mg of dry KBr to the die, followed by pressing in the normal manner. The KBr lid is made in a similar way using a brass ring (13-mm outer diameter, 10-mm inner diameter) having the same conical taper on the inside as does the disk used for forming the cup. The ring can be cut into two semicircular pieces to facilitate removal of the fin- ished KBr lid. The sample is ground with KBr in a ratio of 1/10 to 1/20, added to the KBr cup, and the cup and contents are dried in an oven. After drying, the cup is quickly covered with the lid, and the whole ensemble is compressed into one pellet in the conventional die. Reported spectra [94] of hydroscopic samples using this tech- nique show very pronounced reduction of the broad water band in the 3500 to 3000 cm^{-1} region when compared with conventionally prepared pellets.

E. Sample Decomposition in the Laser Beam

Normal operating procedure in Raman spectroscopy includes match- ing the color of the laser light as closely as possible to that of the sample, thereby minimizing absorption. Absorption at the laser frequency minimizes the detectable Raman-scattered light and can result in thermal decomposition of the sample. Although continuously tunable dye lasers offer hope for the future, present Raman instruments are limited in color matching capability by the number of discrete wavelengths available from the arsenal of lasers at hand. Most gas lasers offer the ability to operate at several wavelengths, but usable intensity is generally confined to one or two lines. The most commonly used exciting sources are He/Ne 6328Å (red), Kr 6471Å (red), Ar 5145Å (green), and Ar 4880Å (blue). Thus it is not uncommon to encounter samples which have appreciable absorption at all the wavelengths of the lasers available to the spectroscopist.

Thermal decomposition from absorption sometimes can be minimized by either defocusing the laser at the sample or reducing the power level of the laser. Either technique will reduce the intensity of the Raman light reaching the spectrometer, and detection will become more difficult. Furthermore, if the sample absorbs at the laser's wavelength it probably also absorbs at the wavelengths of the Raman-scattered light, further reducing the ability to detect the spectrum.

A better solution to the problems of absorption is the technique of sample rotation, developed by Kiefer and Bernstein [95]. A typical apparatus for solids consists of a stainless steel disk which can be spun about its axis of rotation by a small high speed motor ($\sim$1000 - 4000 rpm). The outer flat surface of the disk has a circular groove in which the sample can be compressed over a substrate of potassium bromide. The spinning disk is oriented so that the laser beam strikes the sample in the groove and the Raman light is reflected into the spectrometer. Decomposition is avoided, since each part of the sample spends only a very brief amount of time in the laser beam.

Silver chromate, Ag_2CrO_4, provides an example of a sample which required sample rotation [96]. Powdered Ag_2CrO_4 is very dark brown in color and absorbs most common laser frequencies. Attempts to obtain a Raman spectrum with 90 mW He/Ne laser excitation (6328Å) gave no measurable intensities, due to the low power of the laser and the sample absorption. Rapid sample decomposition occurred with the more powerful Kr laser (6471Å), but it was possible to observe the symmetric CrO_4-stretching band before the sample was incinerated. A successful spectrum was obtained with Kr-laser excitation (200 mW at the sample) when sample rotation at 3000 rpm was used (Fig. 27). Although the sample rotation technique allows the use of an absorbed wavelength for excitation, the laser chosen still must be one which minimizes the absorption. In the case of Ag_2CrO_4, for example, sample rotation used with the green 5145Å light of an argon laser yielded a spectrum with additional broad spurious bands, evidently caused by fluorescence.

Sample rotation can also be used with liquid samples. Here a typical apparatus consists of a small glass cylinder with a hole in one end to allow introduction of the sample. The cylinder is spun about its axis and is oriented so that the laser beam passes up through the bottom, parallel to the axis of rotation. The scattered radiation is observed at 90°. All present manufacturers of laser-Raman systems offer sample rotation devices of varying designs. It is also quite easy to design and construct such equipment for use with any existing spectrometer.

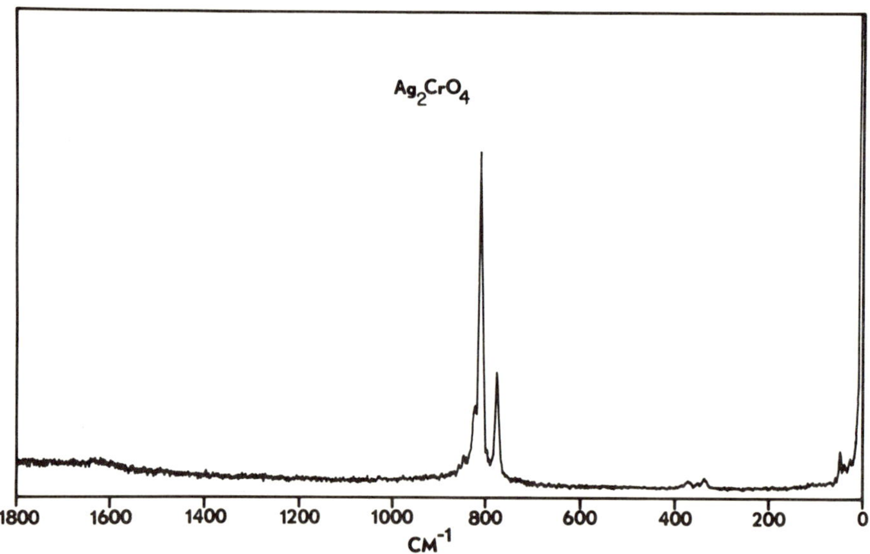

FIG. 27. Raman spectrum of Ag_2CrO_4 obtained by sample rotation (3000 rpm) and 6471Å Kr-laser excitation (200 mW).

APPENDIX A: CHARACTER TABLES

C_s	E	σ		
A'	1	1	x, y; R_z	x^2, y^2, z^2, xy
A''	1	-1	z; R_x, R_y	yz, xz

C_i	E	i		
A_g	1	1	R_x, R_y, R_z	x^2, y^2, z^2, xy, xz, yz
A_u	1	-1	x, y, z	

C_2	E	C_2		
A	1	1	z; R_z	x^2, y^2, z^2, xy
B	1	-1	x, y; R_x, R_y	yz, xz

C_3 table: $\varepsilon = \exp(2\pi i/3)$; $\varepsilon + \varepsilon^* = -1$

C_3	E	C_3	C_3^2		
A	1	1	1	z; R_z	$x^2 + y^2$, z^2, $x^2 - y^2$, xy
E	$\left\{\begin{matrix}1 \\ 1\end{matrix}\right.$	$\begin{matrix}\varepsilon \\ \varepsilon^*\end{matrix}$	$\left.\begin{matrix}\varepsilon^* \\ \varepsilon\end{matrix}\right\}$	x, y; R_x, R_y	xz, yz

C_4	E	C_4	C_2	C_4^3		
A	1	1	1	1	z; R_z	$x^2 + y^2$, z^2
B	1	-1	1	-1		$x^2 - y^2$, xy
E	$\left\{\begin{matrix}1 \\ 1\end{matrix}\right.$	$\begin{matrix}i \\ -i\end{matrix}$	$\begin{matrix}-1 \\ -1\end{matrix}$	$\left.\begin{matrix}-i \\ i\end{matrix}\right\}$	x, y; R_x, R_y	xz, yz

C_5	E	C_5	C_5^2	C_5^3	C_5^4			$\varepsilon = \exp(2\pi i/5);\ \varepsilon^n + \varepsilon^{n*} = 2\cos(2\pi n/5)$
A	1	1	1	1	1	$z;\ R_z$		$x^2 + y^2,\ z^2$
E_1	$\left\{\begin{matrix}1\\1\end{matrix}\right.$	$\begin{matrix}\varepsilon\\\varepsilon^*\end{matrix}$	$\begin{matrix}\varepsilon^2\\\varepsilon^{2*}\end{matrix}$	$\begin{matrix}\varepsilon^{2*}\\\varepsilon^2\end{matrix}$	$\left.\begin{matrix}\varepsilon^*\\\varepsilon\end{matrix}\right\}$	$x,\ y;\ R_x,\ R_y$		$xz,\ yz$
E_2	$\left\{\begin{matrix}1\\1\end{matrix}\right.$	$\begin{matrix}\varepsilon^2\\\varepsilon^{2*}\end{matrix}$	$\begin{matrix}\varepsilon^*\\\varepsilon\end{matrix}$	$\begin{matrix}\varepsilon\\\varepsilon^*\end{matrix}$	$\left.\begin{matrix}\varepsilon^{2*}\\\varepsilon^2\end{matrix}\right\}$			$x^2 - y^2,\ xy$

C_6	E	C_6	C_3	C_2	C_3^2	C_6^5			$\varepsilon = \exp(2\pi i/6);\ \varepsilon + \varepsilon^* = 1$
A	1	1	1	1	1	1	$z;\ R_z$		$x^2 + y^2,\ z^2$
B	1	-1	1	-1	1	-1			
E_1	$\left\{\begin{matrix}1\\1\end{matrix}\right.$	$\begin{matrix}\varepsilon\\\varepsilon^*\end{matrix}$	$\begin{matrix}-\varepsilon^*\\-\varepsilon\end{matrix}$	$\begin{matrix}-1\\-1\end{matrix}$	$\begin{matrix}-\varepsilon\\-\varepsilon^*\end{matrix}$	$\left.\begin{matrix}\varepsilon^*\\\varepsilon\end{matrix}\right\}$	$x,\ y;\ R_x,\ R_y$		$xz,\ yz$
E_2	$\left\{\begin{matrix}1\\1\end{matrix}\right.$	$\begin{matrix}-\varepsilon^*\\-\varepsilon\end{matrix}$	$\begin{matrix}-\varepsilon\\-\varepsilon^*\end{matrix}$	$\begin{matrix}1\\1\end{matrix}$	$\begin{matrix}-\varepsilon^*\\-\varepsilon\end{matrix}$	$\left.\begin{matrix}-\varepsilon\\-\varepsilon^*\end{matrix}\right\}$			$x^2 - y^2,\ xy$

C_7	E	C_7	C_7^2	C_7^3	C_7^4	C_7^5	C_7^6			$\varepsilon = \exp(2\pi i/7);\ \varepsilon^n + \varepsilon^{n*} = 2\cos(2\pi n/7)$
A	1	1	1	1	1	1	1	$z;\ R_z$		$x^2 + y^2,\ z^2$
E_1	$\left\{\begin{matrix}1\\1\end{matrix}\right.$	$\begin{matrix}\varepsilon\\\varepsilon^*\end{matrix}$	$\begin{matrix}\varepsilon^2\\\varepsilon^{2*}\end{matrix}$	$\begin{matrix}\varepsilon^3\\\varepsilon^{3*}\end{matrix}$	$\begin{matrix}\varepsilon^{3*}\\\varepsilon^3\end{matrix}$	$\begin{matrix}\varepsilon^{2*}\\\varepsilon^2\end{matrix}$	$\left.\begin{matrix}\varepsilon^*\\\varepsilon\end{matrix}\right\}$	$x,\ y;\ R_x,\ R_y$		$xz,\ yz$
E_2	$\left\{\begin{matrix}1\\1\end{matrix}\right.$	$\begin{matrix}\varepsilon^2\\\varepsilon^{2*}\end{matrix}$	$\begin{matrix}\varepsilon^{3*}\\\varepsilon^3\end{matrix}$	$\begin{matrix}\varepsilon^*\\\varepsilon\end{matrix}$	$\begin{matrix}\varepsilon\\\varepsilon^*\end{matrix}$	$\begin{matrix}\varepsilon^3\\\varepsilon^{3*}\end{matrix}$	$\left.\begin{matrix}\varepsilon^{2*}\\\varepsilon^2\end{matrix}\right\}$			$x^2 - y^2,\ xy$
E_3	$\left\{\begin{matrix}1\\1\end{matrix}\right.$	$\begin{matrix}\varepsilon^3\\\varepsilon^{3*}\end{matrix}$	$\begin{matrix}\varepsilon^*\\\varepsilon\end{matrix}$	$\begin{matrix}\varepsilon^2\\\varepsilon^{2*}\end{matrix}$	$\begin{matrix}\varepsilon^{2*}\\\varepsilon^2\end{matrix}$	$\begin{matrix}\varepsilon\\\varepsilon^*\end{matrix}$	$\left.\begin{matrix}\varepsilon^{3*}\\\varepsilon^3\end{matrix}\right\}$			

C_8	E	C_8	C_4	C_2	C_4^3	C_8^3	C_8^5	C_8^7			$\varepsilon = \exp(2\pi i/8);\ \varepsilon + \varepsilon^* = \sqrt{2}$
A	1	1	1	1	1	1	1	1	$z;\ R_z$		$x^2 + y^2,\ z^2$
B	1	-1	1	1	1	-1	-1	-1			
E_1	$\left\{\begin{matrix}1\\1\end{matrix}\right.$	$\begin{matrix}\varepsilon\\\varepsilon^*\end{matrix}$	$\begin{matrix}i\\-i\end{matrix}$	$\begin{matrix}-1\\-1\end{matrix}$	$\begin{matrix}-i\\i\end{matrix}$	$\begin{matrix}-\varepsilon^*\\-\varepsilon\end{matrix}$	$\begin{matrix}-\varepsilon\\-\varepsilon^*\end{matrix}$	$\left.\begin{matrix}\varepsilon^*\\\varepsilon\end{matrix}\right\}$	$x,\ y;\ R_x,\ R_y$		$xz,\ yz$
E_2	$\left\{\begin{matrix}1\\1\end{matrix}\right.$	$\begin{matrix}i\\-i\end{matrix}$	$\begin{matrix}-1\\-1\end{matrix}$	$\begin{matrix}1\\1\end{matrix}$	$\begin{matrix}-1\\-1\end{matrix}$	$\begin{matrix}-i\\i\end{matrix}$	$\begin{matrix}i\\-i\end{matrix}$	$\left.\begin{matrix}-i\\i\end{matrix}\right\}$			$x^2 - y^2,\ xy$
E_3	$\left\{\begin{matrix}1\\1\end{matrix}\right.$	$\begin{matrix}-\varepsilon\\-\varepsilon^*\end{matrix}$	$\begin{matrix}i\\-i\end{matrix}$	$\begin{matrix}-1\\-1\end{matrix}$	$\begin{matrix}-i\\i\end{matrix}$	$\begin{matrix}\varepsilon^*\\\varepsilon\end{matrix}$	$\begin{matrix}\varepsilon\\\varepsilon^*\end{matrix}$	$\left.\begin{matrix}-\varepsilon^*\\-\varepsilon\end{matrix}\right\}$			

C_{2v}	E	C_2	$\sigma_v(xz)$	$\sigma_v(yz)$		
A_1	1	1	1	1	z	x^2, y^2, z^2
A_2	1	1	-1	-1	R_z	xy
B_1	1	-1	1	-1	x; R_y	xz
B_2	1	-1	-1	1	y; R_x	yz

C_{3v}	E	$2C_3$	$3\sigma_v$		
A_1	1	1	1	z	$x^2 + y^2$, z^2
A_2	1	1	-1	R_z	
E	2	-1	0	x, y; R_x, R_y	$x^2 - y^2$, xy, xz, yz

C_{4v}	E	$2C_4$	C_2	$2\sigma_v$	$2\sigma_d$		
A_1	1	1	1	1	1	z	$x^2 + y^2$, z^2
A_2	1	1	1	-1	-1	R_z	
B_1	1	-1	1	1	-1		$x^2 - y^2$
B_2	1	-1	1	-1	1		xy
E	2	0	-2	0	0	x, y; R_x, R_y	xz, yz

C_{5v}	E	$2C_5$	$2C_5^2$	$5\sigma_v$		
A_1	1	1	1	1	z	$x^2 + y^2$, z^2
A_2	1	1	1	-1	R_z	
E_1	2	$2\cos 72°$	$2\cos 144°$	0	x, y; R_x, R_y	xz, yz
E_2	2	$2\cos 144°$	$2\cos 72°$	0		$x^2 - y^2$, xy

C_{6v}	E	$2C_6$	$2C_3$	C_2	$3\sigma_v$	$3\sigma_d$		
A_1	1	1	1	1	1	1	z	$x^2 + y^2,\ z^2$
A_2	1	1	1	1	-1	-1	R_z	
B_1	1	-1	1	-1	1	-1		
B_2	1	-1	1	-1	-1	1		
E_1	2	1	-1	-2	0	0	$x,\ y;\ R_x,\ R_y$	$xz,\ yz$
E_2	2	-1	-1	2	0	0		$x^2 - y^2,\ xy$

C_{2h}	E	C_2	i	σ_h		
A_g	1	1	1	1	R_z	$x^2,\ y^2,\ z^2,\ xy$
B_g	1	-1	1	-1	$R_x,\ R_y$	$xz,\ yz$
A_u	1	1	-1	-1	z	
B_u	1	-1	-1	1	x, y	

C_{3h} $\varepsilon = \exp(2\pi i/3);\ \varepsilon + \varepsilon^* = -1$

C_{3h}	E	C_3	C_3^2	σ_h	S_3	S_3^5		
A'	1	1	1	1	1	1	R_z	$x^2 + y^2,\ z^2$
E'	$\begin{cases}1 \\ 1\end{cases}$	$\begin{matrix}\varepsilon \\ \varepsilon^*\end{matrix}$	$\begin{matrix}\varepsilon^* \\ \varepsilon\end{matrix}$	$\begin{matrix}1 \\ 1\end{matrix}$	$\begin{matrix}\varepsilon \\ \varepsilon^*\end{matrix}$	$\begin{matrix}\varepsilon^* \\ \varepsilon\end{matrix}$	$x,\ y$	$x^2 - y^2,\ xy$
A''	1	1	1	-1	-1	-1	z	
E''	$\begin{cases}1 \\ 1\end{cases}$	$\begin{matrix}\varepsilon \\ \varepsilon^*\end{matrix}$	$\begin{matrix}\varepsilon^* \\ \varepsilon\end{matrix}$	$\begin{matrix}-1 \\ -1\end{matrix}$	$\begin{matrix}-\varepsilon \\ -\varepsilon^*\end{matrix}$	$\begin{matrix}-\varepsilon^* \\ -\varepsilon\end{matrix}$	$R_x,\ R_y$	$xz,\ yz$

C_{4h}	E	C_4	C_2	C_4^3	i	S_4^3	σ_h	S_4		
A_g	1	1	1	1	1	1	1	1	R_z	$x^2 + y^2,\ z^2$
B_g	1	-1	1	-1	1	-1	1	-1		$x^2 - y^2,\ xy$
E_g	$\begin{cases}1 \\ 1\end{cases}$	$\begin{matrix}i \\ -i\end{matrix}$	$\begin{matrix}-1 \\ -1\end{matrix}$	$\begin{matrix}-i \\ i\end{matrix}$	$\begin{matrix}1 \\ 1\end{matrix}$	$\begin{matrix}i \\ -i\end{matrix}$	$\begin{matrix}-1 \\ -1\end{matrix}$	$\begin{matrix}-i \\ i\end{matrix}$	$R_x,\ R_y$	$xz,\ yz$
A_u	1	1	1	1	-1	-1	-1	-1	z	
B_u	1	-1	1	-1	-1	1	-1	1		
E_u	$\begin{cases}1 \\ 1\end{cases}$	$\begin{matrix}i \\ -i\end{matrix}$	$\begin{matrix}-1 \\ -1\end{matrix}$	$\begin{matrix}-i \\ i\end{matrix}$	$\begin{matrix}-1 \\ -1\end{matrix}$	$\begin{matrix}-i \\ i\end{matrix}$	$\begin{matrix}1 \\ 1\end{matrix}$	$\begin{matrix}i \\ -i\end{matrix}$	$x,\ y$	

C_{5h}, $\varepsilon = \exp(2\pi i/5)$

C_{5h}	E	C_5	C_5^2	C_5^3	C_5^4	σ_h	S_5	S_5^7	S_5^3	S_5^9		
A'	1	1	1	1	1	1	1	1	1	1	R_z	$x^2+y^2,\ z^2$
E_1'	$\left\{\begin{matrix}1\\1\end{matrix}\right.$	$\begin{matrix}\varepsilon\\\varepsilon^*\end{matrix}$	$\begin{matrix}\varepsilon^2\\\varepsilon^{2*}\end{matrix}$	$\begin{matrix}\varepsilon^{2*}\\\varepsilon^2\end{matrix}$	$\begin{matrix}\varepsilon^*\\\varepsilon\end{matrix}$	$\begin{matrix}1\\1\end{matrix}$	$\begin{matrix}\varepsilon\\\varepsilon^*\end{matrix}$	$\begin{matrix}\varepsilon^2\\\varepsilon^{2*}\end{matrix}$	$\begin{matrix}\varepsilon^{2*}\\\varepsilon^2\end{matrix}$	$\left.\begin{matrix}\varepsilon^*\\\varepsilon\end{matrix}\right\}$	$x,\ y$	
E_2'	$\left\{\begin{matrix}1\\1\end{matrix}\right.$	$\begin{matrix}\varepsilon^2\\\varepsilon^{2*}\end{matrix}$	$\begin{matrix}\varepsilon^*\\\varepsilon\end{matrix}$	$\begin{matrix}\varepsilon\\\varepsilon^*\end{matrix}$	$\begin{matrix}\varepsilon^{2*}\\\varepsilon^2\end{matrix}$	$\begin{matrix}1\\1\end{matrix}$	$\begin{matrix}\varepsilon^2\\\varepsilon^{2*}\end{matrix}$	$\begin{matrix}\varepsilon^*\\\varepsilon\end{matrix}$	$\begin{matrix}\varepsilon\\\varepsilon^*\end{matrix}$	$\left.\begin{matrix}\varepsilon^{2*}\\\varepsilon^2\end{matrix}\right\}$		$x^2-y^2,\ xy$
A''	1	1	1	1	1	-1	-1	-1	-1	-1	z	
E_1''	$\left\{\begin{matrix}1\\1\end{matrix}\right.$	$\begin{matrix}\varepsilon\\\varepsilon^*\end{matrix}$	$\begin{matrix}\varepsilon^2\\\varepsilon^{2*}\end{matrix}$	$\begin{matrix}\varepsilon^{2*}\\\varepsilon^2\end{matrix}$	$\begin{matrix}\varepsilon^*\\\varepsilon\end{matrix}$	$\begin{matrix}-1\\-1\end{matrix}$	$\begin{matrix}-\varepsilon\\-\varepsilon^*\end{matrix}$	$\begin{matrix}-\varepsilon^2\\-\varepsilon^{2*}\end{matrix}$	$\begin{matrix}-\varepsilon^{2*}\\-\varepsilon^2\end{matrix}$	$\left.\begin{matrix}-\varepsilon^*\\-\varepsilon\end{matrix}\right\}$	$R_x,\ R_y$	
E_2''	$\left\{\begin{matrix}1\\1\end{matrix}\right.$	$\begin{matrix}\varepsilon^2\\\varepsilon^{2*}\end{matrix}$	$\begin{matrix}\varepsilon^*\\\varepsilon\end{matrix}$	$\begin{matrix}\varepsilon\\\varepsilon^*\end{matrix}$	$\begin{matrix}\varepsilon^{2*}\\\varepsilon^2\end{matrix}$	$\begin{matrix}-1\\-1\end{matrix}$	$\begin{matrix}-\varepsilon^2\\-\varepsilon^{2*}\end{matrix}$	$\begin{matrix}-\varepsilon^*\\-\varepsilon\end{matrix}$	$\begin{matrix}-\varepsilon\\-\varepsilon^*\end{matrix}$	$\left.\begin{matrix}-\varepsilon^{2*}\\-\varepsilon^2\end{matrix}\right\}$		

C_{6h}, $\varepsilon = \exp(2\pi i/6);\ \varepsilon+\varepsilon^* = 1$

C_{6h}	E	C_6	C_3	C_2	C_3^2	C_6^5	i	S_3^5	S_6^5	σ_h	S_6	S_3		
A_g	1	1	1	1	1	1	1	1	1	1	1	1	R_z	$x^2+y^2,\ z^2$
B_g	1	-1	1	-1	1	-1	1	-1	1	-1	1	-1		
E_{1g}	$\left\{\begin{matrix}1\\1\end{matrix}\right.$	$\begin{matrix}\varepsilon\\\varepsilon^*\end{matrix}$	$\begin{matrix}-\varepsilon^*\\-\varepsilon\end{matrix}$	$\begin{matrix}-1\\-1\end{matrix}$	$\begin{matrix}-\varepsilon\\-\varepsilon^*\end{matrix}$	$\begin{matrix}\varepsilon^*\\\varepsilon\end{matrix}$	$\begin{matrix}1\\1\end{matrix}$	$\begin{matrix}\varepsilon\\\varepsilon^*\end{matrix}$	$\begin{matrix}-\varepsilon^*\\-\varepsilon\end{matrix}$	$\begin{matrix}-1\\-1\end{matrix}$	$\begin{matrix}-\varepsilon\\-\varepsilon^*\end{matrix}$	$\left.\begin{matrix}\varepsilon^*\\\varepsilon\end{matrix}\right\}$	$R_x,\ R_y$	$xz,\ yz$
E_{2g}	$\left\{\begin{matrix}1\\1\end{matrix}\right.$	$\begin{matrix}-\varepsilon^*\\-\varepsilon\end{matrix}$	$\begin{matrix}-\varepsilon\\-\varepsilon^*\end{matrix}$	$\begin{matrix}1\\1\end{matrix}$	$\begin{matrix}-\varepsilon^*\\-\varepsilon\end{matrix}$	$\begin{matrix}-\varepsilon\\-\varepsilon^*\end{matrix}$	$\begin{matrix}1\\1\end{matrix}$	$\begin{matrix}-\varepsilon^*\\-\varepsilon\end{matrix}$	$\begin{matrix}-\varepsilon\\-\varepsilon^*\end{matrix}$	$\begin{matrix}1\\1\end{matrix}$	$\begin{matrix}-\varepsilon^*\\-\varepsilon\end{matrix}$	$\left.\begin{matrix}-\varepsilon\\-\varepsilon^*\end{matrix}\right\}$		$x^2-y^2,\ xy$
A_u	1	1	1	1	1	1	-1	-1	-1	-1	-1	-1	z	
B_u	1	-1	1	-1	1	-1	-1	1	-1	1	-1	1		
E_{1u}	$\left\{\begin{matrix}1\\1\end{matrix}\right.$	$\begin{matrix}\varepsilon\\\varepsilon^*\end{matrix}$	$\begin{matrix}-\varepsilon^*\\-\varepsilon\end{matrix}$	$\begin{matrix}-1\\-1\end{matrix}$	$\begin{matrix}-\varepsilon\\-\varepsilon^*\end{matrix}$	$\begin{matrix}\varepsilon^*\\\varepsilon\end{matrix}$	$\begin{matrix}-1\\-1\end{matrix}$	$\begin{matrix}-\varepsilon\\-\varepsilon^*\end{matrix}$	$\begin{matrix}\varepsilon^*\\\varepsilon\end{matrix}$	$\begin{matrix}1\\1\end{matrix}$	$\begin{matrix}\varepsilon\\\varepsilon^*\end{matrix}$	$\left.\begin{matrix}-\varepsilon^*\\-\varepsilon\end{matrix}\right\}$	$x,\ y$	
E_{2u}	$\left\{\begin{matrix}1\\1\end{matrix}\right.$	$\begin{matrix}-\varepsilon^*\\-\varepsilon\end{matrix}$	$\begin{matrix}-\varepsilon\\-\varepsilon^*\end{matrix}$	$\begin{matrix}1\\1\end{matrix}$	$\begin{matrix}-\varepsilon^*\\-\varepsilon\end{matrix}$	$\begin{matrix}-\varepsilon\\-\varepsilon^*\end{matrix}$	$\begin{matrix}-1\\-1\end{matrix}$	$\begin{matrix}\varepsilon^*\\\varepsilon\end{matrix}$	$\begin{matrix}\varepsilon\\\varepsilon^*\end{matrix}$	$\begin{matrix}-1\\-1\end{matrix}$	$\begin{matrix}\varepsilon^*\\\varepsilon\end{matrix}$	$\left.\begin{matrix}\varepsilon\\\varepsilon^*\end{matrix}\right\}$		

D_2	E	$C_2(x)$	$C_2(y)$	$C_2(z)$		
A	1	1	1	1		$x^2,\ y^2,\ z^2$
B_1	1	1	-1	-1	$z;\ R_z$	xy
B_2	1	-1	1	-1	$y;\ R_y$	xz
B_3	1	-1	-1	1	$x;\ R_x$	yz

D_3	E	$2C_3$	$3C_2$		
A_1	1	1	1		$x^2 + y^2,\ z^2$
A_2	1	1	-1	$z;\ R_z$	
E	2	-1	0	$x,\ y;\ R_x,\ R_y$	$x^2 - y^2,\ xy,\ xz,\ yz$

D_4	E	$2C_4$	$C_4^2 = C_2$	$2C_2'$	$2C_2''$		
A_1	1	1	1	1	1		$x^2 + y^2,\ z^2$
A_2	1	1	1	-1	-1	$z;\ R_z$	
B_1	1	-1	1	1	-1		$x^2 - y^2$
B_2	1	-1	1	-1	1		xy
E	2	0	-2	0	0	$x,\ y;\ R_x,\ R_y$	$xz,\ yz$

D_5	E	$2C_5$	$2C_5^2$	$5C_2$		
A_1	1	1	1	1		$x^2 + y^2,\ z^2$
A_2	1	1	1	-1	$z;\ R_z$	
E_1	2	$2\cos 72°$	$2\cos 144°$	0	$x,\ y;\ R_x,\ R_y$	$xz,\ yz$
E_2	2	$2\cos 144°$	$2\cos 72°$	0		$x^2 - y^2,\ xy$

D_6	E	$2C_6$	$2C_3$	C_2	$3C_2'$	$3C_2''$		
A_1	1	1	1	1	1	1		$x^2 + y^2,\ z^2$
A_2	1	1	1	1	-1	-1	$z;\ R_z$	
B_1	1	-1	1	-1	1	-1		
B_2	1	-1	1	-1	-1	1		
E_1	2	1	-1	-2	0	0	$x,\ y;\ R_x,\ R_y$	$xz,\ yz$
E_2	2	-1	-1	2	0	0		$x^2 - y^2,\ xy$

$D_{2h}=V_h$	E	$C_2(z)$	$C_2(y)$	$C_2(x)$	i	$\sigma(xy)$	$\sigma(zx)$	$\sigma(yz)$		
A_g	1	1	1	1	1	1	1	1		$x^2,\ y^2,\ z^2$
B_{1g}	1	1	-1	-1	1	1	-1	-1	R_z	xy
B_{2g}	1	-1	1	-1	1	-1	1	-1	R_y	xz
B_{3g}	1	-1	-1	1	1	-1	-1	1	R_x	yz
A_u	1	1	1	1	-1	-1	-1	-1		
B_{1u}	1	1	-1	-1	-1	-1	1	1	z	
B_{2u}	1	-1	1	-1	-1	1	-1	1	y	
B_{3u}	1	-1	-1	1	-1	1	1	-1	x	

D_{3h}	E	$2C_3$	$3C_2$	σ_h	$2S_3$	$3\sigma_v$		
A_1'	1	1	1	1	1	1		$x^2+y^2,\ z^2$
A_2'	1	1	-1	1	1	-1	R_z	
E'	2	-1	0	2	-1	0	$x,\ y$	$x^2-y^2,\ xy$
A_1''	1	1	1	-1	-1	-1		
A_2''	1	1	-1	-1	-1	1	z	
E''	2	-1	0	-2	1	0	$R_x,\ R_y$	$xz,\ yz$

D_{4h}	E	$2C_4$	C_2	$2C_2'$	$2C_2''$	i	$2S_4$	σ_h	$2\sigma_v$	$2\sigma_d$		
A_{1g}	1	1	1	1	1	1	1	1	1	1		$x^2+y^2,\ z^2$
A_{2g}	1	1	1	-1	-1	1	1	1	-1	-1	R_z	
B_{1g}	1	-1	1	1	-1	1	-1	1	1	-1		x^2-y^2
B_{2g}	1	-1	1	-1	1	1	-1	1	-1	1		xy
E_g	2	0	-2	0	0	2	0	-2	0	0	$R_x,\ R_y$	$xz,\ yz$
A_{1u}	1	1	1	1	1	-1	-1	-1	-1	-1		
A_{2u}	1	1	1	-1	-1	-1	-1	-1	1	1	z	
B_{1u}	1	-1	1	1	-1	-1	1	-1	-1	1		
B_{2u}	1	-1	1	-1	1	-1	1	-1	1	-1		
E_u	2	0	-2	0	0	-2	0	2	0	0	$x,\ y$	

D_{5h}	E	$2C_5$	$2C_5^2$	$5C_2$	σ_h	$2S_5$	$2S_5^3$	$5\sigma_v$		
A_1'	1	1	1	1	1	1	1	1		x^2+y^2, z^2
A_2'	1	1	1	-1	1	1	1	-1	R_z	
E_1'	2	$2\cos 72°$	$2\cos 144°$	0	2	$2\cos 72°$	$2\cos 144°$	0	x,y	
E_2'	2	$2\cos 144°$	$2\cos 72°$	0	2	$2\cos 144°$	$2\cos 72°$	0		x^2-y^2, xy
A_1''	1	1	1	1	-1	-1	-1	-1		
A_2''	1	1	1	-1	-1	-1	-1	1	z	
E_1''	2	$2\cos 72°$	$2\cos 144°$	0	-2	$-2\cos 72°$	$-2\cos 144°$	0	R_x, R_y	xz, yz
E_2''	2	$2\cos 144°$	$2\cos 72°$	0	-2	$-2\cos 144°$	$-2\cos 72°$	0		

D_{6h}	E	$2C_6$	$2C_3$	C_2	$3C_2'$	$3C_2''$	i	$2S_3$	$2S_6$	σ_h	$3\sigma_e$	$3\sigma_v$		
A_{1g}	1	1	1	1	1	1	1	1	1	1	1	1		$x^2 + y^2, z^2$
A_{2g}	1	1	1	1	-1	-1	1	1	1	1	-1	-1	R_z	
B_{1g}	1	-1	1	-1	1	-1	1	-1	1	-1	1	-1		
B_{2g}	1	-1	1	-1	-1	1	1	-1	1	-1	-1	1		
E_{1g}	2	1	-1	-2	0	0	2	1	-1	-2	0	0	R_x, R_y	xz, yz
E_{2g}	2	-1	-1	2	0	0	2	-1	-1	2	0	0		$x^2 - y^2, xy$
A_{1u}	1	1	1	1	1	1	-1	-1	-1	-1	-1	-1		
A_{2u}	1	1	1	1	-1	-1	-1	-1	-1	-1	1	1	z	
B_{1u}	1	-1	1	-1	1	-1	-1	1	-1	1	-1	1		
B_{2u}	1	-1	1	-1	-1	1	-1	1	-1	1	1	-1		
E_{1u}	2	1	-1	-2	0	0	-2	-1	1	2	0	0	x, y	
E_{2u}	2	-1	-1	2	0	0	-2	1	1	-2	0	0		

$D_{2d}=V_d$	E	$2S_4$	C_2	$2C_2'$	$2\sigma_d$		
A_1	1	1	1	1	1		$x^2 + y^2, z^2$
A_2	1	1	1	-1	-1	R_z	
B_1	1	-1	1	1	-1		$x^2 - y^2$
B_2	1	-1	1	-1	1	z	xy
E	2	0	-2	0	0	$x, y; R_x, R_y$	xz, yz

D_{3d}	E	$2C_3$	$3C_2$	i	$2S_6$	$3\sigma_d$		
A_{1g}	1	1	1	1	1	1		$x^2 + y^2,\ z^2$
A_{2g}	1	1	-1	1	1	-1	R_z	
E_g	2	-1	0	2	-1	0	$R_x,\ R_y$	$x^2 - y^2,\ xy,\ xz,\ yz$
A_{1u}	1	1	1	-1	-1	-1		
A_{2u}	1	1	-1	-1	-1	1	z	
E_u	2	-1	0	-2	1	0	x, y	

D_{4d}	E	$2S_8$	$2C_4$	$2S_8^3$	C_2	$4C_2'$	$4\sigma_d$		
A_1	1	1	1	1	1	1	1		$x^2 + y^2,\ z^2$
A_2	1	1	1	1	1	-1	-1	R_z	
B_1	1	-1	1	-1	1	1	-1		
B_2	1	-1	1	-1	1	-1	1	z	
E_1	2	$\sqrt{2}$	0	$-\sqrt{2}$	-2	0	0	x, y	
E_2	2	0	-2	0	2	0	0		$x^2 - y^2,\ xy$
E_3	2	$-\sqrt{2}$	0	$\sqrt{2}$	-2	0	0	$R_x,\ R_y$	xz, yz

D_{5d}	E	$2C_5$	$2C_5^2$	$5C_2$	i	$2S_{10}^3$	$2S_{10}$	$5\sigma_d$		
A_{1g}	1	1	1	1	1	1	1	1		$x^2+y^2,\ z^2$
A_{2g}	1	1	1	-1	1	1	1	-1	R_z	
E_{1g}	2	$2\cos 72°$	$2\cos 144°$	0	2	$2\cos 72°$	$2\cos 144°$	0	R_x, R_y	xz, yz
E_{2g}	2	$2\cos 144°$	$2\cos 72°$	0	2	$2\cos 144°$	$2\cos 72°$	0		$x^2-y^2,\ xy$
A_{1u}	1	1	1	1	-1	-1	-1	-1		
A_{2u}	1	1	1	-1	-1	-1	-1	1	z	
E_{1u}	2	$2\cos 72°$	$2\cos 144°$	0	-2	$-2\cos 72°$	$-2\cos 144°$	0	x, y	
E_{2u}	2	$2\cos 144°$	$2\cos 72°$	0	-2	$-2\cos 144°$	$-2\cos 72°$	0		

D_{6d}	E	$2S_{12}$	$2C_6$	$2S_4$	$2C_3$	$2S_{12}^5$	C_2	$6C_2'$	$6\sigma_d$		
A_1	1	1	1	1	1	1	1	1	1		$x^2 + y^2,\ z^2$
A_2	1	1	1	1	1	1	1	-1	-1	R_z	
B_1	1	-1	1	-1	1	-1	1	1	-1		
B_2	1	-1	1	-1	1	-1	1	-1	1	z	
E_1	2	$\sqrt{3}$	1	0	-1	$-\sqrt{3}$	-2	0	0	$x,\ y$	
E_2	2	1	-1	-2	-1	1	2	0	0		$x^2 - y^2,\ xy$
E_3	2	0	-2	0	2	0	-2	0	0		
E_4	2	-1	-1	2	-1	-1	2	0	0		
E_5	2	$-\sqrt{3}$	1	0	-1	$\sqrt{3}$	-2	0	0	$R_x,\ R_y$	$xz,\ yz$

S_4	E	S_4	C_2	S_4^3		
A	1	1	1	1	R_z	$x^2 + y^2,\ z^2$
B	1	-1	1	-1	z	$x^2 - y^2,\ xy$
E	$\left\{\begin{matrix}1 \\ 1\end{matrix}\right.$	$\begin{matrix}i \\ -i\end{matrix}$	$\begin{matrix}-1 \\ -1\end{matrix}$	$\left.\begin{matrix}-i \\ i\end{matrix}\right\}$	$x,\ y;\ R_x,\ R_y$	$xz,\ yz$

S_6	E	C_3	C_3^2	i	S_6^5	S_6			$\varepsilon = \exp(2\pi i/3);\ \varepsilon + \varepsilon^* = -1$
A_g	1	1	1	1	1	1	R_z	$x^2 + y^2,\ z^2$	
E_g	$\left\{\begin{matrix}1 \\ 1\end{matrix}\right.$	$\begin{matrix}\varepsilon \\ \varepsilon^*\end{matrix}$	$\begin{matrix}\varepsilon^* \\ \varepsilon\end{matrix}$	$\begin{matrix}1 \\ 1\end{matrix}$	$\begin{matrix}\varepsilon \\ \varepsilon^*\end{matrix}$	$\left.\begin{matrix}\varepsilon^* \\ \varepsilon\end{matrix}\right\}$	$R_x,\ R_y$	$x^2 - y^2,\ xy,\ xz,\ yz$	
A_u	1	1	1	-1	-1	-1	z		
E_u	$\left\{\begin{matrix}1 \\ 1\end{matrix}\right.$	$\begin{matrix}\varepsilon \\ \varepsilon^*\end{matrix}$	$\begin{matrix}\varepsilon^* \\ \varepsilon\end{matrix}$	$\begin{matrix}-1 \\ -1\end{matrix}$	$\begin{matrix}-\varepsilon \\ -\varepsilon^*\end{matrix}$	$\left.\begin{matrix}-\varepsilon^* \\ -\varepsilon\end{matrix}\right\}$	$x,\ y$		

S_8	E	S_8	C_4	S_8^3	C_2	S_8^5	C_4^3	S_8^7			$\varepsilon = \exp(2\pi i/8);\ \varepsilon + \varepsilon^* = \sqrt{2}$
A	1	1	1	1	1	1	1	1	R_z	$x^2 + y^2,\ z^2$	
B	1	-1	1	-1	1	-1	1	-1	z		
E_1	$\left\{\begin{matrix}1 \\ 1\end{matrix}\right.$	$\begin{matrix}\varepsilon \\ \varepsilon^*\end{matrix}$	$\begin{matrix}i \\ -i\end{matrix}$	$\begin{matrix}-\varepsilon^* \\ -\varepsilon\end{matrix}$	$\begin{matrix}-1 \\ -1\end{matrix}$	$\begin{matrix}-\varepsilon \\ -\varepsilon^*\end{matrix}$	$\begin{matrix}-i \\ i\end{matrix}$	$\left.\begin{matrix}\varepsilon^* \\ \varepsilon\end{matrix}\right\}$	$x,\ y;\ R_x,\ R_y$		
E_2	$\left\{\begin{matrix}1 \\ 1\end{matrix}\right.$	$\begin{matrix}i \\ -i\end{matrix}$	$\begin{matrix}-1 \\ -1\end{matrix}$	$\begin{matrix}-i \\ i\end{matrix}$	$\begin{matrix}1 \\ 1\end{matrix}$	$\begin{matrix}i \\ -i\end{matrix}$	$\begin{matrix}-1 \\ -1\end{matrix}$	$\left.\begin{matrix}-i \\ i\end{matrix}\right\}$		$x^2 - y^2,\ xy$	
E_3	$\left\{\begin{matrix}1 \\ 1\end{matrix}\right.$	$\begin{matrix}-\varepsilon^* \\ -\varepsilon\end{matrix}$	$\begin{matrix}-i \\ i\end{matrix}$	$\begin{matrix}\varepsilon \\ \varepsilon^*\end{matrix}$	$\begin{matrix}-1 \\ -1\end{matrix}$	$\begin{matrix}\varepsilon^* \\ \varepsilon\end{matrix}$	$\begin{matrix}i \\ -i\end{matrix}$	$\left.\begin{matrix}-\varepsilon \\ -\varepsilon^*\end{matrix}\right\}$		$xz,\ yz$	

T	E	$4C_3$	$4C_3^2$	$3C_2$			$\varepsilon = \exp(2\pi i/3); \; \varepsilon + \varepsilon^* = -1$
A	1	1	1	1			$x^2 + y^2 + z^2$
E	$\begin{cases}1\\1\end{cases}$	$\begin{matrix}\varepsilon\\\varepsilon^*\end{matrix}$	$\begin{matrix}\varepsilon^*\\\varepsilon\end{matrix}$	$\begin{matrix}1\\1\end{matrix}$			$x^2 + y^2 - 2z^2; \; x^2 - y^2$
T	3	0	0	-1		$\begin{Bmatrix}x,\; y,\; z\\R_x, R_y, R_z\end{Bmatrix}$	$xy,\; xz,\; yz$

T_h	E	$4C_3$	$4C_3^2$	$3C_2$	i	$4S_6$	$4S_6^5$	$3\sigma_h$		$\varepsilon = \exp(2\pi i/3); \; \varepsilon + \varepsilon^* = -1$
A_g	1	1	1	1	1	1	1	1		$x^2+y^2+z^2$
E_g	$\begin{cases}1\\1\end{cases}$	$\begin{matrix}\varepsilon\\\varepsilon^*\end{matrix}$	$\begin{matrix}\varepsilon^*\\\varepsilon\end{matrix}$	$\begin{matrix}1\\1\end{matrix}$	$\begin{matrix}1\\1\end{matrix}$	$\begin{matrix}\varepsilon\\\varepsilon^*\end{matrix}$	$\begin{matrix}\varepsilon^*\\\varepsilon\end{matrix}$	$\begin{matrix}1\\1\end{matrix}$		$x^2+y^2-2z^2, x^2-y^2$
T_g	3	0	0	-1	3	0	0	-1	R_x, R_y, R_z	$xy,\; xz,\; yz$
A_u	1	1	1	1	-1	-1	-1	-1		
E_u	$\begin{cases}1\\1\end{cases}$	$\begin{matrix}\varepsilon\\\varepsilon^*\end{matrix}$	$\begin{matrix}\varepsilon^*\\\varepsilon\end{matrix}$	$\begin{matrix}1\\1\end{matrix}$	$\begin{matrix}-1\\-1\end{matrix}$	$\begin{matrix}-\varepsilon\\-\varepsilon^*\end{matrix}$	$\begin{matrix}-\varepsilon^*\\-\varepsilon\end{matrix}$	$\begin{matrix}-1\\-1\end{matrix}$		
T_u	3	0	0	-1	-3	0	0	1	$x,\; y,\; z$	

T_d	E	$8C_3$	$3C_2$	$6S_4$	$6\sigma_d$		
A_1	1	1	1	1	1		$x^2 + y^2 + z^2$
A_2	1	1	1	-1	-1		
E	2	-1	2	0	0		$x^2 + y^2 - 2z^2, \; x^2 - y^2$
T_1	3	0	-1	1	-1	$R_x, \; R_y, \; R_z$	
T_2	3	0	-1	-1	1	$x, \; y, \; z$	$xy,\; xz,\; yz$

O	E	$8C_3$	$3C_2$	$6C_4$	$6C_2'$		
A_1	1	1	1	1	1		$x^2 + y^2 + z^2$
A_2	1	1	1	-1	-1		
E	2	-1	2	0	0		$x^2 + y^2 - 2z^2,$
T_1	3	0	-1	1	-1	$R_x, \; R_y, \; R_z, \; x, \; y, \; z$	$x^2 - y^2$
T_2	3	0	-1	-1	1		$xy,\; xz,\; yz$

O_h	E	$8C_2$	$6C_2$	$6C_4$	$3C_2(=C_4^2)$	i	$6S_4$	$8S_6$	$3\sigma_h$	$6\sigma_d$		
A_{1g}	1	1	1	1	1	1	1	1	1	1		$x^2+y^2+z^2$
A_{2g}	1	1	-1	-1	1	1	-1	1	1	-1		
E_g	2	-1	0	0	2	2	0	-1	2	0		$x^2+y^2-2z^2,$
T_{1g}	3	0	-1	1	-1	3	1	0	-1	-1	R_x,R_y,R_z	x^2-y^2
T_{2g}	3	0	1	-1	-1	3	-1	0	-1	1		$xy,\ xz,\ yz$
A_{1u}	1	1	1	1	1	-1	-1	-1	-1	-1		
A_{2u}	1	1	-1	-1	1	-1	1	-1	-1	1		
E_u	2	-1	0	0	2	-2	0	1	-2	0		
T_{1u}	3	0	-1	1	-1	-3	-1	0	1	1	x,y,z	
T_{2u}	3	0	1	-1	-1	-3	1	0	1	-1		

I	E	$12C_5$	$12C_5^2$	$20C_3$	$15C_2$		
A	1	1	1	1	1		$x^2+y^2+z^2$
T_1	3	$\dfrac{1+\sqrt{5}}{2}$	$\dfrac{1-\sqrt{5}}{2}$	0	-1	$\left\{\begin{array}{c}x,y,z\\R_x,R_y,R_z\end{array}\right\}$	
T_2	3	$\dfrac{1-\sqrt{5}}{2}$	$\dfrac{1+\sqrt{5}}{2}$	0	-1		
G	4	-1	-1	1	0		
H	5	0	0	-1	1		$\left\{\begin{array}{c}x^2+y^2-2z^2,\\x^2-y^2,\ xy,\ xz,\ yz\end{array}\right\}$

I_h	E	$12C_5$	$12C_5^2$	$20C_3$	$15C_2$	i	$12S_{10}$	$12S_{10}^3$	$20S_6$	15σ		
A_g	1	1	1	1	1	1	1	1	1	1		$x^2+y^2+z^2$
T_{1g}	3	$\dfrac{1+\sqrt{5}}{2}$	$\dfrac{1-\sqrt{5}}{2}$	0	-1	3	$\dfrac{1-\sqrt{5}}{2}$	$\dfrac{1+\sqrt{5}}{2}$	0	-1	$\left\{\begin{array}{c}R_x,R_y,\\R_z\end{array}\right\}$	
T_{2g}	3	$\dfrac{1-\sqrt{5}}{2}$	$\dfrac{1+\sqrt{5}}{2}$	0	-1	3	$\dfrac{1+\sqrt{5}}{2}$	$\dfrac{1-\sqrt{5}}{2}$	0	-1		
G_g	4	-1	-1	1	0	4	-1	-1	1	0		$\left\{\begin{array}{c}x^2+y^2-2z^2\\x^2-y^2,\\xy,xz,yz\end{array}\right\}$
H_g	5	0	0	-1	1	5	0	0	-1	1		
A_u	1	1	1	1	1	-1	-1	-1	-1	-1		
T_{1u}	3	$\dfrac{1+\sqrt{5}}{2}$	$\dfrac{1-\sqrt{5}}{2}$	0	-1	3	$\dfrac{-(1-\sqrt{5})}{2}$	$\dfrac{-(1+\sqrt{5})}{2}$	0	1	x,y,z	
T_{2u}	3	$\dfrac{1-\sqrt{5}}{2}$	$\dfrac{1+\sqrt{5}}{2}$	0	-1	-3	$\dfrac{-(1+\sqrt{5})}{2}$	$\dfrac{-(1-\sqrt{5})}{2}$	0	1		
G_u	4	-1	-1	1	0	-4	1	1	-1	0		
H_u	5	0	0	-1	1	-5	0	0	1	-1		

$C_{\infty v}$	E	$2C_\infty^\phi$	$\cdots$	$\infty\sigma_v$		
$A_1 \equiv \Sigma^+$	1	1	$\cdots$	1	z	$x^2 + y^2,\ z^2$
$A_2 \equiv \Sigma^-$	1	1	$\cdots$	-1	R_z	
$E_1 \equiv \Pi$	2	$2\cos\phi$	$\cdots$	0	$x,\ y;\ R_x,\ R_y$	$xz,\ yz$
$E_2 \equiv \Delta$	2	$2\cos 2\phi$	$\cdots$	0		$x^2 - y^2,\ xy$
$E_3 \equiv \phi$	2	$2\cos 3\phi$	$\cdots$	0		
$\cdots$	$\cdots$	$\cdots$	$\cdots$	$\cdots$		

$D_{\infty h}$	E	$2C_\infty^\phi$	$\cdots$	$\infty\sigma_v$	i	$2S_\infty^\phi$	$\cdots$	∞C_2	σ_h		
Σ_g^+	1	1	$\cdots$	1	1	1	$\cdots$	1	1		x^2+y^2, z^2
Σ_g^-	1	1	$\cdots$	-1	1	1	$\cdots$	-1	1	R_z	
Π_g	2	$2\cos\phi$	$\cdots$	0	2	$-2\cos\phi$	$\cdots$	0	-2	R_x, R_y	xz, yz
Δ_g	2	$2\cos 2\phi$	$\cdots$	0	2	$2\cos 2\phi$	$\cdots$	0	2		x^2-y^2, xy
$\cdots$	$\cdots$	$\cdots$	$\cdots$	$\cdots$	$\cdots$	$\cdots$	$\cdots$	$\cdots$	$\cdots$		
Σ_u^+	1	1	$\cdots$	1	-1	-1	$\cdots$	-1	-1	z	
Σ_u^-	1	1	$\cdots$	-1	-1	-1	$\cdots$	1	-1		
Π_u	2	$2\cos\phi$	$\cdots$	0	-2	$2\cos\phi$	$\cdots$	0	2	$x,\ y$	
Δ_u	2	$2\cos 2\phi$	$\cdots$	0	-2	$-2\cos 2\phi$	$\cdots$	0	-2		
$\cdots$	$\cdots$	$\cdots$	$\cdots$	$\cdots$	$\cdots$	$\cdots$	$\cdots$	$\cdots$	$\cdots$		

APPENDIX B: CORRELATION TABLES

Correlation tables show the relationships between the irreducible representations of a group and its subgroups. This appendix contains tables for the groups most commonly encountered with real molecular problems. More complete tables can be found in Wilson et al. [97] and elsewhere.

More than one correlation exists for some subgroups, depending upon which specific elements of the group are carried over into the subgroup. In such cases the symmetry element of the parent group which becomes a member of the subgroup is noted in the tables. The correlation of smaller groups with their subgroups can be found within the tables for supergroups of both the small group and its subgroup. For example, the correlation of C_{2v} with C_2 may be found in the table for D_{2h} (Table B. 1) by comparing the species under C_{2v} $[C_2(z)]$ with those for C_2 $[C_2(z)]$, or any other listed pair with a common C_2 axis. The correlation of certain of the larger groups with their smaller subgroups may be accomplished through an intermediate subgroup. Thus, to correlate O_h with C_3, correlate O_h with D_{3d}, and then correlate D_{3d} with C_3.

TABLE B.1

D_{2h}	D_2 C_{2v} $C_2(z)$	C_{2v} $C_2(y)$	C_{2v} $C_2(x)$	C_{2h} $C_2(z)$	C_{2h} $C_2(y)$	C_{2h} $C_2(x)$	C_2 $C_2(z)$	C_2 $C_2(y)$	C_2 $C_2(x)$	C_3 $\sigma(xy)$	C_3 $\sigma(zx)$	C_3 $\sigma(yz)$	C_i
A_g	A A_1	A_1	A_1	A_g	A_g	A_g	A	A	A	A'	A'	A'	A_g
B_{1g}	B_1 A_2	B_2	B_1	A_g	B_g	B_g	A	B	B	A'	A''	A''	A_g
B_{2g}	B_2 B_1	A_2	B_2	B_g	A_g	B_g	B	A	B	A''	A'	A''	A_g
B_{3g}	B_3 B_2	B_1	A_2	B_g	B_g	A_g	B	B	A	A''	A''	A'	A_g
A_u	A A_2	A_2	A_2	A_u	A_u	A_u	A	A	A	A''	A''	A''	A_u
B_{1u}	B_1 A_1	B_1	B_2	A_u	B_u	B_u	A	B	B	A''	A'	A'	A_u
B_{2u}	B_2 B_2	A_1	B_1	B_u	A_u	B_u	B	A	B	A'	A''	A'	A_u
B_{3u}	B_3 B_1	B_2	A_1	B_u	B_u	A_u	B	B	A	A'	A'	A''	A_u

TABLE B.2

D_{3h}	C_{3h}	D_3	C_{3v}	C_{2v} $\sigma_h \rightarrow \sigma_v(zy)$	C_3	C_2	C_3 σ_h	C_3 σ_v
A_1'	A'	A_1	A_1	A_1	A	A	A'	A'
A_2'	A'	A_2	A_2	B_2	A	B	A'	A''
E'	E'	E	E	$A_1 + B_2$	E	A + B	2A'	A' + A''
A_1''	A''	A_1	A_2	A_2	A	A	A''	A''
A_2''	A''	A_2	A_1	B_1	A	B	A''	A'
E''	E''	E	E	$A_2 + B_1$	E	A + B	2A''	A' + A''

TABLE B.3

D_{4h}	D_4	D_{2d} $C_2' \to C_2'$	D_{2d} $C_2'' \to C_2'$	C_{4v}	C_{4h}	C_4	S_4	D_{2h} C_2'	D_{2h} C_2''	D_2 C_2'	D_2 C_2''	C_{2v} C_2,σ_v	C_{2v} C_2,σ_d	C_{2v} C_2'	C_{2v} C_2''
A_{1g}	A_1	A_1	A_1	A_1	A_g	A	A	A_g	A_g	A	A	A_1	A_1	A_1	A_1
A_{2g}	A_2	A_2	A_2	A_2	A_g	A	A	B_{1g}	B_{1g}	B_1	B_1	A_2	A_2	B_1	B_1
B_{1g}	B_1	B_1	B_2	B_1	B_g	B	B	A_g	B_{1g}	A	B_1	A_1	A_2	A_1	B_1
B_{2g}	B_2	B_2	B_1	B_2	B_g	B	B	B_{1g}	A_g	B_1	A	A_2	A_1	B_1	A_1
E_g	E	E	E	E	E_g	E	E	$B_{2g}+B_{3g}$	$B_{2g}+B_{3g}$	B_2+B_3	B_2+B_3	B_1+B_2	B_1+B_2	A_2+B_2	A_2+B_2
A_{1u}	A_1	B_1	B_1	A_2	A_u	A	B	A_u	A_u	A	A	A_2	A_2	A_2	A_2
A_{2u}	A_2	B_2	B_2	A_1	A_u	A	B	B_{1u}	B_{1u}	B_1	B_1	A_1	A_1	B_2	B_2
B_{1u}	B_1	A_1	A_2	B_2	B_u	B	A	A_u	B_{1u}	A	B_1	A_2	A_1	A_2	B_2
B_{2u}	B_2	A_2	A_1	B_1	B_u	B	A	B_{1u}	A_u	B_1	A	A_1	A_2	B_2	A_2
E_u	E	E	E	E	E_u	E	E	$B_{2u}+B_{3u}$	$B_{2u}+B_{3u}$	B_2+B_3	B_2+B_3	B_1+B_2	B_1+B_2	A_1+B_1	A_1+B_1

Other subgroups: $3C_{2h}$, $3C_2$, $3C_s$, C_i

TABLE B.4

D_{5h}	D_5	C_{5v}	C_{5h}	C_5	C_{2v} $\sigma_h \rightarrow \sigma(zx)$	C_2	C_3 σ_h	C_3 σ_v
A_1'	A_1	A_1	A'	A	A_1	A	A'	A'
A_2'	A_2	A_2	A'	A	B_1	B	A'	A''
E_1'	E_1	E_1	E_1'	E_1	$A_1 + B_1$	$A + B$	$2A'$	$A' + A''$
E_2'	E_2	E_2	E_2'	E_2	$A_1 + B_1$	$A + B$	$2A'$	$A' + A''$
A_1''	A_1	A_2	A''	A	A_2	A	A''	A''
A_2''	A_2	A_1	A''	A	B_2	B	A''	A'
E_1''	E_1	E_1	E_1''	E_1	$A_2 + B_2$	$A + B$	$2A''$	$A' + A''$
E_2''	E_2	E_2	E_2''	E_2	$A_2 + B_2$	$A + B$	$2A''$	$A' + A''$

TABLE B.5

D_{6h}	D_6	D_{3h} C_2'	D_{3h} C_2''	C_{6v}	C_{6h}	D_{3d} C_2''	D_{3d} C_2'	D_{2h} $\sigma_v\to\sigma(yz)$ $\sigma_h\to\sigma(xy)$	C_6	C_{3h}	D_3 C_2'	D_3 C_2''	C_{3v} σ_v	C_{3v} σ_d	S_6	D_2
A_{1g}	A_1	A_1'	A_1'	A_1	A_g	A_{1g}	A_{1g}	A_g	A	A'	A_1	A_1	A_1	A_1	A_g	A
A_{2g}	A_2	A_2'	A_2'	A_2	A_g	A_{2g}	A_{2g}	B_{1g}	A	A'	A_2	A_2	A_2	A_2	A_g	B_1
B_{1g}	B_1	A_1''	A_2''	B_2	B_g	A_{2g}	A_{1g}	B_{2g}	B	A''	A_1	A_2	A_2	A_1	A_g	B_2
B_{2g}	B_2	A_2''	A_1''	B_1	B_g	A_{1g}	A_{2g}	B_{3g}	B	A''	A_2	A_1	A_1	A_2	A_g	B_3
E_{1g}	E_1	E''	E''	E_1	E_{1g}	E_g	E_g	$B_{2g} + B_{3g}$	E_1	E''	E	E	E	E	E_g	$B_2 + B_3$
E_{2g}	E_2	E'	E'	E_2	E_{2g}	E_g	E_g	$A_g + B_{1g}$	E_2	E'	E	E	E	E	E_g	$A + B_1$
A_{1u}	A_1	A_1''	A_1''	A_2	A_u	A_{1u}	A_{1u}	A_u	A	A''	A_1	A_1	A_2	A_2	A_u	A
A_{2u}	A_2	A_2''	A_2''	A_1	A_u	A_{2u}	A_{2u}	B_{1u}	A	A''	A_2	A_2	A_1	A_1	A_u	B_1
B_{1u}	B_1	A_1'	A_2'	B_1	B_u	A_{2u}	A_{1u}	B_{2u}	B	A'	A_1	A_2	A_1	A_2	A_u	B_2
B_{2u}	B_2	A_2'	A_1'	B_2	B_u	A_{1u}	A_{2u}	B_{3u}	B	A'	A_2	A_1	A_2	A_1	A_u	B_3
E_{1u}	E_1	E'	E'	E_1	E_{1u}	E_u	E_u	$B_{2u} + B_{3u}$	E_1	E'	E	E	E	E	E_u	$B_2 + B_3$
E_{2u}	E_2	E''	E''	E_2	E_{2u}	E_u	E_u	$A_u + B_{1u}$	E_2	E''	E	E	E	E	E_u	$A + B_1$

Other subgroups: $2C_{2v}$, $3C_{2h}$, C_3, $3C_2$, $3C_s$, C_i

TABLE B.6

D_{2d}	S_4	D_2 $C_2 \rightarrow C_2(z)$	C_{2v}	C_2 C_2	C_2 C_2'	C_3
A_1	A	A	A_1	A	A	A'
A_2	A	B_1	A_2	A	B	A''
B_1	B	A	A_2	A	A	A''
B_2	B	B_1	A_1	A	B	A'
E	E	$B_2 + B_3$	$B_1 + B_2$	2B	A + B	A' + A''

TABLE B.7

D_{3d}	D_3	C_{3v}	S_6	C_3	C_{2h}	C_2	C_s	C_i
A_{1g}	A_1	A_1	A_g	A	A_g	A	A'	A_g
A_{2g}	A_2	A_2	A_g	A	B_g	B	A''	A_g
E_g	E	E	E_g	E	$A_g + B_g$	A + B	A' + A''	$2A_g$
A_{1u}	A_1	A_2	A_u	A	A_u	A	A''	A_u
A_{2u}	A_2	A_1	A_u	A	B_u	B	A'	A_u
E_u	E	E	E_u	E	$A_u + B_u$	A + B	A' + A''	$2A_u$

TABLE B.8

D_{4d}	D_4	C_{4v}	S_8	C_4	C_{2v}	C_2 C_2	C_2 C_2'	C_3
A_1	A_1	A_1	A	A	A_1	A	A	A'
A_2	A_2	A_2	A	A	A_2	A	B	A''
B_1	A_1	A_2	B	A	A_2	A	A	A''
B_2	A_2	A_1	B	A	A_1	A	B	A'
E_1	E	E	E_1	E	$B_1 + B_2$	2B	A + B	A' + A''
E_2	$B_1 + B_2$	$B_1 + B_2$	E_2	2B	$A_1 + A_2$	2A	A + B	A' + A''
E_3	E	E	E_3	E	$B_1 + B_2$	2B	A + B	A' + A''

TABLE B.9

D_{5d}	D_5	C_{5v}	C_5	C_2	C_s	C_i
A_{1g}	A_1	A_1	A	A	A'	A_g
A_{2g}	A_2	A_2	A	B	A''	A_g
E_{1g}	E_1	E_1	E_1	$A + B$	$A' + A''$	$2A_g$
E_{2g}	E_2	E_2	E_2	$A + B$	$A' + A''$	$2A_g$
A_{1u}	A_1	A_2	A	A	A''	A_u
A_{2u}	A_2	A_1	A	B	A'	A_u
E_{1u}	E_1	E_1	E_1	$A + B$	$A' + A''$	$2A_u$
E_{2u}	E_2	E_2	E_2	$A + B$	$A' + A''$	$2A_u$

TABLE B.10

D_{6d}	D_6	C_{6v}	C_6	D_3	C_{3v}	D_{2d}
A_1	A_1	A_1	A	A_1	A_1	A_1
A_2	A_2	A_2	A	A_2	A_2	A_2
B_1	A_1	A_2	A	A_1	A_2	B_1
B_2	A_2	A_1	A	A_2	A_1	B_2
E_1	E_1	E_1	E_1	E	E	E
E_2	E_2	E_2	E_2	E	E	$B_1 + B_2$
E_3	$B_1 + B_2$	$B_1 + B_2$	$2B$	$A_1 + A_2$	$A_1 + A_2$	E
E_4	E_2	E_2	E_2	E	E	$A_1 + A_2$
E_5	E_1	E_1	E_1	E	E	E

Other subgroups: D_2, C_{2v}, S_4, C_3, $2C_2$, C_s

TABLE B.11

T_d	T	D_{2d}	C_{3v}	S_4	D_2	C_{2v}	C_3	C_2	C_3
A_1	A	A_1	A_1	A	A	A_1	A	A	A'
A_2	A	B_1	A_2	B	A	A_2	A	A	A''
E	E	$A_1 + B_1$	E	$A + B$	$2A$	$A_1 + A_2$	E	$2A$	$A' + A''$
T_1	T	$A_2 + E$	$A_2 + E$	$A + E$	$B_1 + B_2 + B_3$	$A_2 + B_1 + B_2$	$A + E$	$A + 2B$	$A' + 2A''$
T_2	T	$B_2 + E$	$A_1 + E$	$B + E$	$B_1 + B_2 + B_3$	$A_1 + B_1 + B_2$	$A + E$	$A + 2B$	$2A' + A''$

TABLE B.12

O_h	O	T_d	T_h	D_{4h}	D_{3d}
A_{1g}	A_1	A_1	A_g	A_{1g}	A_{1g}
A_{2g}	A_2	A_2	A_g	B_{1g}	A_{2g}
E_g	E	E	E_g	$A_{1g} + B_{1g}$	E_g
T_{1g}	T_1	T_1	T_g	$A_{2g} + E_g$	$A_{2g} + E_g$
T_{2g}	T_2	T_2	T_g	$B_{2g} + E_g$	$A_{1g} + E_g$
A_{1u}	A_1	A_2	A_u	A_{1u}	A_{1u}
A_{2u}	A_2	A_1	A_u	B_{1u}	B_{1u}
E_u	E	E	E_u	$A_{1u} + B_{1u}$	E_u
T_{1u}	T_1	T_2	T_u	$A_{2u} + E_u$	$A_{2u} + E_u$
T_{2u}	T_2	T_1	T_u	$B_{2u} + E_u$	$A_{1u} + E_u$

Other subgroups: T, D_4, D_{2d}, C_{4h}, C_{4v}, $2D_{2h}$, D_3, C_{3v}, S_6, C_4, S_4, $3C_{2v}$, $2D_2$, $2C_{2h}$, C_3, $2C_2$, S_2, C_s.

ACKNOWLEDGMENTS

This work was supported in part by a Faculty Research Grant of the University of Massachusetts. Special thanks are due to Messrs. John V. Gazzaniga and Rafi Manoukian for critically reading certain portions of the manuscript. The typing of the manuscript by my wife and her continued support through the preparation of this chapter have been greatly appreciated.

REFERENCES

1. F. A. Cotton, "Chemical Applications of Group Theory," Interscience, New York, 1963.

2. H. H. Jaffe and Milton Orchin, "Symmetry in Chemistry," Wiley, New York, 1965.

3. E. B. Wilson, Jr., J. C. Decius, and P. C. Cross, "Molecular Vibrations," McGraw-Hill, New York, 1955.

4. G. Herzberg, "Molecular Spectra and Molecular Structure, vol. 2, Infrared and Raman Spectra of Polyatomic Molecules," Van Nostrand, New Jersey, 1945.

5. E. B. Wilson, Jr., J. C. Decius, and P. C. Cross, "Molecular Vibrations," p. 105, McGraw-Hill, New York, 1955.

6. P. Dawson, *Spectrochim. Acta, 28A,* 715 (1972).

7. D. P. Strommen and E. P. Lippincott, *J. Chem. Educ., 49,* 341 (1972).

8. D. M. Adams, *Annual Reports, 68A,* 47 (1971).

9. H. Stammreich, D. Bassi, O. Sala, and H. Siebert, *Spectrochim. Acta, 13,* 192 (1958).

10. R. L. Carter and C. E. Bricker, *Spectrochim. Acta, 29A,* 253 (1973).

11. I. R. Beattie and J. R. Horder, *J. Chem. Soc. (A),* 2655 (1969).

12. L. M. Schwartz and K. W. Schwartz, *Comput. Programs Biomed., 2,* 257 (1972).

13. E. B. Wilson, Jr., J. C. Decius, and P. C. Cross, "Molecular Vibrations," p. 151, McGraw-Hill, New York, 1955.

14. H. H. Claassen, C. L. Chernick, and J. G. Malm, *J. Am. Chem. Soc., 85,* 1927 (1963).

15. C. P. Courtoy, *Ann. Soc. Sci. Bruxelles,* Ser. 1, *73,* 5 (1959).

16. G. Amat and M. Pimbert, *J. Mol. Spectrosc., 16,* 278 (1965).

17. H. R. Gordon and T. K. McCubbin, Jr., *J. Mol. Spectrosc., 19,* 137 (1966).

18. R. E. Scruby, J. R. Lacher and J. D. Park, *J. Chem. Phys., 19,* 386 (1951).

19. A. H. Nielsen, *J. Chem. Phys., 22,* 659 (1954).

20. J. Vanderryn, *J. Chem. Phys., 30,* 331 (1959).

21. D. A. Dows, *J. Chem. Phys., 31,* 1637 (1959).

22. D. A. Dows and G. Bottger, *J. Chem. Phys., 34,* 689 (1961).

23. D. E. Irish and G. E. Walrafen, *J. Chem. Phys., 46,* 378 (1967).

24. J. P. Mathieu and M. Lounsbury, *Discuss. Faraday Soc., 9,* 196 (1950).

25. R. E. Hester and R. A. Plane, *J. Chem. Phys., 40,* 411 (1964).

26. J. R. Ferraro, *J. Mol. Spectrosc., 4,* 99 (1960).

27. H. Lee and J. K. Wilmhurst, *Aust. J. Chem., 17,* 943 (1964).

28. L. I. Katzin, *J. Inorg. Nucl. Chem., 25,* 1129 (1963).

29. R. E. Hester and R. A. Plane, *Inorg. Chem., 3,* 769 (1964).

30. D. E. Irish and A. R. Davis, *Can. J. Chem., 46,* 943 (1968).

31. R. P. Oertel and R. A. Plane, *Inorg. Chem., 7,* 1192 (1968).

32. J. T. Miller and D. E. Irish, *Can. J. Chem., 45,* 147 (1967).

33. B. Strauch and L. N. Komissarova, *Z. Chem.*, *6*, 474 (1966).

34. (a) S. C. Wait, Jr. and A. T. Ward, *J. Chem. Phys.*, *44*, 448 (1966); (b) G. E. Walrafen and D. E. Irish, *J. Chem. Phys.*, *40*, 911 (1964); (c) R. K. Khanna, J. Lingscheid, and J. C. Decius, *Spectrochim. Acta*, *20*, 1109 (1963); (d) G. J. Janz and D. W. James, *J. Chem. Phys.*, *35*, 739 (1961); (e) J. K. Wilmshurst and S. Senderoff, *J. Chem. Phys.*, *35*, 1078 (1961).

35. G. M. Barrow, "Introduction to Molecular Spectroscopy," Chap. 7, McGraw-Hill, New York, 1962.

36. (a) F. N. Masri and W. H. Fletcher, *J. Chem. Phys.*, *52*, 5759 (1970); (b) R. J. H. Clark and D. M. Rippon, *Chem. Commun.*, 1295 (1971); (c) S. Sportouch, C. Lacoste, and R. Gaufres, *J. Mol. Struct.*, *9*, 119 (1971); (d) R. J. H. Clark and D. M. Rippon, *J. Mol. Spectrosc.*, *44*, 479 (1972).

37. H. H. Hyman, *Science*, *145*, 773 (1964).

38. H. H. Claasen, G. L. Goodman, J. G. Malm, and F. Schreiner, *J. Chem. Phys.*, *42*, 1229 (1965).

39. P. A. Agron, G. M. Begun, H. A. Levy, A. A. Mason, G. G. Jones, and D. F. Smith, *Science*, *139*, 842 (1963).

40. R. D. Burbank, W. E. Falconer, and W. A. Sunder, *Science*, *178*, 1285 (1972).

41. J. L. Huston, *J. Phys. Chem.*, *71*, 3339 (1967).

42. H. H. Claassen, E. L. Gasner, H. Kim, and J. L. Huston, *J. Chem. Phys.*, *49*, 253 (1968).

43. (a) D. F. Smith, *J. Chem. Phys.*, *38*, 270 (1963); (b) H. H. Claassen, C. L. Chernick, and J. G. Malm, "Noble-Gas Compounds," (H. H. Hyman, ed.), p. 287, University of Chicago Press, Chicago, 1963; (c) H. H. Claassen and G. Knapp, *J. Am. Chem. Soc.*, *86*, 2341 (1964); (d) H. Selig, H. H. Claassen, C. L. Chernick, J. G. Malm, and J. L. Huston, *Science*, *143*, 1322 (1964).

44. D. R. Lide and J. J. Comeford, *Spectrochim. Acta*, *21*, 497 (1965).

45. W. E. Hobbs, *J. Chem. Phys.*, *28*, 1220 (1958).

46. R. Paetzold and K. H. Ziegenbalg, *Z. Chem.*, *4*, 461 (1964).

47. (a) R. Gillespie, *J. Chem. Educ.*, *40*, 295 (1963); (b) R. J. Gillespie, "Noble-Gas Compounds," (H. H. Hyman, ed.), p. 333, University of Chicago Press, Chicago, 1963.

48. S. Cohen, J. R. Lacher, and J. D. Park, *J. Am. Chem. Soc.*, *81*, 3480 (1959).

49. M. Ito and R. West, *J. Am. Chem. Soc.*, *85*, 2580 (1963).

50. W. M. Macintyre and M. S. Werkema, *J. Chem. Phys.*, *42*, 3563 (1964).

51. R. L. Carter, *J. Chem. Educ.*, *48*, 297 (1971).

52.   H. Stammreich, D. Bassi, and O. Sala, *Spectrochim. Acta, 12,*
      403 (1958).

53.   L. Glasser, *J. Chem. Educ., 44,* 502 (1967).

54.   "International Tables for X-ray Crystallography," (2nd ed.),
      (N. T. M. Henry and K. Lonsdale, eds.), Kynoch Press, Birming-
      ham, England, 1965.

55.   R. S. Halford, *J. Chem. Phys., 14,* 8 (1946).

56.   L. Couture, *J. Chem. Phys., 15,* 153 (1947).

57.   (a) J. E. Bertie and J. W. Bell, *J. Chem. Phys., 54,* 160 (1971);
      (b) J. E. Bertie and R. Kopelman, *J. Chem. Phys., 55,* 3613
      (1971).

58.   M. Tinkham, "Group Theory and Quantum Mechanics," Chap. 2,
      McGraw-Hill, New York, 1964.

59.   A. Bhagavantam and T. Venkatarayuda, "Theory of Groups and Its
      Application to Physical Problems," (3rd ed.), University of
      Andhra Press, Waltair, India, 1964. (Now published by Academic
      Press, New York.)

60.   S. S. Mitra, Vibrational Selection Rules in Solids in "Solid
      State Physics," (F. Seitz and D. Turnbull, eds.), vol. 13,
      Academic Press, New York, 1962.

61.   D. F. Hornig, *J. Chem. Phys., 16,* 1063 (1948).

62.   H. Winston and R. S. Halford, *J. Chem. Phys., 17,* 607 (1949).

63.   S. S. Mitra, *Z. Kristallogr., 116,* 149 (1961).

64.   R. L. Carter and C. E. Bricker, *Spectrochim. Acta, 27A,* 569
      (1971).

65.   B. D. Saksena, *Proc. Indian Acad. Sci., Sect. A, 11,* 229 (1940).

66.   I. R. Beattie and T. R. Gilson, *Proc. R. Soc. London, Ser. A.,*
      307, 407 (1968).

67.   T. C. Damen, S. P. S. Porto, and B. Tell, *Phys. Rev., 142,* 570
      (1966).

68.   F. D. Bloss, "An Introduction to the Methods of Optical Crystal-
      lography," p. 160 ff., Holt, Rinehart, and Winston, New York,
      1961.

69.   R. Loudon, *Ad. Phys., 13,* 423 (1964); *14,* 621 (1965).

70.   S. S. Mitra, Infrared and Raman Spectra Due to Lattice Vibra-
      tions, in "Optical Properties of Solids," Plenum Press, N.Y.,
      1969, Chapter 14.

71.   J. F. Scott and S. P. S. Porto, *Phys. Rev., 161,* 903 (1967).

72.   R. Newman and R. S. Halford, *J. Chem. Phys., 18,* 1276 (1950).

73.   (a) F. D. Bloss, "An Introduction to Optical Crystallography,"
      Holt, Rinehart and Winston, N.Y., 1961; (b) E. E. Wahlstrom,
      "Optical Crystallography," (3rd ed.), New York, 1960.

74. (a) E. J. Ambrose, A. Elliot, and R. B. Temple, *Proc. R. Soc.,
     Ser. A, 206,* 192 (1951); (b) G. C. Pimentel and A. L. McClellan,
     J. Chem. Phys., 20, 270 (1952).

75. E. Charney, J. Opt. Soc. Am., *45,* 980 (1955).

76. D. L. Wood and S. S. Mitra, *J. Opt. Soc. Am., 48,* 537 (1958).

77. R. L. Carter and C. E. Bricker, *Spectrochim. Acta, 27A,* 825
     (1971).

78. J. A. A. Ketelaar and E. Wegerif, *Rec. Trav. Chim., 57,* 1269
     (1938).

79. F. M. Jaeger and J. E. Zanstra, *Rec. Trav. Chim., 51,* 1013 (1932).

80. I. R. Beattie, K. M. S. Livingston, D. J. Reynolds, and G. A.
     Ozin, *J. Chem. Soc.,* 1970A, 1210.

81. D. E. Sands, A. Zalkin, and R. E. Elson, *Acta Crystallogr., 12,*
     21 (1969).

82. R. S. Drago, "Physical Methods in Inorganic Chemistry," pp. 220-
     222, Rheinhold, New York, 1965.

83. F. A. Miller and C. H. Wilkins, *Anal. Chem., 24,* 1253 (1952).

84. F. A. Miller, G. L. Carlson, F. F. Bentley, and W. H. Jones,
     Spectrochim. Acta, 16, 135 (1960).

85. K. Nakamoto, "Infrared Spectra of Inorganic and Coordination
     Compounds," (2nd ed.), Wiley, New York, 1970.

86. H. Siebert, "Anwendungen der Schwingungsspektroskopie in der
     Anorganischen Chemie," Springer-Verlag, Berlin, 1966.

87. *Spectroscopic Properties of Inorganic and Organometallic Com-
     pounds: A Specialist Periodical Report,* vol. 1-5, The Chemical
     Society, London, 1968-72.

88. H. A. Willis, R. G. J. Miller, D. M. Adams, and H. A. Gebbie,
     Spectrochim. Acta, 19, 1457 (1963).

89. W. B. Barish, G. T. Behnke, and K. Nakamoto, *Appl. Spectrosc.,
     22,* 337 (1968).

90. D. M. Adams, "Metal-Ligand and Related Vibrations," Arnold,
     London, 1967.

91. J. R. Ferraro, "Low-Frequency Vibrations of Inorganic and Co-
     ordination Compounds," Plenum Press, New York, 1971.

92. (a) K. Nakamoto, K. Shobatake, and B. Hutchinson, *Chem. Commun.,*
     1451 (1969); (b) K. Shobatake and K. Nakamoto, *J. Am. Chem.
     Soc., 92,* 3332, 3339 (1970); (c) B. Hutchinson, J. Takemoto, and
     K. Nakamoto, *J. Am. Chem. Soc., 92,* 3335 (1970); (d) K. Nakamoto,
     C. Udovich, and J. Takemoto, *J. Am. Chem. Soc., 92,* 3973 (1970);
     (e) J. Takemoto and K. Nakamoto, *Chem. Commun.,* 1017 (1970);
     (f) N. Ohkaku and K. Nakamoto, *Inorg. Chem., 10,* 798 (1971);
     (g) Y. Saito, J. Takemoto, B. Hutchinson, and K. Nakamoto,
     Inorg. Chem., 11, 2003 (1972); (h) Y. Saito, M. Cordes, and

    K. Nakamoto, *Spectrochim. Acta, 28A,* 1459 (1972); (i) A. Müller, F. Königer, K. Nakamoto, and N. Ohkaku, *Spectrochim. Acta, 28A,* 1933 (1972); (j) A. Müller, N. Weinstock, K. H. Schmidt, K. Nakamoto, and C. W. Schläpfer, *Spectrochim. Acta, 28A,* 2293 (1972).

93. J. H. van der Maas and A. Tolk, *Spectrochim. Acta, 18,* 235 (1962).

94. T. Nortia and E. Kontas, *Spectrochim. Acta, 29A,* 1493 (1973).

95. W. Kiefer and H. J. Bernstein, *Appl. Spectrosc., 25,* 609 (1971).

96. R. L. Carter, *Spectrosc. Lett., 5,* 401 (1972).

97. E. B. Wilson, Jr., J. C. Decius, and P. C. Cross, "Molecular Vibrations," pp. 333-340, McGraw-Hill, New York, 1955.

Chapter 3

ORGANOMETALLIC COMPOUNDS:
VIBRATIONAL ANALYSIS

Walter F. Edgell

Department of Chemistry
Purdue University
West Lafayette, Indiana

I. INTRODUCTION

More than 2,000 papers are published each year on the vibrational
spectra of organometallic compounds and related molecular systems.
With few exceptions, each of these papers advances our knowledge of
these important chemical systems. It is with temerity, then, that
the author sits down to write this chapter. A comprehensive treatment of the work carried out in this area during recent years is a
herculean task and beyond the present confines of time and space.

This chapter concerns itself with an exposition of some of the
principal types of spectroscopic studies being made in the field.
It is hoped that the reader will come away with some feeling for the
methods which are used to study this aspect of organometallic chemistry. Basic treatments of molecular vibration and the spectra
which they produce are found in many places. Chapter 1 of this book
is an excellent starting place. The book by Wilson et al. [1] provides a foremost exposition of the fundamental aspects of the subject. This chapter deals with the vibrational spectra of organometallic compounds. Virtually all of the papers currently being
published fall into this category. Recent developments have provided
new spectroscopic ways of probing the microscopic character of the
ionic solutions involved in organometallic chemistry. This has
sparked a growing number of such studies, which promise to increase
the knowledge of this important aspect of this field of chemistry.
These studies form the topic of Chapter 4. The emphasis in both
chapters is on the methods used in these researches.

In this chapter, the role of vibrational spectroscopy in chemical analysis is illustrated. These spectra are an index to the
nature and behavior of these compounds. They have their origin in
the molecular motion which takes place, and the relation between
this motion and the spectra which appear will be illustrated. In
some cases, a comprehensive study of the infrared and Raman spectra
of an organometallic compound yields all its frequencies of vibration.
These provide the basis for a detailed treatment of structure and
bonding in the molecule. Fortunately, it is possible to obtain some

very useful information about molecular structure and chemical bonding from more abbreviated studies. Examples of both kinds of research are given. One tool for a detailed study of molecular structure and bonding is a molecular force field study. An example demonstrates the type of spectroscopic data required and the results which are obtained. Here also one can make some very useful and needed computations without resort to a complete vibrational assignment.

II. MOLECULAR POINT GROUPS AND SELECTION RULES

Infrared and Raman spectra contain a wealth of information about the microscopic nature and behavior of the substances which give rise to them. This is one reason that this type of study of organometallic compounds is so popular. To obtain this knowledge, one must have some understanding of how these spectra are generated by the molecules. In the gas phase, the center of mass of a molecule moves with time (translates) while the molecule tumbles (rotates) and undergoes structural deformation (vibrates). The infrared and Raman bands of concern to us here arise from the excitation of vibrational motion in a molecule. At the same time, there is a change in the rotational (and translational) state of the system which provides the wealth of structure in the bands. Since we are not interested here in high resolution studies, we need concern ourselves only with the vibrational motion.

The first question one might ask is "How many vibrational bands will appear in the infrared or Raman spectrum of a molecule?". There are $3n - 6$ different (independent) ways in which a nonlinear molecule of n atoms can vibrate. In the general case, each of these modes of vibration has a different frequency and can give a band in both the infrared and Raman spectrum. However, some molecules have symmetry. It is a general consequence of the existence of molecular symmetry that some of the different modes of vibration have the same frequency of vibration. A further result of symmetry is that the intensity of the infrared or Raman band for some modes of vibration must have zero intensity.

The primary source of symmetry is the geometry of the molecule.
This symmetry is expressed by a *point group*. For example, the
$Ni(CO)_4$ molecule has four CO groups arranged tetrahedrally in space
about the Ni atom--a geometric arrangement which corresponds to the
point group T_d. The consequences of this symmetry are collected in
Table 1. First, this table indicates that a molecule of this kind
may have vibrations of symmetry type A_1, A_2, E, F_1, and F_2. Columns
2, 3, and 4 indicate that a vibration of type A_1 does not appear in
the infrared spectrum (zero intensity) and gives rise to a band in
the Raman spectrum which is polarized. Further, one sees that an F_2
frequency appears in both kinds of spectra and that the Raman band
will have a depolarization ratio of 3/4. Column 5 gives the number
of different modes which give rise to a single frequency of vibration
of the symmetry class.

All of the above information applies to any molecule with T_d
symmetry. The corresponding information for a molecule of any other
symmetry is found in tables collected in basic texts which treat the
symmetry aspects of molecular vibration [1-3]. It is also possible
to say how many frequencies of each symmetry type occur for any specific molecule [1-3]. This is collected in the last column of Table
1 for $Ni(CO)_4$. One symbol appears here for each frequency. Thus,
there occur two A_1, two E, one F_1, and four F_2 frequencies. This is
often expressed in the language of group theory as

TABLE 1

Consequences of Geometric Symmetry

for the $Ni(CO)_4$ Molecule

T_d	IR	Raman	ρ	d	Frequency
A_1	0	a	p	1	r,q
A_2	0	0	-	1	-
E	0	a	3/4	2	α,β
F_1	0	0	-	3	β
F_2	a	a	3/4	3	r,q,α,β

0 = zero intensity; a = intensity >0; ρ = depolarization ratio; p = polarized range; d = degeneracy

$$\Gamma = 2A_1 + 2E + F_1 + 4F_2$$

It is also possible to say something about the types of molecular
distortion which occur in the vibrations of each symmetry class from
symmetry alone. Thus, only bond deformations can occur in the A_1
modes, only angle distortions in the E and F_1 modes, and both in F_2
modes. One obtains this knowledge through the formation of symmetry
coordinates (or the equivalent)[1-3].

This is as far as one can go on the basis of symmetry alone.
However, experience gained from the general study of the vibrational
dynamics and aided by any experimental and dynamical results for
molecules with similar structural units will frequently permit one
to be rather specific about the character expected for the several
modes of vibration. For example, the basic distortions which are
found in metal carbonyl vibrations are known. Thus, CO stretching
vibrations occur with the basic q distortion shown in Fig. 1. Starting with these displacements and using standard group theory methods
[1-3], one can form the modes with q type distortions for each symmetry class. The result is that CO stretching vibrations with this
basic distortion at the CO units exist as class A_1 and F_2 vibrations
for $Ni(CO)_4$. This is indicated by the symbol q in the A_1 and F_2
lines of column 6 of Table 1. The "metal-carbon" stretching vibra-

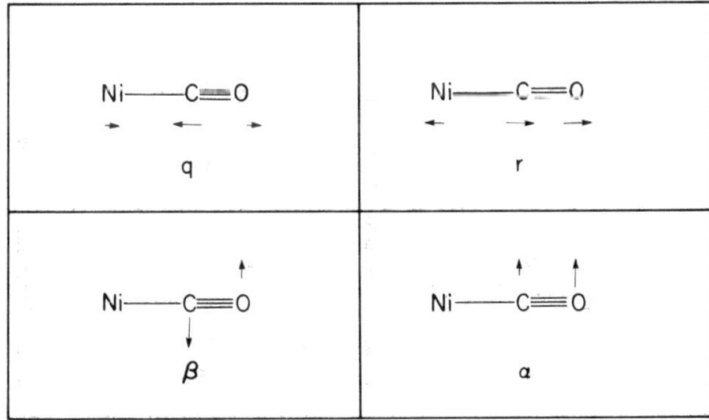

FIG. 1. Basic vibrational distortions in $Ni(CO)_4$.

tions occur with the basic r-type distortion of Fig. 1 and they also give rise to A_1 and F_2 modes. Here the CO group moves (virtually) as a rigid unit. Metal carbonyl vibrations take place with the basic angle distortions shown in Fig. 1. In the α distortion, only the CMC angles are deformed; in the β bend, both CMC and MCO angles are deformed. Bending modes can be formed with these distortions, as indicated in Table 1. Column 6 provides a good description of the modes of vibration to be expected for $Ni(CO)_4$. Thus, the two A_1 modes are expected to be r and q type stretching vibrations, the E frequencies to be α and β bending vibrations, the F_1 frequency to come from β motion, and the four F_2 frequencies to be q and r stretches and α and β bends.

This section provides an example of how one can use the geometric symmetry of an organometallic molecule to provide detailed expectations for its infrared and Raman spectra. It is possible to determine how many frequencies of vibration exist, if these frequencies appear in the spectra, and the depolarization ratio of the Raman bands, and to obtain some notion of the type of distortions which occur in the vibrations of each symmetry class. When information is available from earlier work on related molecules, one can often provide a surprisingly accurate description of the motion expected in each mode of vibration. Many spectroscopic studies of organometallic compounds seek to make a detailed assignment of the observed bands to the expected fundamental frequencies. These studies use the basic methods outlined above.

Before leaving this section some comments should be made about the applicability of such selection rules. They are derived under the assumption that the atoms move under forces which have the molecular symmetry. This is strictly true for molecules in the gas phase. But it is also the case for some liquid organometallic compounds and for both liquid and solid organometallic compounds when dissolved in a variety of solvents. What is required here is that there be no strong interaction of the molecule with one in its near-neighbor environment. Matrix isolation studies, which surround a

molecule with an inert substrate, also provide an environment for
molecular selection rules. In contrast, this situation usually is
not obtained for crystalline organometallic compounds, where the
field of the neighbors causes the molecular symmetry to give way to
site symmetry. When coupling between the molecule and its neighbors
occurs, the crystal factor group governs the expectations from symmetry [4].

III. HIDDEN SYMMETRY AND SELECTION RULES

The last section showed how the point group of a molecule specifies
its symmetry and influences the spectra that it generates. It will
be the contention of this section that other sources of symmetry are
also present in some molecules. This "hidden symmetry" then also
exerts an influence on the spectra and must be reckoned with in order
to obtain a satisfactory understanding of the observations. A method of obtaining the selection rules in the presence of hidden symmetry is indicated [5].

The nature of the vibrations which occur for a molecule is
fundamentally a dynamical problem. The atomic displacements in each
mode result from the kinetic and potential energy. These displacements determine the dipole moment generated in the vibration, which
fixes the intensity of the infrared band and the polarizability
change, which determine the Raman band intensity. Thus, the fundamental quantities are the kinetic and potential energies. Symmetry
in these quantities is the proximate cause of the symmetry in the
molecular vibrations. Of course, the geometry of a molecule is the
primary source of symmetry in the kinetic and potential energy. But
it is not the only source.

Molecules are made up of a number of structural units. Common
ones which occur in organometallic compounds are methyl and phenyl
groups, CO, NO, metal-hydrogen, and metal-oxygen groups, and a variety of organic ligands. It is common to find the same structural
unit repeated in the same organometallic molecule. For example,
metal carbonyls contain a number of CO groups, and even a single

methyl group contains three C-H bonds. In such a case, two different situations can occur, which are illustrated in Fig. 2. In
$Ni(CO)_4$, all the CO groups are equivalent in the group theory sense
of the term, i.e., one CO is carried into each one of the others by
the operations of the point group T_d. All the CO groups of $Mn(CO)_5Br$,
however, are not equivalent, because no operation of the point group
C_{4v} will carry the axial CO into a planar CO, or vice-versa. The
group theoretical procedures of the C_{4v} point group treat the planar
and axial CO's as different units. The vibrational consequences of
the presence of the planar CO groups are devised as though the axial
CO unit were not there--it might as well be a PF_3 group. In addition, the consequences of the axial CO unit are obtained as though
the planar units were not there. Thus, the point group C_{4v} recog-
nizes the geometric difference between the axial and planar CO units
in $Mn(CO)_5Br$ and treats them as fundamentally different units.

   To make the discussion easier to follow, let us consider the
specific case of the CO stretching vibrations in $Mn(CO)_5Br$. The
atomic displacements associated with the distortion of a single CO
unit are illustrated by q in Fig. 1. Let q be the coordinate governing this distortion at CO group number 1 (see Fig. 2) and k be the

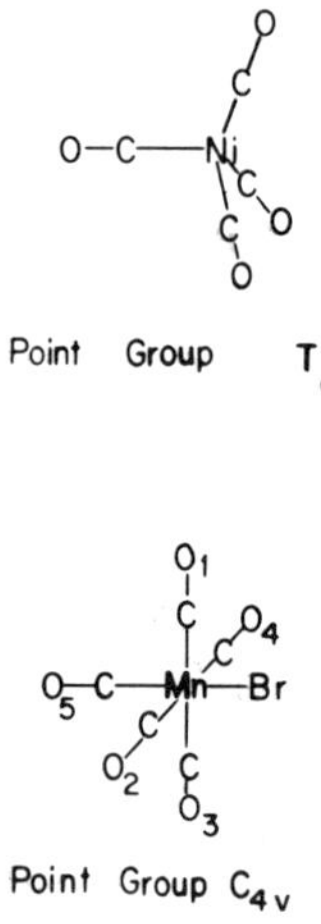

FIG. 2. Equivalent (T_d) and nonequivalent (C_{4v}) CO groups in
metal carbonyl compounds.

constant defining the force which resists its distortion. Then the potential energy involved in the CO stretching modes of $Mn(CO)_5Br$ is

$$2V = \underline{q}\underline{K}\tilde{q} \tag{1}$$

where $\underline{q}$ is the row matrix $(q_1 q_2 q_3 q_4 q_5)$, $\tilde{q}$ is its transpose, and $\underline{K}$ is the "force constant" matrix

$$
\begin{array}{c|ccccc}
 & q_1 & q_2 & q_3 & q_4 & q_5 \\
\hline
q_1 & k & k' & k'' & k' & k''' \\
q_2 &   & k & k' & k'' & k''' \\
q_3 &   &   & k & k' & k''' \\
q_4 &   &   &   & k   & k''' \\
q_5 &   &   &   &   & k^o \\
\end{array} \tag{2}
$$

Only the upper triangular part of $\underline{K}$ is shown because it is symmetrical. Here k is the constant for the planar CO units and k^o that for the axial unit. Three interaction constants k', k'', and k''' occur because of the geometry. This potential energy is invariant under the operations of the C_{4v} group and the detailed expression of this symmetry is given by the character table (see Table 2). Its use in the standard ways [1-3] leads to the following CO stretching frequencies

$$\Gamma = 2A_1 + B_1 + E$$

whose spectral expectations are collected in Table 3.

TABLE 2

Character Table for the Point Group C_{4v}

C_{4v}	E	$2C_4$	C_2	$2\sigma_v$	$2\sigma_d$
A_1	1	1	1	1	1
A_2	1	1	1	-1	-1
B_1	1	-1	1	1	-1
B_2	1	-1	1	-1	1
E	2	0	-2	0	0

TABLE 3

Consequences of Geometric Symmetry for the

CO Stretching Frequencies of $Mn(CO)_5Br$

C_{4v}	IR	Raman	ρ	d	Number CO Frequencies
A_1	a	a	p	1	2
A_2	0	0	-	1	0
B_1	0	a	3/4	1	1
B_2	0	a	3/4	1	0
E	a	a	3/4	2	1

But how different are the axial and planar CO units in $Mn(CO)_5Br$? What would be the spectroscopic consequences if the five CO groups were the same in every way but their position in space? The potential energy which defines this situation is given by the $\underline{K}$ matrix

$$
\begin{array}{c|ccccc}
 & q_1 & q_2 & q_3 & q_4 & q_5 \\
\hline
q_1 & k & k' & k' & k' & k' \\
q_2 &   & k & k' & k' & k' \\
q_3 &   &   & k & k' & k' \\
q_4 &   &   &   & k & k' \\
q_5 &   &   &   &   & k
\end{array}
\qquad (3)
$$

The potential energy in this situation would be invarient to all the operations which permute the five CO groups. There are 120 such operations and they form the permutation group P_5, whose character table appears as Table 4. The permutations defined by the table symbols are

$$
\begin{aligned}
E: &\quad (12345) \rightarrow (12345) \\
R_5: &\quad (12345) \rightarrow (51234) \\
R_4: &\quad (12345) \rightarrow (41235) \\
R_3: &\quad (12345) \rightarrow (31245) \\
R_2: &\quad (12345) \rightarrow (34125) \\
m: &\quad (12345) \rightarrow (12354) \\
R_3m: &\quad (12345) \rightarrow (31254)
\end{aligned}
$$

TABLE 4

Character Table of the Permutation Group P_5

P_5	E	$24R_5$	$30R_4$	$20R_3$	$15R_2$	$10m$	$20R_3m$
A_1	1	1	1	1	1	1	1
A_2	1	1	-1	1	1	-1	-1
G_1	4	-1	0	1	0	2	-1
G_2	4	-1	0	1	0	-2	1
H_1	5	0	-1	-1	1	1	1
H_2	5	0	1	-1	1	-1	-1
I	6	1	0	0	-2	0	0

The nomenclature used here is readily apparent from the specification that R_3 means that CO number 3 has been moved to the number 1 position, 1 to the 2 position, 3 to the 1 position, and 4 and 5 left where they were. The seven irreducible representations of this group are designated by the sumbols A_1, A_2, G_1, G_2, H_1, H_2 and I in an obvious extension of the Schoenflies nomenclature. The degeneracy of a frequency of vibration of each of these seven symmetry types is 1, 1, 4, 4, 5, 5, and 6, as shown by the character of the unity operation E.

The use of this table with the standard procedures shows the following CO stretching frequencies for $Mn(CO)_5Br$,

$$\Gamma = A_1 + G_1$$

with the spectral expectations found in Table 5 [5]. If the five CO groups were in fact the same, these considerations show that only two CO stretching frequencies would occur, one nondegenerate and one with fourfold degeneracy, that both of these would appear in the infrared and Raman spectrum, and that the A_1 Raman band would be polarized while the ρ value for G_1 band would be 3/4. These expectations from P_5 symmetry differ from those which result from C_{4v} symmetry. Which situation actually occurs in nature can be determined by observing the infrared and Raman spectrum of $Mn(CO)_5Br$.

TABLE 5

Consequences of P_5 Symmetry

for the CO Stretching Frequencies of $Mn(CO)_5Br$

P_5	IR	Raman	ρ	d	Number CO Frequencies
A_1	a	a	p	1	1
G_1	a	a	3/4	4	1

Before making the appeal to experiment, there is still another possibility which must be considered. Let us carry out the following hypothetical sequence. Suppose one starts with an $Mn(CO)_5Br$ molecule in which all five CO units are the same. Then, make small changes in each CO group in a way which reflects C_{4v} symmetry. This will produce small deviations from the P_5 expectations outlined above. As long as the distortion of the CO groups from their symmetrical P_5 condition is small, the deviations which occur will vary linearly with the distortion. We call this range of difference between axial and planar CO groups the *linear deviation range* [5]. As the distortion of the CO groups grows into the large category, the deviations from P_5 behavior become substantial and no longer vary linearly with distortion.

Let us consider the spectral expectations for the linear deviation range. It has been shown that the atom displacements of the modes of vibration which occur in this range satisfy the symmetry requirements of the point group C_{4v} and, at the same time, satisfy the symmetry requirements of the P_5 permutation group [5]. The CO stretching frequencies have the description

$$\Gamma = A_1/A_1 + A_1/G_1 + B_1/G_1 + E/G_1$$

Thus, there is a frequency whose mode of vibration transforms as A_1 under the operations of C_{4v} and A_1 under the operations of P_5, a frequency whose mode transforms as A_1 under C_{4v} and G_1 under P_5, a frequency whose mode shows both B_1 and G_1 symmetry, and lastly, one which shows both E and G_1 symmetry. This symmetry situation is re-

presented by the symbol C_{4v}/P_5. The spectral expectations for the
CO stretching vibration in this eventuality have been obtained [5]
and are found in Table 6.

We can summarize these results as follows. The *geometry* of
the $Mn(CO)_5Br$ molecule is described by the point group C_{4v}. This,
however, does not uniquely determine the symmetry of the kinetic
and potential energy associated with the vibrations. If there is a
pronounced difference between the axial and planar CO groups, the
CO stretching frequencies will follow C_{4v} selection rules. On the
other hand, if the CO groups are the same except for the spatial
positions, P_5 selection rules are expected. Finally, if the differences between axial and planar CO groups are small enough to fall
into the linear deviation range, the CO stretching frequencies follow C_{4v}/P_5 selection rules. The expectations are different for the
three cases. The comparison with experiment is made in the next
section.

In this section, we have introduced a fundamental principle:
the proximate cause of symmetry in molecular vibrations is the symmetry present in the kinetic and potential energy that is relevant
[5,6]. It is this symmetry which yields the spectral expectations.
Of course, the geometry of the molecule, expressed by its point
group, shapes this symmetry. In the last section, vibrational selection rules were derived for $Ni(CO)_4$ from the geometric symmetry. But

TABLE 6

Consequences of C_{4v}/P_5 Symmetry
for the CO Stretching Frequencies of $Mn(CO)_5Br$

C_{4v}/P_5	IR	Raman	ρ	d	Number CO Frequencies
A_1/A_1	a	a	p	1	1
A_1/G_1	a	a	~3/4	1	1
B_1/G_1	0	a	3/4	1	1
E/G_1	a	~0	–	2	1

geometry is not the only source of symmetry in the relevant kinetic and potential energy. In this section, it is seen that three different sets of selection rules are possible for the $Mn(CO)_5Br$ molecule, whose geometric symmetry is described by the C_{4v} point group. The physical feature which determines which of the three applies is the difference between an axial and a planar CO group in this molecule. This hidden symmetry occurs frequently in organometallic compounds.

There is another way in which the kinetic and potential energy governing vibrations may show symmetry which differs from the symmetry of the molecular geometry. This can, and generally does, occur when "localization" is present in the modes of vibration of a molecule. Since localization is the source of "group frequencies" in vibrational spectra, one sees that its occurrence is also common in organometallic molecules. It generally appears in the "ligand" bands. A general treatment of "localization symmetry" and its spectral consequences has appeared in Ref. 6.

IV. VIBRATIONAL SPECTRA AND ASSIGNMENTS
IN $Ni(CO)_4$ AND $Mn(CO)_5Br$

To make an assignment of an observed band is to give a description of the mode of vibration from which it arises. This description may be minimal. For example, one may be able to say only that a band arises from metal-carbon stretching motion, or it may specifically include the symmetry aspects of the motion, as discussed in the previous two sections. An example here would be the assignment of the 336 cm^{-1} infrared band in RuO_4 to the F_2 metal-oxygen bending mode in the T_d point group [7]. The ultimate in an assignment is to be specific about the atom displacements in the mode as a consequence of computations of the nature of the vibrations. Thus, one can say that the carbon atom moves 0.14 Å while the oxygen atom moves 0.10 Å in the A_1 CO stretching mode of $Ni(CO)_4$.

Finding the frequencies for all the modes of vibration of an organometallic compound is usually an arduous spectroscopic task. This often involves studies of the spectra of the gas, liquid, and

solution phases of the compound. In recent years, such data have
been supplemented by studies of the compound in low-temperature matrix isolation and in the pure crystal phase. Even this is sometimes not enough to "coax out" the last frequency or two. A useful
technique, frequently resorted to in such cases, is to study derivatives of the molecule which lowers its symmetry. This works best
when one can form a derivative by replacing only a few atoms of the
molecule. The general procedure of making such an assignment is the
same whether the molecule is simple or complex. The vibrational assignment of the infrared and Raman spectra for $Ni(CO)_4$ will be given
as an example of a complete assignment for an organometallic compound. The assignment of the spectral data for $Mn(CO)_5Br$ in the
2000 cm^{-1} region will serve as an example of a compound assignment
for a limited spectral region.

Early studies of the infrared [8] and Raman [9] spectra of
$Ni(CO)_4$ were made by Crawford and Cross and their co-workers. These
studies established, in a rough way, the nature of the modes of vibration and the general spectral region in which each occurred. Fewer
bands were observed than called for by the T_d selection rules discussed above and the assignment that was made is imperfect. Nevertheless, a sound foundation had been laid--and not being perfect always accompanies the proud distinction of being first. The task of
establishing the remaining frequencies was taken up by Jones [10],
then by Jones and his co-workers [11], and by Bigorgne and his associates [12]. These workers extended the infrared studies into both
the far and near regions of the spectrum and obtained both Raman and
infrared spectra of the substance both in the gas phase and in the
solid state at low temperatures. Some of the gas phase measurements
were made at several meters of sample path length in folded-path
cells. Studies were made not only on the compound of normal isotopic
content but also with the substance containing heavy carbon ^{13}C atoms
and the substance with heavy oxygen ^{18}O atoms. Then, in a massive
effort to "flush out" the elusive β bending mode of type E symmetry,
the spectral studies were extended to substituted nickel carbonyl

compounds. The sum of all this work is given in Table 7, whose
data are taken from the paper of Jones, McDowell, and Goldblatt [13].
The frequencies listed are for the normal compound in CCl_4 solution.

With each advance in the spectroscopic data, there was an advance in the frequency assignment. We shall not be concerned here
with the various stages of this development, although they provide
a classic example of such work, but will present only the end result
[13]. The selection rules of Table 1 provide the guide to the assignment. This is supplemented by the general knowledge that the CO
stretches (the q modes of Fig. 1) will occur at high wave numbers,
the metal-carbon stretches (the r modes) and the NiCO bends (the β
modes) at middle wave numbers, and the CNiC bends (the α modes) at
low wave numbers.

The two frequencies observed in the 2000 cm^{-1} region are clearly
the CO stretching modes. The 2125 cm^{-1} frequency is assigned to the
A_1 mode because it gives a polarized Raman band and does not occur
in the infrared spectrum. The F_2 modes have the 2045 cm^{-1} value and,
as expected, appear as strong bands in both infrared and Raman spec-

TABLE 7

Spectral Data for $Ni(CO)_4$[a]

Infrared	Raman	ρ
	2125.0 w	p
2044.5 s	2044.5 s	
	598 w	p
455 m	461 w[b]	
422.5 s		
	379.8 m	p
91 w		
	78 s	

[a] In 5-10% solution in CCl_4; units are cm^{-1}.

[b] Seen in spectrum of pure liquid but not that
of dilute solution.

tra. The weakness of the Raman band arising from the A_1 mode has long been a puzzle, but Kettle, Paul, and Stamper [14] have recently offered an explanation in terms of the basic polarizability terms involved.

Two α bends are expected and two low frequencies are found. The F_2 modes can give a band in both infrared and Raman spectra, while the band arising from the E modes can appear only in the Raman spectrum. The observed band at 91 cm^{-1} is assigned to the F_2 frequency because it appears in the infrared spectrum. The Raman band at 78 cm^{-1} is then the E frequency.

The place to start in the middle frequency region is with the A_1 Ni-C stretch, because a polarized Raman band is expected from this mode. Two polarized Raman bands are found--the weak band at 598 cm^{-1} and the band of medium intensity at 380 cm^{-1}. The latter band must be assigned to this mode because of its intensity. The two F_2 frequencies are expected in both infrared and Raman spectra while the E frequency is allowed only in the Raman spectrum. It follows, therefore, that the infrared bands at 423 and 455 cm^{-1} must be the F_2 bands. While allowed in the Raman spectrum, they are too weak to appear in the spectrum of the dilute solution and only 455 cm^{-1} is seen in the spectrum of the pure liquid. It would be desirable to be able to say something about the form of the distortion of the molecule in each of these F_2 modes. One's intuition, sharpened from experience with other molecules, suggests that the stronger band involves the Ni-C stretching motion rather than the β bending. This consideration would assign the 423 cm^{-1} band to the Ni-C stretch and the 455 cm^{-1} band to β bending. Such an assignment is supported by vibration frequency calculations on two counts. First, this assignment gives "better looking" values of force constants but, more important, the calculation predicts the correct relative shifts of the bands on ^{13}C and ^{18}O substitution with the above assignment. There remains the E and F_1 frequencies without assigned values and only the weak, polarized Raman band at 598 cm^{-1} which has not been assigned. Before a polarization measurement was available for the 598 cm^{-1} band, it was usually assigned to the E frequency, since F_1 bands are not allowed in either Raman or infrared spectra. However, such an assignment is no longer tenable.

We have not discussed selection rules for overtone or combination bands but they may be obtained from the basic symmetry properties of the T_d point group [1-3]. These considerations show that the overtone of the F_1 frequency is allowed in the Raman spectrum and may be polarized. The 598 cm^{-1} band is assigned to this overtone, which places the F_1 β bend at about 300 cm^{-1}. Credence is given to such an assignment by the appearance of a band at 303 cm^{-1} in the Raman spectrum of solid $Ni(CO)_4$, where the T_d selection rules can be expected to break down as discussed in a previous section. Finally, we see that there is no observed band which can be assigned to the E type β bend. Nevertheless, it is still possible to estimate a frequency value for these modes. One can prepare derivatives of $Ni(CO)_4$ by replacing one or more CO groups by another group L. The vibrational spectra of these compounds must correlate with the $Ni(CO)_4$ spectra. Compounds of this type have been prepared, studied, and the correlations made. Out of this has come an estimated value of 380 cm^{-1} for the β bend of E summetry. The assignment is collected in Table 8.

Before leaving this topic, it might be pointed out that the early data consisted of infrared spectra for the gas and Raman spectra for the liquid. Shifts in band positions occur for this compound with a change in phase. They produce coincidences between gas phase frequencies and liquid state values which disappear when infrared and Raman spectra of the same phase are compared. A glance at the selection rules shows how this leads to error in the early assignments. This case is not an isolated instance of this difficulty. It is a common occurrence not only in metal carbonyl compounds and their derivatives but also in other organometallic compounds. It is such a hazard that it must be directly reckoned with in making vibrational assignments for compounds of interest to us in this chapter.

Now let us turn to a consideration of the CO stretching frequencies of $Mn(CO)_5Br$. This compound is an example of closely re-

TABLE 8

Vibrational Assignment for $Ni(CO)_4$

Designation	Type	Symmetry	ν
ν_1	q	A_1	2125.0 cm^{-1}
ν_2	r	A_1	379.8
ν_3	β	E	380[a]
ν_4	α	E	78
ν_5	q	F_2	2044.5
ν_6	β	F_2	455
ν_7	r	F_2	422.5
ν_8	α	F_2	91
ν_9	β	F_1	300[b]

[a]Estimated by comparison to related compounds, see text.

[b]From the assignment of 598 cm^{-1} to $2\nu_g$.

lated substances of the type $M(CO)_5X$, where X = Cl, Br, I and M = Mn, Tc, Re. The infrared spectra of these compounds in the 2000 cm^{-1} region have been established in a series of studies by Abel and co-workers [15], by Wilson [16], and by Kaesz and his associates [17]. The early work on the first few members of this class of compounds showed a variation in the number of bands which appeared and, of course, lead to uncertainty about which of them should be regarded as fundamentals. The situation was clarified by the extensive study of all members of the series which showed a pronounced variation of the intensity of certain bands with changes in X and M. Further, the study of the infrared spectrum of ^{13}C enriched $Mn(CO)_5Br$ establishes experimentally what often had been asserted from theory; namely, that certain of the bands in this region of the spectrum arise from the isotopic compounds present as a result of the natural abundance of ^{13}C. Thus, the frequencies of the isotopic species were established. The infrared frequencies arising from the fundamental modes of vibration of the normal $Mn(CO)_5Br$ molecule are listed in Table 9.

TABLE 9

Spectral Data for $Mn(CO)_5Br$[a]

Infrared	Raman	ρ
2135 m	2136 vs	0.03
	2082 s	0.77
2053 vs		
2003 s	2001 s	0.75

[a]CCl_4 solutions; units are cm^{-1}.

$Mn(CO)_5Br$ decomposes rapidly under the radiation of a Hg arc. As a result, repeated efforts to obtain its Raman spectrum failed, because Hg arcs were the sources used in these scattering experiments. With the advent of laser sources with output in the red end of the spectrum, it became feasible to obtain the Raman spectrum of this compound. Kaesz et al. [17b] report that the strongest band for $Mn(CO)_5Br$ occurs at 2079.5 cm^{-1} for the molecule dissolved in cyclohexane. The full Raman spectrum in the 2000 cm^{-1} region of $Mn(CO)_5Br$ was obtained in this laboratory, including the all important depolarization ratios [5]. These results are also found in Table 9. A photoelectric recording of this region may be seen in Fig. 3.

In the last section, three different sets of selection rules were deduced for these modes of $Mn(CO)_5Br$: the P_5 set, which would be obtained if all five CO groups were the same, except for their position in space; the C_{4v} set, which would be obtained if there were a pronounced difference between a planar and an axial CO group; and the C_{4v}/P_5 set, which would apply if the axial-planar difference is small enough to fall into the linear deviation range. An examination of Table 5 shows that two CO stretching frequencies would appear in the spectra if P_5 selection rules obtain. The spectra contain four fundamental bands (Table 9). One concludes, therefore, that the spectral results show that a difference does exist between axial and planar CO groups.

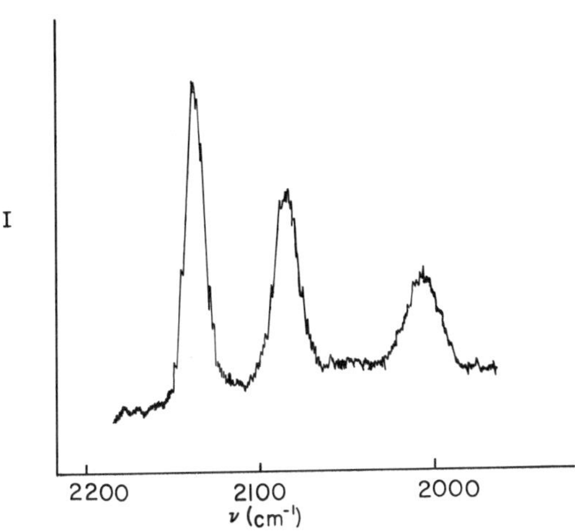

FIG. 3.   The Raman spectrum of $Mn(CO)_5Br$ near 2000 cm^{-1}.

The accepted principle in treating spectra is that the vibrational selection rules are fixed by the point group, which details the geometry of the molecule, and the spectra of the $M(CO)_5X$ molecules have been interpreted in terms of the C_{4v} selection rules [17-20].   These rules (Table 3) call for two A_1 bands and one E band in the infrared spectrum.   Three bands are indeed found and all agree that the bands at 2135 and 2003 cm^{-1} arise from the A_1 modes, while 2053 cm^{-1} is the E frequency.   This assignment is supported by intensity considerations and vibrational frequency calculations.   All of these vibrational assignment studies have worked solely with infrared data.   Only recently, have a little Raman spectral data become available for these molecules without, apparently, any depolarization ratio values.

Let us see how the Raman spectral data for $Mn(CO)_5Br$ fit the C_{4v} rules.   First, Table 3 shows that four Raman bands are expected-- only three are seen.   Second, the band at 2001 cm^{-1}, which is established from the infrared spectra and other studies as arising from an A_1 mode, does not have the low value of the depolarization ratio ρ expected (see Table 3).

Now let us examine these data from the point of view of the C_{4v}/P_5 expectations. Table 6 calls for three Raman bands with more than essentially zero intensity--three are found. Table 6 calls for only one Raman band with a low ρ value and for two with values essentially at 3/4. This is exactly what is observed. Further, the infrared spectral data fit the expectations of Table 6 equally well.

It is seen that the spectral data for $Mn(CO)_5Br$ in the 2000 cm^{-1} region do not agree with the expectations based upon the shape (geometry) of the molecule (C_{4v}) but, in fact, correspond to those of the C_{4v}/P_5 symmetry. The difference between the axial and a planar CO group is seen to be pronounced, as shown by the frequency values of the observed bands. However, this difference is not sufficient to establish the C_{4v} selection rules. The assignment of the spectral data is made in terms of the C_{4v}/P_5 rules and is given in Table 10 [5].

$Mn(CO)_5Br$ is not an isolated case. The hidden symmetry implied by the C_{4v}/P_5 selection rules for this compound is also found in the spectra of other organometallic compounds [5]. We should replace the idea that vibrational selection rules are determined by the point group which expresses the molecular geometry with the principle that they arise from the symmetry in the modes of motion. This latter symmetry, whose proximate cause is the nature of the kinetic and

TABLE 10

Vibrational Assignment for $Mn(CO)_5Br$

under C_{4v}/P_5 Symmetry

Designation	Symmetry	ν
ν_1	A_1/A_1	2135 cm^{-1}
ν_2	A_1/G_1	2003
ν_9	B_1/G_1	2082
ν_{15}	E/G_1	2053

potential energy, has a broader base than the geometry of the molecule. One should look for hidden symmetry whenever one finds several of the same atoms (or other chemical units) in a molecule in
spatial positions which are not equivalent under the covering operations of the point group of the molecule.

In this section, it is seen that one can understand the infrared and Raman spectra of a molecule and assign its features to the
expectations of the selection rules. A somewhat realistic consideration of what is required to collect spectral data for such an assignment is presented.

V. VIBRATIONAL SPECTRA AND ASSIGNMENTS FOR DIMETHYL TIN(IV) CHLORIDE

Many organometallic compounds fall into the category of larger molecules. Their spectra show the presence of "group" and "framework"
frequencies. The vibrational assignment for a compound of this kind
is considered in this section.

The dimethyl tin (IV) moiety forms a number of compounds in
which the tin atom has a coordination number of six. Among these
are the anions $(CH_3)_2SnX_4^{2-}$, with X = F, Cl, Br, NCS. Besides their
chemistry, these anions are of interest for their structure and for
the effect which the substitution of two methyl groups for X atoms
in SnX_6^{2-} has on the remaining Sn-X bonds. Infrared and Raman spectra of $(CH_3)_2SnX_4^{2-}$ salts have been studied by Beattie and McQuillan
[21], Wilkins and Haendler [22], Clark and Wilkins [23], Wada and
Okawara [24], and by Hobbs and Tobias [25]. The spectral data of
Hobbs and Tobias for $(CH_3)_2SnCl_4^{2-}$ are found in Table 11. Both infrared and Raman spectra were obtained from powdered samples of
$Cs_2(CH_3)_2SnCl_4$--the powder being suspended in mulls for the infrared
studies.

Figure 4 shows the trans model for the anion. The elements of
symmetry for this geometry are: a center of symmetry, a twofold
axis, and a plane of symmetry perpendicular to the axis. These
elements constitute the point group C_{2h} and the spectral expectations

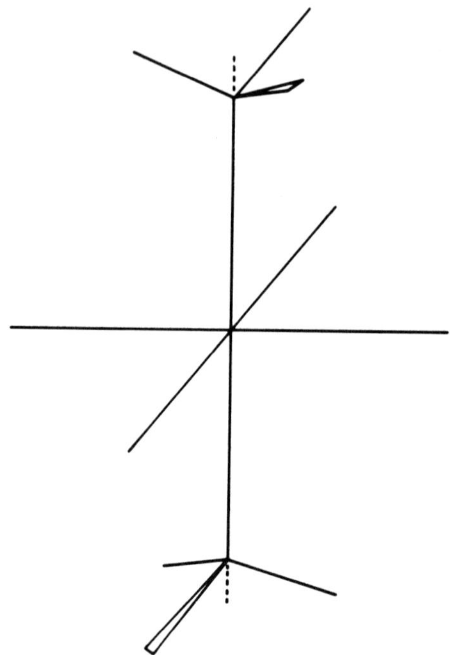

FIG. 4. The spatial arrangement of the atoms of the $(CH_3)_2SnCl_4^{2-}$ ion in the C_{2h} conformation.

which result from it are found in Table 12. The symbols therein
stand for the following types of deformation:

    ν CH_3 stretching
    δ CH_3 deformation
    r CH_3 rocking
    t CH_3 twisting
    R $SnCl_4$ stretching
    R' SnC_2 stretching
    β ClSnCl bending
    α CSnCl bending

It is at once seen that 33 frequencies are expected in the vibrational spectra, but less than half this number actually appear. Finding
fewer frequencies than one predicts from the geometric point group
is a common occurrence with larger molecules.

TABLE 11

Spectral Data[a] for $Cs_2(CH_3)_2SnCl_4$

	Infrared	Raman
		139 m
		197 m
		207 s
	235 s	
		508 vs
	508 s	
	793 m	
	1192 vw	
		1203 m
	1410 vw	
	1438 vw	
	2930 vvw	2926 m
	3020 vvw	3023 w

[a]Units are cm^{-1}.

TABLE 12

Consequences of C_{2h} Symmetry
for the $(CH_3)_2SnCl_4^{2-}$ Ion

C_{2v}	IR	Raman	ρ	d	Frequency
A_g	0	a	p	1	$2\nu,\delta,r,2R,R',\alpha$
B_g	0	a	3/4	1	$\nu,\delta,r,t,\beta,\alpha$
A_u	a	0	-	1	$\nu,\delta,r,t,R,\beta,\alpha$
B_u	a	0	-	1	$2\nu,3\delta,r,R,R',\beta,3\alpha$

Since a difficulty exists with larger molecules, one might start
its resolution by characterizing such a molecule. The symmetry operations of a small molecule move atoms to equivalent atoms. A new
feature appears with larger molecules--these operations result in

two kinds of transport phenomena: the transport of atom to atom within a functional group (CH_3, C_6H_5, C_5H_5, etc.) and the transport of one functional group to another [6]. The second dominant feature of larger molecules is the existence of group frequencies in their spectra, i.e., of frequencies which are associated with the presence of functional groups in the molecule.

The frequencies which characterize the functional groups that appear in larger molecules have been the subject of extended study and are collected in basic reference works [26,27]. For an excellent treatment of the theoretical basis for group frequencies, see the discussion of King and Crawford [28]. It has long been known that the presence in a molecule of functional groups, which could be re-garded as "tops" attached to a "framework," result in the occurrence of fewer fundamental frequencies than predicted by the molecular point group [29].

The existence of group frequencies for a molecule implies the existence of modes of vibration which confine the displacements (virtually) to the atoms of the functional group. A recent discussion of localization in the vibration of larger molecules emphasizes the "extra" symmetry which results [6]. Localization of the displacements in a mode of vibration to a functional group involves an exclusion of some parts of the molecule from deformation in the vibration. If there are modes of vibration which confine the motion to atoms of a group, then the orthogonality condition requires that there be other modes in which the functional group moves as a rigid unit. These are "framework" modes. Yet framework modes also involve exclusions, since they do not distort the functional group in question. Both kinds of exclusions enter the vibrational analysis in the same abstract way and localization is here defined to include both group and framework modes.

A vibration is localized when the number of displacement coordinates required for its description is less than the number required to express the kinetic and potential energy describing all vibrations of the molecule. Let q be the set of coordinates which span the whole vibrational domain. The vibrational kinetic and potential

energies for the molecule are $V(q)$ and $T(\dot{q})$, respectively. Let q_i

be the subset of the coordinates which are involved in the ith group

of localized modes. By definition, there are terms in $V(q)$ and $T(\dot{q})$

which do not involve the coordinates in q_i. These terms have no

connection with or influence on the vibrations of the ith group--and

can, therefore, be ignored in treating them. The truncated potential

and kinetic energies

$$V_{eff} = V(q_i)$$

$$T_{eff} = T(\dot{q}_i)$$

yield the frequencies and the normal coordinates for the modes of

vibration of the ith group through the secular equation

$$|\underline{V}_{eff} - \underline{T}_{eff}\, E| = 0$$

or its computational equivalent. Thus, localization is seen to be

the basis for factoring the kinetic and potential energies and, hence,

the vibrational problem. Here $\underline{V}_{eff}$ and $\underline{T}_{eff}$ are the matrices of

the coefficients of the coordinates in $V(q_i)$ and $T(\dot{q}_i)$. The symmetry

shown by the vibrations of the ith group will be the symmetry con-

tained in V_{eff} and T_{eff}. This symmetry is expressed by a group G

formed from those operations which permute the coordinates q_i and

leave V_{eff} and T_{eff} invariant. The group G for each set of localized

modes is characteristic of that set and is generally different from

the point group which details the geometry of the molecule. While G

is frequently isomorphic with one of the 32 point groups, it need

not be, in which case, it is a permutation group.

The group G depends upon the nature of V_{eff} and T_{eff}. It is

well-known that the presence of a methyl group in a molecule gives

rise to group frequencies [26-28]. But will the modes of the two

methyl groups in $(CH_3)_2SnCl_4^{2-}$ couple? Only experience with this or

similar systems can give the answer. We consider both possibilities

and obtain the selection rules for each. First, suppose that the

displacements in these group frequencies are localized in a single

methyl group. The vibrating moiety m is then the CH_3 group [6].
Suppose all C-H bonds were equivalent. Then

$$G = C_{3v} \qquad m = CH_3$$

and the selection rules would be those listed in Table 13. Thus,
one expects five methyl group frequencies in the spectrum of
$(CH_3)_2SnCl_4^{2-}$--two CH_3 stretches, two CH_3 deformations, and one CH_3
rock--if this were the symmetry of V_{eff} and T_{eff} for these modes.
Each mode of vibration requires one coordinate to describe the extent
of the distortion which occurs in it. The degeneracy of a given frequency is the number of modes which vibrate with that frequency. It
is also the number of coordinates which describe distortions of that
frequency. Since there are two methyl groups in the anion, it takes
one coordinate to describe the a_1 stretching motion in the first
methyl group and a second coordinate to describe the a_1 stretching
motion in the second methyl group. Thus, the a_1 stretching frequency
is twofold degenerate. In the same way, each type e frequency is
fourfold degenerate.

But G need not be C_{3v} for the methyl group. Note that one CH
bond is different from the other two in the model of Fig. 4. Then
the geometric site symmetry is C_s, and if this were reflected in
V_{eff} and T_{eff}, G would be C_s, or perhaps C_s/C_{3v}. These two possibilities yield expectations which do not correspond to observation
and are not considered further here.

TABLE 13

Consequences of C_{3v} Symmetry

on the Methyl Modes of $(CH_3)_2SnCl_4^{2-}$

when the Vibrating Moiety is CH_3

$G = C_{3v}$	IR	Raman	ρ	d	Frequency
a_1	a	a	p	2	ν, δ
a_2	0	0	-	2	t
e	a	a	3/4	4	ν, δ, r

Now let us consider the selection rules if the two methyl groups are coupled. Since a single mode involves distortion of both methyl groups, the vibrating moiety is now $m = (CH_3)_2$. A consideration of Fig. 4 suggests that the symmetry of V_{eff} and T_{eff} in the case of coupling is likely to be [6]

$$G = C_{3v}^{\ell} \times C_2^{i} \qquad m = (CH_3)_2$$

The ith operation of G group consists of two parts. The first part is a local operation L_j which exchanges atoms within the methyl group. The second part is an interchange operation I_k which moves one methyl group to another. Thus, the ith operation may be written as

$$G_i = L_j I_k$$

The local operations L_j form the group $L = C_{3v}^{\ell}$ (all C-H bonds the same) and the interchange operations I_k form the group $I = C_2^{i}$. Then the group formed by the operations G_i which leave V_{eff} and T_{eff} invariant is the *direct product* group [6]

$$G = L \times I = C_{3v}^{\ell} \times C_2^{i}$$

The selection rules for this case are easily obtained by the methods of Ref. 6 and give the results found in Table 14. A methyl group mode will have the symmetry of one of the six symmetry types, a_1A, a_2A, ..., eB. The lower case symbol gives the local symmetry of the mode, while the upper case symbol gives the symmetry under interchange of the methyl groups. For example, a vibration with A character is symmetrical to the interchange of the two methyl groups by rotation about the C_2^{i} axis, while a mode with B character is antisymmetrical to this exchange. Moreover, a mode with a_1 character is symmetrical to all operations of C_{3v}^{ℓ} at each CH_3 group, and two modes with e character transform locally as e modes under C_{3v}^{ℓ}. It follows then that a mode of type a_1B shows a_1 local symmetry and is antisymmetric to the C_2^{i} interchange of the two CH_3 groups.

TABLE 14

Consequences of $G = C_{3v}^{\ell} \times C_2^{i}$ Symmetry

on the Methyl Modes of $(CH_3)_2SnCl^{2-}$

when they are Coupled

$G = C_{3v}^{\ell} \times C_2^{i}$	IR	Raman	ρ	d	Frequency
a_1A	0	a	p	1	ν,δ
a_2A	0	0	-	1	t
eA	0	a	3/4	2	ν,δ,r
a_1B	a	0	-	1	ν,δ
a_2B	0	0	-	1	t
eB	a	0	-	2	ν,δ,r

With this coupling, Table 14 shows that one expects 10 methyl
group frequencies to appear--five as Raman bands and five as infrared bands. The specific form of these modes is given in its last
column. No single frequency can appear in both Raman and infrared
spectra. Note that the group G is not required to be the same for
each type of methyl group vibration. Thus, G for the stretching
modes may be different from G for the deformation modes which can be
different from the G for the rocking modes.

We turn now to the interpretation of the experimental results.
In this, the assignment of Hobbs and Tobias is followed, although the
language used in describing the modes varies somewhat from theirs.
Consider first the CH_3 stretching modes. The symmetric a_1 motion
gives a frequency near 2900 cm^{-1} while the asymmetric e frequency
appears near 3000 cm^{-1} in a wide variety of molecules [26,27]. Therefore, the bands observed at 2928 ± 2 cm^{-1} and 3021 ± 2 cm^{-1} arise
from a_1 and e type motions, respectively. The remaining question is
whether $G = C_{3v}$ or $G = C_{3v}^{\ell} \times C_2^{i}$. This hinges on whether the infrared
and Raman frequencies are the same. Since the experimental uncer-
tainties are ±2 cm^{-1} for each value, it follows that the Raman and
infrared values are the same within the experimental value. Thus,
$G = C_{3v}$ and any coupling present produces no spectral consequences
greater than the experimental uncertainties.

The CH_3 deformation modes come in the middle frequency region. The e type frequency is found between 1400 and 1500 cm^{-1} in other molecules, depending upon the atom to which the CH_3 group is attached [26,27]. Similarly, the a_1 type motion appears between 1200 and 1400 cm^{-1}. When the methyl group is bonded to the $SnCl_3$ moiety, the values are 1402 and 1196 cm^{-1}, respectively [30]. A comparison of these values with the entries in Table 11 permits one to assign the 1410 cm^{-1} frequency to e type modes and the 1192, 1203 cm^{-1} values to a_1 type modes. The difference between the infrared band at 1192 cm^{-1} and the Raman band at 1203 cm^{-1} is outside the stated limits for experimental error and therefore must arise from different modes of vibration. This implies coupling between the two methyl groups for the deformation modes and leads to the assignment of 1192, 1203, and 1410 cm^{-1} to the a_1B, a_1A, and eB frequencies of the group $G = C_{3v}^{\ell} \times C_2^i$, respectively. That the eA frequency, which is allowed in the Raman spectrum and expected near 1400 cm^{-1}, does not appear in the observed spectrum is in complete harmony with the fact that the e type motion also does not appear in the Raman spectrum of CH_3SnCl_3, where it is also allowed [30].

The CH_3 rocking frequency has been observed in a number of simple CH_3 derivatives at about 300 to 350 cm^{-1} below the symmetric deformation frequency [31]. This would place it at about 850 to 900 cm^{-1} in these compounds. The methyl rocking frequency is observed for CH_3SnCl_3 as a strong infrared band at 787 cm^{-1}. No Raman band is seen for the same fundamental, although it is allowed by the selection rules. It follows then that one should assign the infrared band observed at 793 cm^{-1} for $(CH_3)_2SnCl_4^{2-}$ to the e type rocking frequency. As expected from the CH_3SnCl_3 results, no corresponding Raman band is seen for the anion and one can not determine if coupling occurs for this motion. The simplest assignment is made: $G = C_{3v}$.

These assignments for the methyl group frequencies are collected in Table 15. The results are in good correspondence with the findings for CH_3SnCl_3 and one may conclude that the major forces on the methyl group are essentially the same in both chemical species.

TABLE 15

Vibrational Assignment of the Methyl
Frequencies from $(CH_3)_2SnCl_4^{2-}$

Mode	Symmetry	G	m	Frequency[a]
ν	e	C_{3v}	CH_3	3021
	a_1			2928
δ	eA	$C_{3v}^{\ell} \times C_2^{i}$	$(CH_3)_2$	
	eB			1410
	$a_1 A$			1203
	$a_1 B$			1192
r	e	C_{3v}	CH_3	793

[a] Units in cm^{-1}.

Turn now to the skeletal or framework modes. The first task is
to consider reasonable forms for V_{eff}, T_{eff}, and hence G. The methyl
groups move (virtually) as rigid units in these modes and the bonds
which are being distorted are the Sn-C and Sn-Cl bonds. The main
problem is whether there is a difference between the Sn-Cl bonds, as
expected with the geometric point group C_{2h}, or whether the bonds are
effectively the same, as chemical intuition suggests. The simplest
assumption is made; namely, that they are equal--an assumption which
is in harmony with the experimental findings for the methyl group
modes. Then the group of operations which leaves V_{eff} and T_{eff} for
these modes invariant is $G = D_{4v}$. The spectral consequences for the
skeletal modes are found by the standard methods and are collected
in Table 16.

The highest frequencies expected among the skeletal modes are
from the Sn-C stretching vibrations. McGrady and Tobias find the
corresponding bands at 529 and 582 cm^{-1} in the $(CH_3)_2Sn(aq)^{2+}$ ion
[32]. On that basis, Hobbs and Tobias assigned the Raman band at
508 and the infrared band at 580 cm^{-1} to the A_{1g} and A_{2u} R' modes,
respectively. Sn-Cl stretching frequencies occur at 363 cm^{-1} for

TABLE 16

Consequences of $G = D_{4h}$ Symmetry

on the Skeletal Modes of $(CH_3)_2SnCl_4^{2-}$

$G = C_{4v}$	IR	Raman	ρ	d	Frequency
A_{1g}	0	a	p	1	R,R'
A_{2g}	0	0	-	1	
B_{1g}	0	a	3/4	1	R
B_{2g}	0	a	3/4	1	β
E_g	0	a	3/4	2	α
A_{1u}	0	0	-	1	
A_{2u}	a	0	-	1	R',α
B_{1u}	0	0	-	1	
B_{2u}	0	0	-	1	α
E_u	a	0	-	2	R,β,α

CH_3SnCl_3, at 344 cm^{-1} for $(CH_3)_2SnCl_2$, and at 318 cm^{-1} for $(CH_3)_3SnCl$ [30]. This type of vibration occurs at 310 cm^{-1} for the $SnCl_6^{2-}$ anion [33,34,35]. The selection rules call for two Sn-Cl stretching frequencies, A_{1g} and B_{1g}, to be active in the Raman spectrum and one, E_u, to be active in the infrared spectrum. By comparison with other molecules with metal-halogen bonds, these bands in $(CH_3)_2SnCl_4^{2-}$ are expected to be strong in both kinds of spectra. An examination of the experimental data in Table 11 shows that no bands are present in the region just above 300 cm^{-1}. On the other hand, three strong bands are present in a region centered just above 200 cm^{-1}. If these are not the R stretching frequencies, they must then come from the α and β bending modes. However, α and β type bending vibrations do not have such high frequency values in other molecules with similar structures. Therefore, these must be the Sn-Cl stretching modes. Hobbs and Tobias assign 197, 207, and 235 cm^{-1} to the B_{1g}, A_{1g}, and E_u frequencies, respectively.

Only one frequency remains, the Raman band at 139 cm^{-1}, which has not been assigned to a mode of vibration. Two Raman active fundamentals, the B_{2g} ClSnCl (β) bend or the E_g CSnCl (α) bend, could be the source of this band. The β bends occur at about one-half the Sn-Cl stretching frequencies in $SnCl_6^{2-}$. Further, the β bend has been assigned to a Raman band at 112 cm^{-1} for CH_3SnCl_3 [30], which is less than one-half the Sn-Cl stretching frequency of 363 cm^{-1} for this compound. On this basis, one predicts the β bend for $(CH_3)_2SnCl_4^{2-}$ to occur at ≤ 100 cm^{-1}. The α bends are expected to occur at higher frequency than the β motions. For example, the α bend in CH_3SnCl_3 has been assigned to a Raman band at 142 cm^{-1} [30]. It would appear then that 139 cm^{-1} for the anion comes from the α-bending fundamental. Hobbs and Tobias support their assignment by noting that McGrady and Tobias find a Raman band at 185 cm^{-1} for the $(CH_3)_2Sn(aq)^{2+}$ ion which is also assigned to α-bending motion.

These assignments are collected in Table 17. No values are available for the A_{2u} and E_u, α-bending fundamentals and the E_u β-bending vibration, all of which are active in the infrared spectrum, because the infrared measurements do not extend to long enough wavelengths to reach their low-frequency values. Of course the B_{2u} α frequency is not active in either spectra. All in all, one sees that while the spectral observations fall far short of what one expects for the $(CH_3)_2SnCl_4^{2-}$ anion on the basis of its geometric point group, they are in good agreement with the expectations of localized-mode selection rules.

Now, let us see what these measurements tell us about the $(CH_3)_2SnCl_4^{2-}$ ion. If the methyl groups were cis to each other, both Sn-C stretching vibrations would be active in the infrared spectrum and also in the Raman spectrum. The expectations for the trans conformation (Table 16) lead to a rule of mutual exclusion. Both Sn-C stretches are observed--one in the Raman spectrum and one in the infrared spectrum. Thus, the trans conformation has been deduced for the anion on the basis of its vibrational spectroscopy. One can also make some comments about the bonding. The Sn-Cl stretching frequencies occur in $SnCl_6^{2-}$ at 317 and 312 cm^{-1}. Substitution of two methyl groups for two chlorine atoms drops the Sn-Cl frequencies to 235, 207, and 197 cm^{-1}. This corresponds to a pronounced weakening of the

TABLE 17

Vibrational Assignment of the Skeletal
Frequencies[a] of $(CH_3)_2SnCl_4^{2-}$ $(G = D_{4h})$

Mode	Symmetry	Frequency	Calculated
R'	A_{2u}	580	571
	A_{1g}	508	516
R	E_u	235	238
	A_{1g}	207	203
	B_{1g}	197	190
α	E_g	139	139
	E_u		80
	A_{2u}		98
β	E_u		127
	B_{2g}		

[a]Units are cm^{-1}.

Sn-Cl bond--a bond which is predominantly ionic in character. Note
that this bond weakening is a cis effect, while similar phenomena
in transition metal compounds occur in a trans fashion. These effects are discussed in greater detail by Hobbs and Tobias [25] and
the references cited by them.

Finally, the spectral data for the CH_3 deformation vibrations--
the δ-bonding modes--show coupling between the two methyl groups.
Some deductions can be made about the nature of the likely origin of
this coupling and the consequences on the atomic displacements in
the several modes of vibration. While the carbon atom is moderately
displaced along the C-Sn bond in the a_1 δ motion, the response of
the Sn atom is quite small, because of its large mass. As a consequence, there is little splitting of the two a_1 type δ frequencies
as a result of kinetic energy coupling. Further, the two methyl
groups are on opposite sides of the anion. Thus, they are rather
far apart and one methyl group is "shielded" from the other by the

intervening $SnCl_4$ moiety. Hence, it is unlikely that any methyl-
methyl interaction terms in the potential energy are large enough to
cause substantial splitting of the two a_1 δ frequencies. These two
considerations leave only the interaction of the methyl modes with
the framework modes as the origin of the observed a_1A and a_1B splitting shown in Table 15. Now the basic act of separating the methyl
modes (ν, δ, r) from the framework motion stretching the C-Sn bond
removes any coupling terms between these motions. Thus, the primary
methyl group-framework interaction can be expected to involve potential energy terms which involve the distortion of the $SnCl_4$ moiety.
A glance at a model of the anion shows that the existence of such
interaction terms are eminently reasonable. Further, they may involve $SnCl_4$ stretching motion as well as bending. Thus, these methyl-
$SnCl_4$ interaction terms in the potential energy are the most likely
origin of much of the splitting shown between the a_1A and A_1B δ frequencies. This means that there must be some $SnCl_4$ deformation in
these two δ vibrations as well as some δ motion of the CH_3 groups in
the $SnCl_4$-framework vibrations.

VI. SPECTRA AND STRUCTURAL UNITS

The last section considered the spectra of an ion in which localization takes place in the modes of vibration. This effect produced
vibrations localized in the framework and those localized to the
methyl groups. That vibrations of the latter kind occur in larger
molecules has been recognized almost from the beginning of vibrational spectroscopy, e.g., the classic researches of Coblentz. Vibrations localized to a particular structural unit are characteristic
of that unit and their presence in the spectrum of a molecule indicates the presence of the unit in the molecule. This result is, obviously, of major importance, and a large number of spectroscopic
studies throughout the years have been devoted to the elucidation of
this relationship and its utilization in structure determinations.
Thus, one can recognize the presence of alkyl groups by the bands
arising from CH_3 and $(CH_2)_n$ groups. Quantitative measurements of

CH_3 band intensities indicate how many CH_3 groups are present and contribute to an understanding of the extent of branching and the length of the chains.

The $>C=C<$ and $-C\equiv C-$ units are recognizable not only from the stretching frequency of the C_2 unit--strong in the Raman spectrum-- but also from $\equiv CH$ and $=CH_2$ frequencies. Moreover, these spectra differ from those which arise from aromatic moieties like the phenyl group, the cyclopentadienyl group, etc. In the same way, the common ligands of organometallic chemistry such as CO, NO, CNS, NH_3, PF_3, pyridine, ethylenediamine, acetylacetonate, ethers, sulfoxides, etc., show their presence in a molecule by their "fingerprint" bands in the spectra. Further, one can say something about the "environment" of the structural unit in a particular molecule. For a detailed consideration of the spectroscopic features which characterize the common structural units of organic moieties, one is referred to extensive treatments of this subject [26,27]. The original literature is a vast storehouse of data of this type for molecular moieties which occur in organometallic compounds. We will consider here several examples, rather particular to the organometallic area, which illustrate the workings of this spectroscopic method.

The first step is to determine the frequencies which arise from a specific molecular structural unit. Sometimes one is faced with the task of simultaneously establishing the structure of the molecule and the characteristic frequencies of the structural units. The undertaking then frequently becomes a bootstrap operation. This is the case for the first example. Hieber and Leutert prepared the first transition metal compound containing a hydrogen atom in 1932 [36]. This unstable and toxic gas was $H_2Fe(CO)_4$. It was soon followed by the preparation of the almost equally unstable and noxious $HCo(CO)_4$ [37,38]. The first structures proposed for these compounds had the hydrogen bound as a proton within the core of the iron or cobalt atom to form a "pseudo-nickel" atom or had it attached to an oxygen forming an O-H bond. The first infrared band associated with hydrogen atom motion was found at 703 cm^{-1} in $HCo(CO)_4$ [39]. In-

sufficient data were available to determine whether this band came
from a stretching or bending type of motion, but it clearly indicated
that the H atom was bound to a Co atom. A model was proposed of C_{3v}
symmetry which placed the Co-H link opposite, and linear with, a
Co-CO bond. At the same time, it was felt that the H atom could also
interact with three adjacent CO groups--the relative amounts of Co-H
and CO-H bonding to be established by further study. Cotton and
Wilkinson examined the infrared spectrum of both $HCo(CO)_4$ and $DCo(CO)_4$
in the 1400 cm^{-1} region, where a Co-D stretching frequency might well
be anticipated [40]. Finding none, they assigned the 703 cm^{-1} band
to a Co-H stretching vibration in a model in which the H atom was
bound only to Co. Shortly, both groups located a Mn-H stretching
frequency at 1782 cm^{-1} in the newly prepared $HMn(CO)_5$ and established
that the hydrogen bending mode, near 660 cm^{-1}, coupled with MnCO
bending modes [41,42]. Thus, it was established that the H atom was
bonded to the Mn atom. It remained to settle the $HCo(CO)_4$ question.
In the same communication [41], it was reported that the 703 cm^{-1}
band of $HCo(CO)_4$ came from a mode of vibration in which the hydrogen
motion was coupled with CoCO bending motion, and the hydrogen motion
in this mode was assigned as Co-H bending with the H atom bound to
Co. Then a detailed study of the infrared spectrum of $HCo(CO)_4$ and
$DCo(CO)_4$ found the Co-H stretch at 1934 cm^{-1} and the Co-D stretch at
1396 cm^{-1} [42] to clearly establish that the bonding of the hydrogen
in $HCo(CO)_4$ was like that in $HMn(CO)_5$. It also established that the
chemical and thermal stability of these hydrido compounds do not, of
necessity, parallel the mechanical strength of the metal-hydrogen
bond, as revealed by the stretching frequency. Thus, the Co-H bond
is stronger than the Mn-H bond (1934 cm^{-1} versus 1782 cm^{-1}) but
$HCo(CO)_4$ is chemically unstable in contrast to $HMn(CO)_5$.

In the early 1960's a series of transition metal compounds were
prepared with hydrido bonds. Thus, Chatt and Shaw prepared compounds
of the type $PtHClL_2$, with $L = PEt_3$ and similar compounds of Ir, Ru,
and Rh [44]. Lewis, Nyholm, and Reddy made octahedral $RhHX_2L_3$ compounds, with $L = AsPh_2Me$ and $X = Cl, Br, I$ [45]. Vaska prepared

$IrHX_2L_3$, IrH_2XL_3, and $OsHX(CO)L_3$, where X = Cl, Br and L = PPh_3, $AsPh_3$, $SbPh_3$ [46]. In all these cases, the presence of the metal-hydride bond was established by the presence of an infrared band near 2000 cm^{-1}, which arises from M-H stretching motion. For some molecules, M-H bending frequencies were also observed, as well as the corresponding deuterium atom vibrations. Table 18 is a collection of typical examples of some of these frequencies. In the decade since these papers were published, a large volume of work has been done on the preparation, chemistry, and spectroscopy of compounds of this type. In this work, the metal-hydrogen frequencies have been a guide to structure.

Many organometallic compounds have metal-metal bonds. Modes of vibration occur in these molecules which distort the cluster of metal atoms and which are localized in the cluster. This results, for the most part, from the disparity between the mass of the metal atoms and the mass of the atoms (C, O, etc.) to which they are bound, and is aided by bond-strength differences. As a result, these vibrations are characteristic of the metal cluster. Covalent bonding between metal atoms results in highly polarizable bonds. The consequence is that these frequencies generally yield intense bands in the Raman spectrum. In contrast, the intensity of the infrared bands is highly variable. The metal-metal frequency is forbidden in the infrared by the selection rules when each metal atom has the same atomic groups bound to it (e.g., R_3M-MR_3). The selection rules allow for frequencies in the infrared spectrum when the two metal atoms are different or when the groups attached to each metal atom differ. But we have seen in previous sections that the fact that a frequency is allowed, i.e., not forbidden, does not guarantee it an observable intensity.

The first assignment of a frequency to a metal-metal vibration was made by Woodward in 1934, who attributed the strong Raman band at 169 cm^{-1} in the spectrum of aqueous mercurous nitrate solutions to the stretching mode of the Hg_2^{2+} ion [47]. Indeed, this classic observation furnished an early proof of the structure of this cation. The corresponding Cd_2^{2+} frequency was found some years later at

TABLE 18

Metal-Hydrogen Frequencies[a]

in Some Hydrido Compounds

Compound	ν_H	ν_D	δ_H	δ_D	Reference
$RhHCl_2(AsPh_2Me)_3$	2077	1485			45
$RhHBr_2(AsPh_2Me)_3$	2073	1483			
$RhHI_2(AsPh_2Me)_3$	2058				
$IrHCl_2(PPh_3)_3$	2200	1580			46(a)
$IrHCl_2(AsPh_3)_3$	2170				
$IrHCl_2(SbPh_3)_3$	2100				
$IrH_2Cl(PPh_3)_3$	2215	1584			46(a)
	2100	1515			
$IrH_2Br(PPh_3)_3$	2240				
	2090				
$OsHCl(CO)(PPh_3)_3$	2097	1506	825	645	46(b)
			801	612	
$OsHBr(CO)(PPh_3)_3$	2100	1508	827	642	
			796	609	

[a]Units are cm^{-1}.

183 cm^{-1} in molten salts [48]. In early studies, the Sn-Sn frequency has been located near 200 cm^{-1} by both Raman and infrared spectroscopy for a series of molecules [49,50]. On the other hand, R. J. H. Clark and his associates found the Pb-Pb frequency as low as 110 cm^{-1} in a definitive study of the infrared and Raman spectra of Pb_2Me_6 and Pb_2Ph_6 [51,52]. Usually the MM' frequency lies between the MM and M'M' values [50]. The metal-metal frequencies for selected molecules are collected in Table 19.

TABLE 19

Metal-Metal Frequencies[a]

in Some Organometallic Compounds

Compound	ν_M	Spectrum[b]	Reference
$Me_3Sn-SnMe_3$	190	Ram	49
$Ph_3Sn-SnPh_3$	208	Ram	49
$Ph_3Sn-SnMe_3$	194	IR	50
$Me_3Pb-PbMe_3$	~116[c]	Ram	51
$Ph_3Pb-Pb-Ph_3$	109	Ram	52
$Me_3Sn-GePh_3$	225	IR	50
$Me_3Sn-Mn(CO)_5$	182	IR	50
$Me_3Sn-Co(CO)_4$	176	IR	50
$Mn_2(CO)_{10}$	159	Ram	49
$Re_2(CO)_{10}$	120	Ram	49
$MnRe(CO)_{10}$	154	Ram	53

[a]Units are cm^{-1}.

[b]Ram = observed in Raman spectrum. IR = observed in infrared spectrum.

[c]Overlapped in spectrum.

Metal carbonyl compounds exist with a variety of metal clusters. The frequencies for the Mn_2, Re_2, and MnRe bonds have been located in Raman studies by Gager, Lewis, and Ware [49] and by Sheline and co-workers [53]. The values are listed in Table 19. More complicated metal clusters can exist with these compounds. For example, the mercuric ion reacts with metal carbonyl anions to form compounds such as $Hg[Co(CO)_4]_2$ and $Hg(Mn(CO)_5]_2$. These compounds have a linear M-Hg-M cluster which gives rise to two metal-metal stretching frequencies. The symmetric vibration ν_s is Raman active and appears near 165 cm^{-1}; the antisymmetric stretch ν_a is allowed only in the infrared spectrum and occurs near 190 cm^{-1} [54]. The exact values for these compounds are listed in Table 20. Another example is the linear polymetallic carbonyls, $Fe(CO)_4[Mn(CO)_5]_2$ and $Fe(CO)_4[Re(CO)_5]_2$, recently prepared by Sheline and co-workers [55]. The values for ν_s and ν_a of the FeM_2 cluster are listed in Table 20.

As another example of the ability to recognize basic structural arrangements in organometallic compounds from vibrational spectroscopy, consider the case of the B-H bond. Price, in studying the infrared spectrum of diborane, B_2H_6, was lead to the conclusion that it had a bridge-type structure [56]. Lord and Nielson made a detailed study of B_2H_6 and B_2D_6 and assigned most of the fundamentals [57]. This work has been verified and extended by the studies of Smith and Mills [58] and Freund and Halford [59]. The result is that terminal B-H stretching vibrations occur at 2525 and 2615 cm^{-1} in the bridged

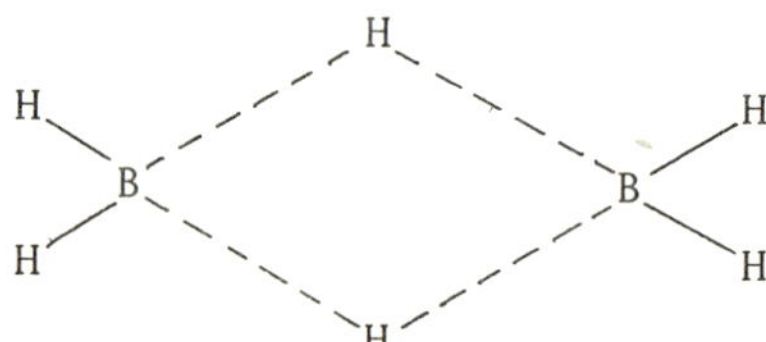

molecule while the BH_2B bridging unit gives B-H stretching modes at 1887 and 2104 cm^{-1} and a B-H deformation (bridge stretch) mode near 1600 cm^{-1}.

This basic frequency pattern has been used to recognize this double hydrogen bridge structure in other molecules. One example is cited here. The properties of alkyl aluminum systems are of interest. The lower alkyl aluminums are dimeric with the alkyl groups in a bridging position. Higher alkyl aluminums, $AlPr_3$, $AlBu_3$, are monomeric, while alkyl aluminum halides generally form species with bridging halogen atoms. Further, there is a lively current interest in the chemistry of metal borohydrides (hydroboration). Aluminum chloride reacts with lithium borohydride to form aluminum hydroborate [60], which in turn reacts with triethylaluminum to form diethylaluminum hydroborate [61].

TABLE 20

Metal-Metal Frequencies[a]

for Some Metal Clusters

Compound	ν_s	ν_a	Reference
$Hg[Co(CO)_4]_2$	161	196	54
$Hg[Mn(CO)_5]_2$	167	188	54
$Fe(CO)_4[Mn(CO)_5]_2$	143	$\begin{cases} 214^b \\ 220 \end{cases}$	55
$Fe(CO)_4[Re(CO)_5]_2$	107		55

[a]Units are cm^{-1}.

[b]X-ray analysis shows two crystallographically distinct molecules per unit cell.

$$3LiBH_4 + AlCl_3 \rightarrow Al(BH_4)_3 + 3LiCl$$

$$Al(BH_4)_3 + Al_2Et_6 \rightarrow 3Et_2AlBH_4$$

The infrared spectrum of Et_2AlBH_4 reveals its basic structure. The alkyl aluminum part of the spectrum is similar to that of Et_2AlX compounds. Further, the spectrum shows strong bands at 2490 and 2420 cm^{-1}, near 2100 cm^{-1}, and near 1500 cm^{-1} which must arise from motion of the four "hydride" hydrogen atoms. The spectrum does not show the single strong band near 2300 cm^{-1} which is characteristic of the tetrahedral BH_4^- ion. Instead, in analogy with diborane, the bands at 2490 and 2420 arise from terminal BH_2 stretching vibrations, and the strong absorption bands near 2100 and 1500 cm^{-1} come from B-H stretching and deformation, respectively, in a double hydrogen bridge. The structure is thus [61]

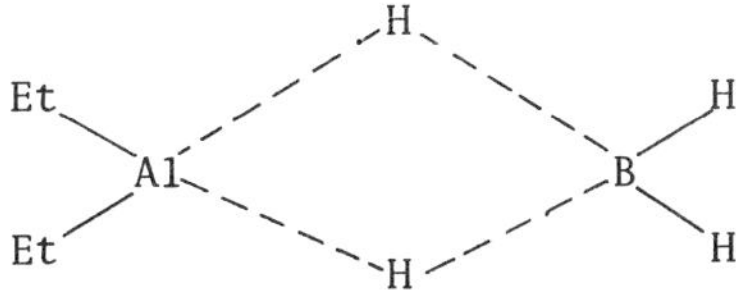

in analogy with that found similarly for the $Al(BH_4)_3$ [62-64].

VII. VIBRATIONAL FREQUENCY CALCULATIONS--
FORCE CONSTANT DETERMINATION

In the harmonic oscillator approximation, the frequency of the radiation absorbed by a molecule in making a vibrational transition is
equal to the classical frequency of vibration of the molecule in the
particular fundamental mode involved. This relation has lead to the
computation of the classical frequencies of vibration of molecules
almost from the beginnings of vibrational spectroscopy.

Two kinds of computations are involved. In the first, one starts
from the experimental values of the vibrational frequencies and seeks
the force constants for the molecule which reproduce these values.
Since a fundamental mode generally involves the distortion of several
bonds and angles of a molecule, this computational process serves to
"uncouple" the molecular data present in the frequencies to furnish
the values of the specific bond and angle parameters. These latter
quantities are directly related to the nature of the chemical bonds
and, hence, such a calculation often preceeds a discussion of bonding
based upon vibrational frequencies. In the second type of calculation, one is usually seeking some help in making a vibrational assignment. If one can make an estimate of the force constants, one can
calculate an approximate frequency spectrum. If a good estimate of
force constants can be made from previous studies of related molecules, the resulting spectrum will be surprisingly valid. Even when
the estimate is less accurate, one generally is able to recognize
basic patterns which are helpful in making an assignment. The prac-
ticalities of the situation demand that one proceed in the same way
for both types of computation--namely, from a set of force constants
to a set of calculated frequencies. When one seeks the force constants which reproduce the experimental frequencies, some kind of an
iteration process is used to drive the computed values to the experimental ones.

In this section, several examples will be given which illustrate
how such computations are used in the studies of the vibrational
spectra of organometallic compounds. For an introduction to the cal-

culation of vibration frequencies, one is referred to the excellent
treatment of Professor Crawford in Chap. 1 of this book. An extensive and detailed treatment of the subject is found in the basic reference book of Wilson et al. [1]. Only those equations which specifically define the quantities referred to here are written in this
section.

Twice the potential energy for the distorted molecule is

$$2V = \tilde{r}kr \tag{4}$$

where r is the column matrix formed from the bond and angle distortions, $\tilde{r}$ is its transpose, and k is the matrix of the force constants. The elements of the k matrix are the basic bond and angle
parameters which describe the resistance of the molecule to distortion. In practice, one does not start the computations from the
elements of r, but from linear combinations of them which reflect
the symmetry. In terms of the symmetry coordinates S,

$$2V = \tilde{S}FS \tag{5}$$

The symmetry force constants of F are linear combinations of the bond
and angle constants of k. Both kinds (k and F) of force constants
are evaluated in practice.

The kinetic energy also enters the computation of vibration frequencies. Usually one calculates the so-called *inverse* kinetic energy matrix G from the atomic masses and the "s" vectors, which convert
atomic displacements to coordinate values. Then the frequencies of
vibration of the molecule appear as the roots, $\lambda - (2\pi c\nu)^2$, of the
secular equation

$$|FG - E\lambda| = 0 \tag{6}$$

E being the unit matrix. This is the "Wilson FG matrix method",
which is widely used today. Virtually all vibrational frequency calculations are now made on digital computers. The best known program
for such a computation is that of Schachtschneider [65].

It is not an easy task to establish the exact force field for
an organometallic molecule. The primary difficulty stems from the
fact that there are more force constants than there are vibrational

frequencies. One way to close this gap is to obtain the frequencies
for a molecule with isotopic atoms, because such a replacement does
not alter the intramolecular forces. A second source of independent
data for gaseous molecules is the Coriolis constants ζ for degenerate
frequencies. For typical organometallic compounds, the Coriolis effect influences the contour of the unresolved infrared or Raman band
[66]. Thus, a contour measurement for the band yields the Coriolis
constant. For example, Edgell and Moynihan showed that the separation between the maximum of the P and R branches of an infrared
fundamental band for a spherical top is given by

$$\Delta\nu = 4\left(\frac{kTA}{hc}\right)^{1/2} (1 - \zeta) \tag{7}$$

where $A = h/8\pi^2 cI$ is the rotational constant for the molecule and ζ
is the Coriolis constant for the modes of vibration generating the
band [66]. This equation was later verified by McDowell [67]. It
is not possible to obtain a closed expression like Eq. (7) for molecules which are symmetric tops. However, an atlas of band shapes
have been computed showing the dependence of the contour on ζ [68].
A comparison of the experimental band with the atlas then yields ζ.
Advances in digital computers also make it possible today to compute band shapes directly for each molecule--with fewer approxima-
tions--to obtain values for ζ and other dynamical parameters [69].

There is a different kind of factor which also enters these
calculations. The fundamental equations of motion, e.g., Eq. (6),
are derived on the assumption that the vibrations are harmonic.
Actually molecular vibrations are slightly anharmonic and force constant calculations of the highest accuracy correct for the anharmonicity.

The first example is the use of vibrational frequencies to evaluate the force field for an organometallic molecule. The force field
which is best understood is that for $Ni(CO)_4$, computed by Jones,
McDowell, and Goldblatt [13]. The primary experimental data available for the computation are the frequencies for $Ni(CO)_4$, $Ni(^{13}CO)_4$,
and $Ni(C^{18}O)_4$ plus the Coriolis ζ constants for the ν_5 and ν_9 bands

(Table 8) as evaluated from band contours by Eq. (7). Corrections
for anharmonicity were made for the CO stretching frequencies. The
principal force constants obtained are listed in the second column
of Table 21. One notes, immediately, the strength of the CO bond
and the weakness of the NiC bond. This result, which is general for
neutral metal carbonyls, forms the basis of any discussion of the
bonding in these molecules. The stretch-stretch interaction constants are listed in the fourth column of Table 21. Of particular
interest here is the large value for $k_{(NiC,CO)}$, which testifies to
the strong synergistic relation between the MC and CO bonds in a
metal carbonyl.

Seldom are this much data available for a force constant calculation. Usually the constants which result are less unique and
less accurate. Yet they are of fundamental importance. It is this
author's opinion that one should always reduce the frequency data
to force constant values. The purpose is to separate the mass effect
from the bond strength effect and to decouple the distortion of the
several structural units of a molecule which generally takes place
in any mode of vibration.

The second example illustrates the use of frequency calculations
as an aid to vibration assignments. Hobbs and Tobias [25] made frequency calculations for the $(CH_3)_2SnCl_4^{2-}$ ion. These were used as

TABLE 21

Some Force Constants for $Ni(CO)_4$ [a]

Type	k	Type	k
CO	17.68	CO,C'O'	0.13
NiC	2.09	NiC,NiC'	0.12
NiCO	0.31	NiC,CO	0.55
CNiC	0.25	NiC,C'O'	-0.10

[a]Based upon solution frequencies; units are mdyn/Å for stretch
constants, mdyn-Å/rad² for bending constants; C,C' and O,O' refer
to atoms of different CO groups.

an aid to the frequency assignment and an index to the bonding. These authors started with trial values arrived at from constants computed from the $(CH_3)_2Sn(aq)^{2+}$ and $SnCl_6^{2-}$ spectra. The computed frequency pattern supported the assignment which is listed in Table 17. Then an iterative procedure was used to adjust the trial values to minimize the weighted difference between observed and computed frequencies. The resulting values, expressed as valency force constants, are 2.44 mdyn/Å, 0.81 mdyn/Å, 0.31 mdyn.Å/rad^2, and 0.30 mdyn.Å/rad^2 for the SnC, SnX, CSnX, and XSnX parameters. They give the computed frequencies for the $(CH_3)_2SnCl_4^{2-}$ anion listed in the last column of Table 17. These results lend a strong air of certitude to the vibrational assignment.

Force constant estimates have been made from a single observed frequency for vibrations which are well localized. Typical examples are the M-M, M-O, M-H, C≡O, C≡N, and C-O bonds. As an example, consider the well-known Cotton-Kraihanzel model for the CO stretching vibrations in octahedral metal carbonyl frequencies [70]. The model is simplicity itself: separate CO oscillators of reduced mass $\mu = m_c m_o/(m_c + m_o)$ and force constant k, coupled through a potential energy coupling constant k_c or k_t, depending on whether the coupled oscillators are cis or trans to each other. Simple expressions result for the frequencies--e.g., for the cis-$L_3M(CO)_3$ molecule,

$$\lambda_{A_1} = \left(\frac{k + 2k_c}{\mu}\right)$$
$$\lambda_E = \left(\frac{k - k_c}{\mu}\right) \tag{8}$$

(Note that Eq. (8) uses the usual symbolism of μ for reduced mass instead of its inverse as used in the original C-K paper.) These so-called "exact" expressions are frequently reduced further by writing

$$k_c = \frac{k_t}{2} = k_i$$

The results of applying this model to the CO stretching frequencies

TABLE 22

Cotton-Kraihanzel Force Constants for $CrL_x(CO)_{6-x}$ [a]

Compound	k_1	k_2	k_i
$Cr(CO)_6$		16.49	0.22
$Cr(PPh_3)(CO)_5$	15.50	15.88	0.33
Trans $Cr(PPh_3)_2(CO)_4$		15.28	0.34
Cis $Cr(Ph_2P(CH_2)_2PPh_2)(CO)_4$	14.49	15.32	0.36
Cis $Cr(H_3CC(CH_2PPh_2)_3)(CO)_3$	13.91		0.38
Cis $Cr(C_2H_4(PPh_2)_2)_2(CO)_2$	13.33		0.47

[a] Units are mdyn/Å; k_1 is CO trans to P, k_2 is CO cis to P.

of some phosphine substituted chromium carbonyls are found in Table
22 [70]. It is at once clear that replacing a CO group by a phosphine ligand results in a drop in the C-K force constant for the CO
bond. This suggests an increase in the bonding of the CO groups to
Cr via the d orbitals of Cr and the antibonding π orbitals of CO.
One sees that this increased bonding is greater for the CO group
trans to a phosphine ligand.

This model is simple and hence is readily used. It has been
widely utilized to guide vibrational assignments and to detect changes
in CO bond strengths which accompany various kinds of ligand substitution. However, C-K force constants should be compared directly
with those computed from a full vibrational treatment with considerable caution because they neglect the coupling of the distortion of
the MC bond with that of the CO bond which occurs in the CO stretching vibrations.

The basic dilemma in making these abbreviated force constant
calculations can be illustrated for the tin-tin bond in $Me_3Sn-SnMe_3$.
The observed metal-metal frequency in this molecule is 190 cm^{-1} and
the Sn-Sn force constant is taken as F = k. But, in calculating the
corresponding inverse kinetic energy term G, does one use the mass

of the Sn atom, the mass of the $SnMe_3$ group, or what? In practice,
"one usually takes a deep breath and makes a choice," or one sets
about collecting additional spectroscopic data for a more complete
$\underline{FG}$ matrix treatment of the vibrating system.

There is another way: take advantage of what one knows in advance about the vibration or may readily deduce about it. An illus-
tration will demonstrate the power of this approach. Consider a CO
stretching vibration in a metal carbonyl. Is only the CO bond deformed as the name might imply? Or is the metal-carbon bond compressed as the CO bond is stretched? If so, how much? Experimental
evidence exists which directly relates to this point. The atomic
displacements determine the Coriolis constant for degenerate vibrations. Edgell and Moynihan [66] measured the separation of the P
and R maxima in ν_5 for $Ni(CO)_4$ and used Eq. (7) to find a value of
$1 - \zeta$ near 1.0 for this mode. If the atoms move in this vibration
in such a way that only the CO bonds are distorted, $1 - \zeta$ would be
1.5 [66]--a value far from the experimental one. However, if the
carbon atom displacement is m_o/m_c times that of the oxygen atom (to
conserve momentum at each CO group), the value of $1 - \zeta$ would be
1.0 [66]-- in good agreement with the experimental finding. One
concludes that the CO stretching vibrations in $Ni(CO)_4$ take place
in the latter manner and is very much like this in all other neutral
metal carbonyl derivatives.

How is one to apply the constraints which define a vibration or
a type of vibration? Constraints shrink the size of the $\underline{F}$ matrix
which must be calculated. Unfortunately, the corresponding truncated
$\underline{G}$ matrix requires all the elements of the full $\underline{G}$ matrix for its eval-
uation. Equations have been given for such a reduction [71]. How-
ever, it is easier to treat the constrained vibration problem in a
somewhat different manner. The constraints are introduced by what
might be called descriptive coordinates. Let R_i be the coordinate
which describes the distortion of a molecule when it is constrained
to move in the ith mode of motion. The displacement of the nth atom
in this mode of motion is $\underline{\tau}_{ni}R_i$. Note that the $\underline{\tau}_{ni}$ define the dis-

tortion which is specified by the coordinate R_i. When there are several different modes of distortion of a molecule, each described by its own coordinate, the vector displacement of the nth atom is

$$\rho_n = \sum_i \tau_{ni} R_i \tag{9}$$

For a molecule limited to move in the manner permitted by the several R_i of Eq. (9), the kinetic energy is

$$2T = \dot{\tilde{R}} T \dot{R}$$

where $\dot{R}$ is the column matrix of the time derivatives $\dot{R}_i$ and $\dot{\tilde{R}}$ is its transpose. The elements of the kinetic energy matrix $\underline{T}$ are

$$T_{ij} = \sum_n m_n \tau_{ni} \cdot \tau_{nj} \tag{10}$$

The corresponding truncated potential energy is

$$2V = \tilde{R} K R$$

The full potential energy in terms of the bond and angle distortions is

$$2V = \sum_{qr} k_{qr} r_q r_r \tag{11}$$

where the internal coordinate r_q is given in terms of Wilson "$\underline{s}$" vectors by

$$r_q = \sum_n s_{qn} \cdot \rho_n$$

For the constrained motions,

$$r_q = \sum_i r_{qi} R_i \tag{12}$$

with

$$r_{qi} = \sum_n s_{qn} \cdot \tau_{ni}$$

By using Eq. (2) in (11), one finds that the elements of the truncated potential energy matrix $\underline{K}$ may be expressed in terms of the bond constants by

$$K_{ij} = \sum_{qr} k_{qr} r_{qi} r_{rj} \tag{13}$$

The frequencies of vibration of the constrained system are the roots λ of the secular equation

$$|\underline{K} - \underline{T}\lambda| = 0 \tag{14}$$

When there is one or two R_i it is practical to solve Eq. (14). Otherwise, one may use a computer in an iterative process to obtain a matrix $\underline{B}$ which drives $\underline{T}$ to a unit matrix

$$\underline{BTB} = \underline{E}$$

and $\underline{K}$ to a diagonal matrix $\underline{\Lambda}$

$$\underline{\tilde{B}KB} = \underline{\Lambda}$$

whose diagonal elements give the frequencies

$$\Lambda_{ii} = (2\pi c \nu_i)^2$$

We now have the expressions needed to apply our knowledge of displacements to a simple yet accurate calculation of CO stretching frequencies from force constants, or vice versa. As an example, consider the trans-$M(CO)_4 L_2$ molecule. CO stretching vibrations occur with A_{1g}, B_{1g}, and E_u symmetry. These may be computed directly from the descriptive coordinates for these modes shown in Fig. 5. The vector displacements at each atom are the $\underline{\tau}_{ni}$ for that coordinate. These were formed from the knowledge of the displacements of the C and O atoms obtained above plus the requirements of symmetry. It will be convenient to normalize the displacements for each coordinate of Fig. 5 by multiplying by N_i. Then, Eqs. (10), (13), and (14) give

$$A_{1g} \text{ Mode:} \quad N_i = \frac{1}{2}(1 + a) \qquad a = \frac{m_o}{m_c}$$

$$T_{11} = 4[m_o N_1^2 + m_c (N_1 a)^2]$$

$$= \mu$$

$$= \frac{m_c m_o}{m_c + m_o}$$

$$K_{11} = 4[k(1 + a)^2 + \hat{k} a^2 - 2\hat{\hat{k}}(1 + a)a + \cdots$$

$$= K + K_t + 2K_c$$

$$\lambda_1 = \frac{K_{11}}{T_{11}}$$

B_{1g} Mode: $\quad N_2 = \frac{1}{2}(1 + a)$

$$T_{22} = \mu$$

$$K_{22} = K + K_t - 2K_c$$

$$\lambda_2 = \frac{K_{22}}{T_{22}}$$

E_u Modes: $\quad N_3 = \frac{1}{\sqrt{2}}(1 + a)$

$$T_{33} = \mu$$

$$K_{33} = K - K_t$$

$$\lambda_3 = \frac{K_{33}}{T_{33}}$$

where

$$K \equiv K + b^2 \hat{k} - 2b\hat{\hat{k}}$$

$$K_t \equiv k_t + b^2\hat{k}_t - 2b\hat{\hat{k}}_t$$

$$K_c \equiv k_c + b^2\hat{k}_c - 2b\hat{\hat{k}}_c$$

$$b \equiv \frac{m_o}{m_c + m_o}$$

Thus, the principal force constant K is a linear combination of the CO bond stretching constant k, the MC bond stretching constant $\hat{k}$, and the interaction constant $\hat{\hat{k}}$ for the stretching of the MC and CO bonds of the same MCO unit. The trans interaction constant K_t depends upon the interaction constant k_t between trans CO bonds, the

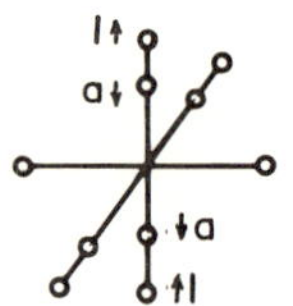

A_{1g} Mode: R_1 B_{1g} Mode: R_2

E_u Mode: R_3

FIG. 5.   CO stretching distortions in trans-$M(CO)_4L_2$. $a = m_0/m_c$.

interaction constant $\hat{k}_t$ between trans MC bonds, and the interaction constant $\acute{k}_t$ between an MC bond and the CO bond trans to it. The cis interaction constant K_c depends upon the corresponding cis interaction bond constants.

It is seen that the three experimental frequencies depend upon the three force constants K, K_t, and K_c. But these are composite or net force constants in that each one depends upon several bond force constants. This is inescapable because more than one kind of bond is distorted in each of the modes of vibration. If one has only the three frequencies corresponding to λ_1, λ_2, and λ_3 to work with, then one may compute the composite constants K, K_t and K_c. These are useful for discussing bond changes in molecules of this type because K is dominated by k and will reflect its changes. This follows from the fact that k ~ 16, $\hat{k}$ ~ 2, $\acute{k}$ ~ .5, and b = 4/7.

The accuracy of this model for CO stretching frequencies in metal carbonyls was tested on the $Mn(CO)_5I$ molecule with the results collected in the upper part of Table 23 [72]. The second column gives the frequencies computed with the model from the bond force constants.

TABLE 23

Descriptive Coordinate Calculations[a] for $Mn(CO)_5I$

Vibration	Descriptive Coordinate Model	Exact
	CO Stretch Calculation	
A_1	2110.1	2110.6
A_1	2008.4	2009.4
E	2052.2	2052.9
	Middle Frequency Calculation	
E	607.3	605.8
E	505.3	506.3
E	477.9	478.0
E	438.3	438.0
A_1	580.3	577.7
A_1	464.1	462.4
A_1	377.8	376.0
A_1	197.0	197.5

[a] Units are cm^{-1}.

The third column gives the exact frequencies calculated by the full
$\underline{F}\ \underline{G}$ matrix method. The two columns are essentially identical and
the model is seen to be an accurate representation of reality.

The results of applying this model to the CO stretching frequencies for all the octahedral species $ML_x(CO)_{6-x}$ are to be found
in Table 24. These equations are analogous to those obtained earlier
by Cotton and Kraihanzel [70]. They differ from the earlier equations in that wherever the Cotton-Kraihanzel equations show bond
force constants k, k_t, k_c, etc., the equations of Table 24 show
composite constants K, K_t, K_c, etc. Thus constants already computed
by the Cotton-Kraihanzel model may be reinterpreted. While they are
only approximate values for the bond force constants, they are, in
fact, accurate values of composite force constants.

The MC stretching motion in a metal carbonyl must be orthogonal
to the CO stretch. When this requirement is combined with the dis-

TABLE 24

CO Stretching Frequencies for $ML_x^a(CO)_{6-x}$ Molecules

Molecule	Symmetry	Type	Spectral Activity	$\lambda = (2\pi c\nu)^2$ [b]
$M(CO)_6$	O_h	A_{1g}	R	$(K + K_t + 4K_c)/\mu$
		E_g	R	$(K + K_t - 2K_c)/\mu$
		F_{1u}	IR	$(K - K_t)/\mu$
$M(CO)_5L$	C_{4v}	A_1	IR/R	$K_{11} = K_1$ [c]
				$K_{22} = K_2 + K_t + 2K_c'$
				$K_{12} = 2K_c$
		B_1	R	$(K_2 + K_t - 2K_c')/\mu$
		E	IR/R	$(K_2 - K_t)/\mu$
trans-$ML_2(CO)_4$	D_{4h}	A_{1g}	R	$(K + K_t + 2K_c)/\mu$
		B_{1g}	R	$(K + K_t - 2K_c)/\mu$
		E_u	IR	$(K - K_t)/\mu$
cis-$ML_2(CO)_4$	C_{2v}	A_1	IR/R	$K_{11} = K_1 + K_c'$ [c]
				$K_{22} = K_2 + K_t$
				$K_{12} = 2K_c$
		B_1	IR/R	$(K_2 - K_t)/\mu$
		B_2	IR/R	$(K_1 - K_c')/\mu$
cis-$ML_3(CO)_3$	C_{3v}	A_1	IR/R	$(K + 2K_c)/\mu$
		E	IR/R	$(K - K_c)/\mu$
trans-$ML_3(CO)_3$	C_{2v}	A_1	IR/R	$K_{11} = K_2 + K_t$ [c]
				$K_{22} = K_1$
				$K_{12} = \sqrt{2}\, K_c$
		B_2	IR/R	$(K_2 - K_t)/\mu$

TABLE 24 (Continued)

Molecule	Symmetry	Type	Spectral Activity	$\lambda = (2\pi c \nu)^2$ [b]
cis-$ML_4(CO)_2$	C_{2v}	A_1	IR/R	$(K + K_c)/\mu$
		B_1	IR/R	$(K - K_c)/\mu$
trans-$ML_4(CO)_2$	D_{4h}	A_{1g}	R	$(K + K_t)/\mu$
		A_{2u}	IR	$(K - K_t)/\mu$
ML_5CO	C_{4v}	A_1	IR/R	K/μ

$$K = k + b^2\hat{k} - 2b\acute{k}$$

$$K_1 = k_1 + b^2\hat{k}_1 - 2b\acute{k}_1$$

$$K_2 = k_2 + b^2\hat{k}_2 - 2b\acute{k}_2$$

$$K_t = k_t + b^2\hat{k}_t - 2b\acute{k}_t$$

$$K_c = k_c + b^2\hat{k}_c - 2b\acute{k}_c$$

$$K_c' = k_c' + b^2\hat{k}_c' - 2b\acute{k}_c'$$

Symbol	Type
k	CO
$\hat{k}$	MC
$\acute{k}$	MC,CO
k_c	CO,C'O'
$\acute{k}_t$	CO,C''O''

[a] A ligand (L) is a group other than CO.

[b] When there are two kinds of CO groups in the same molecule, the subscript 1 refers to a CO with a ligand opposite to it, while the subscript 2 refers to one with another CO opposite. C'O' is cis to CO; C''O'' is trans to CO. When there are two kinds of cis CO groups in the same molecule, the primed constants express the interaction between CO groups neither of which have opposite ligands.

[c] The frequencies here are given by the roots of the secular equation:

$$\begin{vmatrix} K_{11} - \mu\lambda & K_{12} \\ K_{12} & K_{22} - \mu\lambda \end{vmatrix} = 0$$

placements found for the CO stretch, one sees that the CO group must move as a rigid unit in the MC stretch. The β-bending motion (see Fig. 1) must be orthogonal to both the MC and the CO stretch. Thus, the C and O atoms must move perpendicular to the MCO bond in this bending motion. We have found that an excellent model for this bending gives the C and O atoms displacements such that angular momentum about the M atom is conserved at each MCO unit. Thus, we know enough to compute these frequencies separate from all the others in a metal carbonyl. The MC and β-bending frequencies can not be computed separate from each other because these motions couple with each other in the actual vibrations which occur. In metal carbonyl derivatives, these motions also tend to couple to some extent with metal-ligand stretch and it is wise to treat all of these together. This creates no problem, for orthogonality serves to furnish the form of the descriptive coordinate for this motion.

The $Mn(CO)_5I$ molecule provides a test of the accuracy of this model for the MC-stretch and β-bending motion. These motions give four of the eight type E frequencies. These four frequencies may be calculated separately from the others by using the above model distortions as descriptive coordinates. The resulting frequencies are found in the second column in the lower part of Table 23. The accuracy achieved with this model is seen by comparing the model frequencies with the exact values computed for the same modes from the full $\underline{F}\ \underline{G}$ matrix computation (column 3). A further test is obtained from the middle frequencies of Type A_1 where MC stretch and β bending occurs with the Mn-I stretch. The excellent agreement of the truncated, model calculation with the full $\underline{F}\ \underline{G}$ matrix values testifies again to the accuracy of the simple model for the MC and β-bending motions.

The α-bending modes are just as readily dealt with but will not be treated here. Note that orthogonality to the above modes furnished the form of this distortion.

The metal carbonyls are not isolated examples of the ability to treat modes of vibration separately. One does not need Coriolis constants to define atomic displacements with sufficient accuracy for the purpose at hand. In fact, this same information about the character of CO stretching motion is readily obtained from a vibrational calculation on a simple model for a metal carbonyl. Simple models furnish the information for other systems. It is a rule though, and not the exception, that composite force constants appear when separate frequencies are treated to this level of accuracy.

VIII. SPECTRA AND BONDING

What has been said above serves to stress the existence of a relation between vibration frequencies and the mechanical strength of chemical bonds. Since the strength of bonds is sensitive to the nature of the bonding, it follows, of course, that the vibrational spectrum of a molecule relates directly to the character of its bonds. We have seen that a frequency of vibration, in general, arises from the distortion of more than one kind of bond and contains a factor which is dependent upon the masses of the atoms which move in the mode of motion. The computation of force constants serves to undo this coupling and removes the effect of the atomic masses. Thus, the best prelude to a discussion of the bonding in a molecule as revealed by its vibration spectrum is a calculation of its force constants. However, trends in the frequency of a vibration localized in a given structural unit as other parts of the molecule are changed indicate trends in the bonding in this unit because coupling with the rest of the molecule is reduced (i.e., localization) and the mass factor is constant. Both of those kinds of bonding studies are popular in organometallic chemistry and these methods are illustrated here.

The SO and SeO bonds in molecules containing these structural units are composed of both a σ bond and a π bond. The latter is formed from the 2p electrons of the oxygen atom and 3d orbitals of the S atom or 4d orbitals of the Se atom. The nature of the σ and π bonding depend upon the S or Se electronegativity, its oxidation number in the compound, and the total charge on the So or SeO moiety. These factors change with changes in the other part of the molecule.

The SO or SeO stretching vibration is well localized in this structural unit and its variation from molecule to molecule has been studied as an index to the SO or SeO bonding. The studies in the literature have generally used a diatomic molecule model--two atoms and one bond--to compute the "force constant" for the SO or SeO bond from the observed frequency. The considerations of the last section, however, show that what is computed in such a model, is in general, a composite force constant even when the atom displacements are limited to the S and O or Se and O atoms. But the force required to distort the SO or SeO bond is larger than that involved in the distortion of the other structural entities which occur in this mode. It follows from this that changes in the SO or SeO-bond force constant are reflected in the changes which occur in the composite force constant for this mode of vibration. Thus, the composite force constant is an index to the bonding. But it is sometimes useful to keep in mind that all the changes in the composite force constant do not come from the SO or SeO-bond force constant.

When there are more than one SO or SeO bond in a molecule, the situation becomes more complicated in that there are several SO or SeO frequencies. The most correct solution short of a complete treatment of the full set of frequencies for the molecule is to introduce both the localization and the truncation into the vibration problem with descriptive coordinates as illustrated above for the octahedral metal carbonyl derivatives. This has not been done. An approach which is commonly found in current practice is to compute an "isolated" SO or SeO frequency by taking an appropriate geometric or arithmetic mean of the observed SO or SeO values. This "isolated bond" frequency is then used in the diatomic molecule model to calculate the "bond force constant."

Gillespie and Robinson [73] have calculated an approximate force constant for the SO bond in 27 chemical species from the SO frequencies using the diatomic molecule model. These (composite) force constants were found to vary with the SO bond length r in a manner which could be represented by the equation

$$\log K = -7.41 \log r + 7.15$$

This relation is of the general form

$$kr^n = \text{constant}$$

suggested earlier by Linnett [74]. That bond lengths change with bond order is well known. Thus, this frequency variation is fundamentally an index to the variation of the bonding in these species. To establish this relation, Gillespie and Robinson ascribed all the change to variation in the π bond. The longest SO bonds are in the molecules F_5SOOSF_5 and F_5SOF where $r = 1.65 \pm .01$ Å. Here the relevant $3d_{x^2-y^2}$ and $3d_{z^2}$ orbitals are presumed to be primarily used in σ bonding to the F atoms. If this were completely the case, the SO π bonding could be taken as zero. However, some competition for these orbitals by the lone pair electrons of the oxygen atoms is expected even in this limiting case. Thus, a small amount of SO π bonding is expected in these two molecules. These authors then assume an SO bond length of $r = 1.70$ Å for a hypothetical pure σ SO bond and use the above equation to compute that the force constant for this case would be $K = 2.7 \times 10^5$ dyn/cm. The shortest SO bond lengths are in the SO_2F_2 and SOF_4 molecules where $r = 1.40 \pm .01$ Å. The bond order is assumed to have reached its limiting value of $n = 2$ in these molecules. With this value of r, one finds $K = 11.7 \times 10^5$ dyn/cm. Having established limiting values of K, it remains to determine the basic form of the relation between K and the bond order n. The bond order in the SO_4^{2-} ion is taken as 1.5 and if one assumes that fluorine causes the SO bond to achieve its maximum bond order, then $n = 1.67$ in FSO_3^-. This gives two additional points for the K versus n curve. It is found that these two points lie on the straight line established by the $n = 1$ and $n = 2$ limiting points. Thus, Gillespie and Robinson expressed this straight line by the expression

$$n = (1.11 \times 10^{-6})K + 0.7$$

Using the force constant values evaluated from the spectra and this equation one obtains bond order values for molecules with SO bonds. Some examples from this work are SO_2FBr, $n = 1.96$; $SO_2(CH_3)_2$. $n = 1.77$; $C_2H_5OSO_3^-$, $n = 1.65$; and $(CH_3)_2SO$, $n = 1.56$.

Paetzold has computed approximate force constants for 21 compounds with SeO bonds using the diatomic molecule model [75]. He finds that the SeO (composite) force constant is proportional to the SO constant in the compound obtained by replacing the Se with a S atom. This relation is written as

$$K_{SeO} = 0.593 \ K_{SO} + 1.37$$

As in the sulfur compounds, he finds that the force constant is related to the SeO bond length. The equation is

$$\log K = -4.551 \ \log r + 1.785$$

where K is expressed in mdyn/Å and r in Å. He concludes that the bonding in molecules with SeO bonds is similar to that of the SO containing molecules.

Zingaro and his associates [76] have synthesized and studied the infrared spectrum of more than 60 chalcogenides of the general formula R_3MX, where R is an alkyl group, M is a Group Va element, and X is a VIa atom. The R groups are CH_3, C_2H_5, $n\text{-}C_3H_7$, $n\text{-}C_4H_9$, $n\text{-}C_5H_{11}$, cyclo-C_6H_{11}, and $n\text{-}C_8H_{17}$. The M atoms are P, As, and Sb, while X is O, S, Se, and Te. The emphasis was on the assignment of the MX frequency and its correlation with the properties of the M and X atoms. A basic diatomic molecule model is assumed. Using an argument based upon the idea that changing the X atom was similar to an isotope substitution, the dependence of the MX frequency upon masses was found to be similar to that given by the Teller-Redlich product rule. The effect of the variation of the MX force constant was written as proportional to the product of the Pauling electro-negatives, E_M and E_X, for the M and X atoms. Thus, the MX frequency ν is given in this correlation by

$$\nu = AE_M E_X \left(\frac{S}{m_X \, m_M} \right)^{1/2} = AZ$$

where m_X and m_M are the mass of the X and M atoms, S is the mass of the whole molecule, and A is a constant for all molecules with the same R group. Such a correlation for the propyl group is shown in Fig. 6. The constant A varies with the R group. Several examples from this study are

$$CH_3: \quad \nu = -19.62 + 376.72 \, Z$$
$$n\text{-}C_3H_7: \quad \nu = 107.34 + 237.98 \, Z$$
$$cyclo\text{-}C_6H_{11}: \quad \nu = 173.82 + 172.82 \, Z$$

with

$$Z = E_X E_M \left(\frac{S}{m_X \, m_M} \right)^{1/2}$$

The units for ν are cm^{-1}.

The use of vibrational spectroscopy to study bonding in substances of interest to organometallic chemistry was illustrated in this section. There is a wide range in the nature of the current studies of this type, but they all reduce to two basic alternates: one must evaluate the force constant for the bond or one must restrict the compounds studied in such a way that the frequencies themselves become the meaningful index to the bond strength. The bond strength is correlated with a variety of bond properties and indices. In the examples considered in this section, the bond order and the electronegativity of the atoms forming the bond were used. Other properties which have been used in such correlations include the acid-base character of a ligand, steric factors in the ligand and the organometallic compound, electrical charge on a relevant moiety of the molecule, oxidation number of metal atoms, etc. Meaningful studies of this type have been qualitative as well as quantitative.

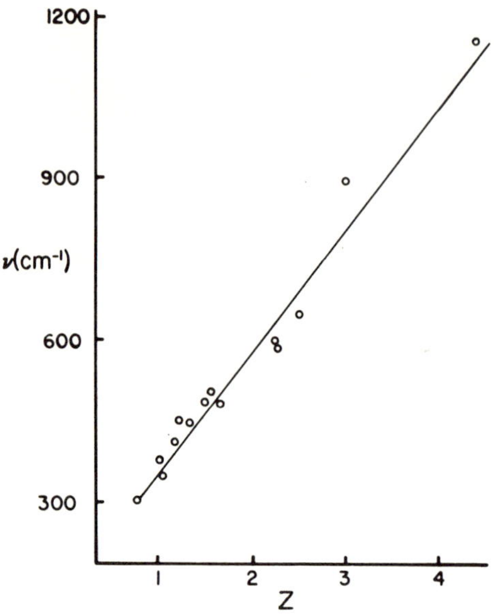

FIG. 6. Correlation of ν for M = X vibrations with Z = $E_M E_X (S/m_X m_M)^{1/2}$ for $(C_3H_7)_3$MX, where M = P, As, Sb; X = O, S, Se, Te (see Sec. VIII).

IX. VIBRATIONAL SPECTRA
AND CHEMICAL ANALYSIS

We turn now to a consideration of chemical analysis. Both quantitative and qualitative aspects of spectroscopic analysis are treated
in basic references (e.g., Bellamy, Colthrup, Rao, Bauman, Potts) and
the reader is referred to these. Nevertheless, one should not complete a treatment of the methods of vibrational spectroscopy, however sparse, without some reference to this important field. We
consider here some aspects of qualitative analysis which are relevant to organometallic chemistry.

A primary task in chemical analysis is to answer the question,
"What is it?" Infrared spectroscopy plays a fundamental role in
this task while Raman spectroscopy is making an increasing contribution. The infrared spectrum of a compound is characteristic of
that material. Thus, it is a physical property of the substance

like a boiling point, density, or index of refraction. However,
the infrared spectrum is far more definitive than the latter type
of physical property and hence far more useful in identifying a substance. Perhaps there is no other single physical property of a
compound which is as useful for this purpose.

The identification of an unknown substance from its infrared
spectrum is a matter of comparison of the spectrum with that of
known compounds. To make the identification is a matter of skill
and experience. It is at once clear that one should have access to
the infrared spectrum of a number of known compounds. The collection of infrared spectra for this purpose has taken place with increasing rapidity over the last two decades. Catalogues of such
spectra include:

1. Sadtler Standard Infrared Spectra, Sadtler Research Laboratories, 3316 Spring Garden Street, Philadelphia, Pennsylvania 19104.

2. The Coblentz Society Evaluated Infrared Reference Spectra.
Available through Sadtler Reserach Laboratories.

3. American Petroleum Institute Catalogue of Infrared Spectra
Data Distribution Offices, A and M Research Foundation, FE Box 130,
College Station, Texas 77843.

4. "Infrared Spectra of Selected Chemical Compounds," R. Mecke
and F. Langenbucher, Hayden and Sons, Ltd., London, 1965.

5. "Infrared Spectra and Characteristic Frequencies, 700-300
cm^{-1}," F. F. Bentley, L. L. Smithson, and A. L. Rozek, Wiley (Interscience), New York, 1968.

6. "The Aldrich Library of Infrared Spectra," Charles Pouchert,
Aldrich Chemical Co., Inc., 1970.

7. "Infrared Spectra of Inorganic Compounds," Richard A.
Nyquist and Ronald A. Kagel, Academic Press, New York, 1971.

8. "An Atlas of Spectral Data and Physical Constants for
Organic Compounds," Jeanette G. Grasselli, Chemical Rubber Co.,
Cleveland, Ohio 1970.

The basic data set now comprises thousands upon thousands of spectra. The Sadtler catalogue contains 45,000 spectra, the Coblentz Society evaluated spectra number 8,000, while the A.P.I. and Mecke-Langenbucher catalogues extend to seven volumes. Most of these spectra are of organic compounds. They are of interest to organometallic chemistry as reactants, as products of reactions, as ligands, and a few as solvents. The library of spectra of inorganic compounds are equally relevant. The only collections of spectra of organometallic compounds of which this author is aware are a small subset of the Sadtler catalogue and the far-infrared spectra of Bentley et al. Of course, a number of private libraries of organometallic spectra must exist in organizations with a particular interest in this class of compounds, and the several thousand spectra of organometallic molecules published each year add to a basic pool of such data.

It is, of course, a verity that the results of an identification match are only as good as the spectra in the library. Instrumentation and experimental techniques have made vast strides in recent years. As a consequence, recent catalogue spectra are almost always of good quality; this is not true of all of the early spectra in these collections. Moreover, one must exercise caution in the use of literature spectra of organometallic compounds. Most are good spectra, but some are of such poor quality that they are of marginal value for identification purposes, and, occasionally, a published spectrum is so badly contaminated with decomposition products it is worse than worthless. Attention should be called to the evaluated spectra of the Coblentz Society. The Class III spectra can be relied upon for high quality. Modern identification techniques sometimes place great demands upon spectral accuracy. The Class II spectra of the Coblentz Society is a response to this requirement. They have the improved accuracy (both wavelength and intensity) resulting from stringent requirements for the recording of the spectra, better sample preparation, and reliable, independent, identification of the substance.

Localized vibrations are discussed in a separate section above. It is seen that they give rise to frequencies which are characteristic of various structural groups in a molecule. The first step in compound identification usually involves a scan of the spectrum for group frequencies. Here is where the skill and knowledge of the experienced spectroscopist comes into play. He will be acquainted with the basic spectral patterns and their subtle relation to structure for the major kinds of compounds important to his work. This examination of the spectrum, coupled with a knowledge of the source of the material, is generally sufficient to reduce the field to several possibilities. The identification is completed by a comparison of various physical properties of the unknown with those of the possible candidates. Of course, the infrared spectrum of the unknown should be compared with that of the several candidates. But this should always be supplemented with other comparisons such as u.v. spectra, nmr spectra, mass spectra, x-ray patterns, chemical analysis, melting point, index of refraction, etc.

Sometimes, the above procedure does not produce a match. Then a more extensive search of the basic spectral data set is required. This search is often carried out today in private organizations with the aid of a computer. The identification match is then made as indicated above with one of the additional candidates suggested by the computer search. The more experienced the spectroscopist, the more likely he is to be successful with the computer search routine. A typical search for an organic compound produces a "hit" in the list of ten to twenty possibilities suggested by the computer routine. It is seldom in the top three to five compounds. It is, perhaps, also proper here to extend the caveat that computer searching in the hands of the inexperienced can be misleading.

Three computer systems in use today are:

1. The Singer System, Chemir Laboratories, 719 W. Kirkham Avenue, Glendale, Missouri 63122.

2. The Sadtler System, Sadtler Research Laboratories, 3316 Spring Garden Street, Philadelphia, Pennsylvania 19104.

3. The Duncan-Erley System, Don Nichols Associates, Saginaw, Michigan.

The first two systems are available on an interactive, time-share basis via telephone lines. The third system is the basis of some private computer systems.

REFERENCES

1. E. B. Wilson, Jr., J. Decius, and P. Cross, "Molecular Vibra-tions," McGraw-Hill, New York, 1955.

2. J. R. Ferraro and J. S. Ziomek, "Introductory Group Theory," Plenum Press, New York, 1969.

3. H. Jaffe and M. Orchin, "Symmetry and Chemistry," Wiley, New York, 1965.

4. W. G. Fateley, F. R. Dollish, N. T. McDevitt, and F. F. Bentley, "Infrared and Raman Selection Rules for Molecular and Lattice Vibrations," Wiley (Interscience), New York, 1972.

5. W. F. Edgell, *Spectrochim. Acta, 31A,* 1623 (1975).

6. W. F. Edgell, *J. Phys. Chem., 75,* 1343 (1971).

7. R. S. McDowell, L. B. Asprey, and L. C. Hoskins, *J. Chem. Phys., 56,* 5712 (1972).

8. B. L. Crawford, Jr. and P. C. Cross, *J. Chem. Phys., 6,* 525 (1938).

9. B. L. Crawford, Jr. and W. Horwitz, *J. Chem. Phys., 16,* 147 (1948).

10. (a) L. H. Jones, *J. Chem. Phys., 23,* 2448 (1955); (b) *ibid., 28,* 1215 (1958); (c) *ibid., Spectrochim. Acta., 19,* 1899 (1963).

11. (a) R. S. McDowell and L. H. Jones, *J. Chem. Phys., 36,* 3321 (1962); (b) L. H. Jones and R. S. McDowell, *J. Chem. Phys., 46,* 1536 (1967); (c) L. H. Jones, *J. Chem. Phys., 47,* 1196 (1967).

12. (a) M. Bigorgne, *Compt. Rend., 251,* 355 (1960); (b) M. Bigorgne and A. Chelkowski, *Compt. Rend., 251,* 538 (1960); (c) M. Bigorgne, Proceedings of the 9th International Conference on Coordination Chemistry, (W. Schneider, ed.) *Verlag Helvetiia Chimica. Acta,* Basel, 1966, p. 140; (d) M. Bigorgne and G. Bouquet, *Compt. Rend., 264,* 1485 (1967).

13. L. H. Jones, R. S. McDowell, and M. Goldblatt, *J. Chem. Phys., 48,* 2663 (1968).

14. S. F. A. Kettle, I. Paul, and P. J. Stamper, *Chem. Commun.,* 1724 (1970).

15. (a) E. W. Abel, G. B. Hargraves, and G. Wilkinson, *J. Chem. Soc., 1958,* 3149; (b) E. W. Abel and G. Wilkinson, *J. Chem. Soc., 1959,* 1501; (c) E. W. Abel, M. A. Bennet, and G. Wilkinson, *Chem. and Ind. London,* 442, 1960.

16. W. E. Wilson, *Z. Naturforsch., 136,* 349 (1958).

17. (a) M. A. El-Sayed and H. D. Kaesz, *J. Mol. Spectrosc. 9,* 310 (1962); (b) H. D. Kaesz, R. Bau, D. Hendrickson, and J. M. Smith, *J. Amer. Chem. Soc., 89,* 2844 (1967).

18. L. E. Orgel, *Inorg. Chem., 1,* 25 (1962).

19. R. Poilblanc and M. Bigorgne, *Bull. Soc. Chim. France,* 1303 (1962).

20. F. A. Cotton and C. S. Kraihanzel, *J. Amer. Chem. Soc., 84,* 4432 (1962).

21. I. R. Beattie and G. P. McQuillan, *J. Chem. Soc.,* 1519 (1963).

22. C. J. Wilkins and H. M. Haendler, *J. Chem. Soc.,* 3174 (1965).

23. J. P. Clark and C. J. Wilkins, *J. Chem. Soc. A,* 871 (1966).

24. M. Wada and R. Okawara, *J. Organometal. Chem. Amsterdam, 8,* 261 (1967).

25. C. W. Hobbs and R. S. Tobias, *Inorg. Chem., 9,* 1037 (1970).

26. L. J. Bellamy, "The Infrared Spectra of Complex Molecules" (2nd ed.), Methuen, London, 1958.

27. C. R. Rao, "Chemical Applications of Infrared Spectroscopy," Academic Press, New York, 1963.

28. W. King and B. Crawford, Jr., *J. Mol. Spectrosc., 5,* 421 (1960).

29. B. Crawford, Jr. and E. Bright Wilson, Jr., *J. Chem. Phys., 9,* 323 (1941).

30. P. W. Moore, Ph.D. Thesis, Purdue University, June 1961.

31. G. Herzberg, "Infrared and Raman Spectra of Polyatomic Molecules," D. Van Nostrand, New York, 1945.

32. M. M. McGrady and R. S. Tobias, *Inorg. Chem., 3,* 1157 (1964).

33. I. Wharf and D. F. Shriver, *Inorg. Chem., 8,* 914 (1969).

34. T. L. Brown, W. G. McDugle, Jr., and L. G. Kent, *J. Amer. Chem. Soc., 92,* 3645 (1970).

35. H. A. Carter, A. M. Qureshi, J. R. Sams, and F. Aubke, *Canad. J. Chem., 48,* 2853 (1970).

36. W. Hieber and F. Leutert, *Z. Anorg. Chemie., 204,* 745 (1932).

37. G. W. Coleman and A. A. Blanchard, *J. Amer. Chem. Soc., 58* 2160 (1936).

38. W. Hieber, *Angew. Chemie., 49,* 463 (1936).

39. W. F. Edgell, C. Magee, and G. Gallup, *J. Amer. Chem. Soc.*, *78*, 4185 (1956).

40. F. A. Cotton and G. Wilkinson, *Chem. and Ind.*, 1305 (1956).

41. W. F. Edgell, G. Asato, W. Wilson, and C. Angell, *J. Amer. Chem. Soc.*, *81*, 2022 (1959).

42. F. A. Cotton, J. L. Down, and G. Wilkinson, *J. Chem. Soc.*, 833 (1959).

43. W. F. Edgell and R. Summitt, *J. Amer. Chem. Soc.*, *83*, 1772 (1961).

44. J. Chatt and B. L. Shaw, *Chem. and Ind.*, 931 (1960).

45. J. Lewis, R. S. Nyholm, and G. K. N. Reddy, *Chem. and Ind.*, 1386 (1960).

46. (a) L. Vaska, *J. Amer. Chem. Soc.*, *83*, 756 (1961); (b) *ibid.*, *86*, 1943 (1964).

47. L. A. Woodward, *Phil. Mag.*, *18*, 823 (1934).

48. J. D. Corbett, *Inorg. Chem.*, *1*, 700 (1962).

49. H. M. Gager, J. Lewis, and M. J. Ware, *Chem. Comm.*, 616 (1966).

50. N. A. P. Carey and H. C. Clark, *Chem. Comm.*, 292 (1967).

51. R. J. H. Clark, A. G. Davies, R. J. Puddephatt, and W. McFarlane, *J. Amer. Chem. Soc.*, *91*, 1334 (1969).

52. R. J. H. Clark, A. G. Davies, and R. J. Puddephatt, *Inorg. Chem.*, *8*, 457 (1969).

53. G. O. Evans, W. T. Wozniak, and R. K. Sheline, *Inorg. Chem.*, *9*, 979 (1970).

54. D. M. Adams, J. B. Cornell, J. L. Davies, and R. D. W. Kemmitt, *Inorg. Nucl. Chem. Letters*, *3*, 437 (1967).

55. (a) E. H. Schubert and R. K. Sheline, *Z. Naturforsch.*, *20b*, 1306 (1965); (b) G. O. Evans, J. P. Hargaden, and R. K. Sheline, *Chem. Comm.*, 186 (1967); (c) G. O. Evans and R. K. Sheline, *J. Inorg. Nucl. Chem.*, *30*, 2863 (1968).

56. W. C. Price, *J. Chem. Phys.*, *16*, 814 (1948).

57. R. C. Lord and E. Nielsen, *J. Chem. Phys.*, *19*, 1 (1951).

58. W. L. Smith and I. M. Mills, *J. Chem. Phys.*, *41*, 1479 (1964).

59. I. Freund and R. S. Halford, *J. Chem. Phys.*, *43*, 3795 (1965).

60. H. I. Schlesinger, H. C. Brown, and E. K. Hyde, *J. Amer. Chem. Soc.*, *75*, 209 (1953).

61. N. Davies, C. A. Smith, and M. G. H. Wallbridge, *J. Chem. Soc. A*, 342 (1970).

62. W. C. Price, *J. Chem. Phys.*, *17*, 1044 (1949).

63. A. Almenningen, G. Gundersen, and A. Haaland, *Acta Chem. Scand.* *22*, 328 (1968).

64. N. Davies, P. H. Bird, and M. G. H. Wallbridge, *J. Chem. Soc.* *A,* 2269 (1968).

65. J. H. Schachtschneider, "Vibrational Analysis of Polyatomic Molecules (V and VI), Technical Report," Shell Development Co., Emeryville, California, 1964.

66. W. F. Edgell and R. E. Moynihan, *J. Chem. Phys., 27,* 155 (1957).

67. R. S. McDowell, *J. Chem. Phys., 43,* 319 (1965).

68. W. F. Edgell and R. E. Moynihan, *J. Chem. Phys., 45,* 1205 (1966); (b) R. E. Moynihan, Ph.D. Thesis, Purdue University, July 1954; (c) R. W. Valentine, Ph.D. Thesis, Purdue University, August 1957.

69. W. F. Edgell, R. W. Weiner, and B. H. Kleinstein, unpublished; (b) B. H. Kleinstein, Ph.D. Thesis, Purdue University, August 1969; (c) R. S. Weiner, Ph.D. Thesis, Purdue University, August 1969.

70. F. A. Cotton and C. S. Kraihanzel, *J. Amer. Chem. Soc., 84,* 4432 (1962).

71. (a) B. L. Crawford, Jr. and J. T. Edsall, *J. Chem. Phys., 7,* 223 (1939); (b) E. B. Wilson, Jr., *J. Chem. Phys., 7,* 1047 (1939); *ibid., 9,* 76 (1941).

72. W. F. Edgell and R. S. Spencer, "Vibrational Factoring in Metal Carbonyls," Eighth European Conference on Molecular Spectroscopy, Copenhagen, Denmark, August 1965. See abstracts of meeting and R. S. Spencer, Ph.D. Thesis, Purdue University, May 1973.

73. R. J. Gillespie and E. A. Robinson, *Can. J. Chem., 41,* 2074 (1963).

74. J. W. Linnett, *Quart. Rev. Chem. Soc., 1,* 73 (1947).

75. R. Paetzold, *Spectrochim. Acta., 24A,* 717 (1968); *ibid., 26A,* 577 (1968).

76. (a) R. A. Zingaro and R. M. Hedges, *J. Phys. Chem., 65,* 1132 (1961); (b) R. A. Zingaro, *Inorg. Chem., 2,* 192 (1963); (c) R. A. Zingaro and R. E. McGlothlia, *J. Chem. Eng. Data, 8,* 226 (1963); (d) R. A. Zingaro, R. E. McGlothlia, and R. M. Hedges, *Trans. Fara. Soc., 59,* 798 (1963); (e) R. A. Zingaro and A. Merijanian, *Inorg. Chem., 3,* 580 (1964); (f) G. N. Chremos and R. A. Zingaro, *J. Organometallic Chem., 22,* 637 (1970); (g) G. N. Chremos and R. A. Zingaro, *J. Organometallic Chem., 22,* 647 (1970).

Chapter 4

IONIC ORGANOMETALLIC SOLUTIONS

Walter F. Edgell

Department of Chemistry
Purdue University
West Lafayette, Indiana

I. INTRODUCTION

Chapter 3 has been concerned with spectra resulting from the excitation of vibrational states of organometallic compounds. This kind
of spectrum has been observed for many years and has been the source
of much knowledge of organometallic compounds. In contrast, infrared

bands arising from the translational motion of ions in organometallic
solution were first reported in 1965 [1]. The states which are being
excited in this case are truly quantum states of the solution itself,
for they cease to exist when the components of the solution environment of an ion are removed to infinity. Consequently, these spectra
are a direct source of information about the microscopic character
of electrolytic organometallic solutions.

The frequencies of the (intramolecular) vibrations of a polyatomic ion in solution are fixed primarily by the intramolecular
forces resisting its distortion and less by the forces of interaction
of the ion with its environment.* This situation has tended to mask
the fact that one is exciting, here also, quantum states of the solution rather than those of some "isolated" species. The solution
characteristics of these states are most apparent in the frequency
distribution of the band intensity. Thus, the study of the contour
of a band arising from the vibration of an ion in solution is also a
primary source of information about the microscopic character of the
solution.

The kind of information potentially available from studies of
this type has broad relevance. The physical properties of solutions
(and glasses) stem from the microscopic behavior of the solution elements and the chemical reactions that take place in them are just
the sum of the microscopic events. Because solutions of metal ions
and of metal containing ions in organic solvents are a concern of
metal-organic chemistry, these translation and vibration spectra are
of interest here.

In this chapter, the translation bands are examined and some
conclusions are drawn about the structure of the solutions, the motion in them, and the size and nature of the short-range forces at
the ions. Then some intramolecular vibration bands are examined.

---

*Hydrogen bonding, in all but its weakest manifestation, is
considered here as a force between two parts of the same species,
rather than a force between a species in its environment.

This leads to a major improvement in characterizing solution structures. A direct relation between solution structure and chemical
reactivity is demonstrated. Finally, a return is made to the question of the nature of the phenomena. A quantum model for an electrolytic solution is given and some results for the time behavior
of the ions are presented.

II. THE ALKALI ION TRANSLATION BAND*

The translation band of the alkali ion appears as a broad band of
medium intensity in the far infrared spectrum. It lies near 400 cm^{-1}
for Li^+ salts, 200 cm^{-1} for Na^+ salts, 150 cm^{-1} for K^+ salts, 125
cm^{-1} for Rb^+ salts, and 110 cm^{-1} for Cs^+ salts [1-3]. The sharp inverse dependence of the band frequency on the mass of the cation
shows that the motion being excited is predominantly that of the
alkali ion. Figure 1 shows the Li^+, Na^+, and K^+ bands for the salts
$LiCo(CO)_4$, $NaCo(CO)_4$, and $KCo(CO)_4$ in THF solutions taken from the
early spectra of Dr. A. I. Watts and Dr. J. Lyford [1,2]. These are
typical of the bands found for a variety of salts in a variety of
solvents. Table 1 gives the frequency of the alkali ion band for a
number of salts dissolved in THF [1,2]. One notes a pronounced dependence of the frequency upon the salt anion. The largest range is
for the Li^+ salts, but the effect is also clearly present with the
Na^+ salts, even though fewer of them are soluble in this solvent.

That the solvent also influences the frequency of the translation band is shown by the data collected in Table 2. The shifts are
pronounced for both Li^+ and Na^+ salts. The solubility of salts of
the heavier alkali ions in these solvents is low, so little data are
available.

---

*There is a semantic problem here. It has been the practice to
refer to the infrared band involved here as arising from the *intermolecular vibration* of the ion in the solution. However, it is the
motion of the center of mass of the ion which is involved, and such
motion is traditionally spoken of as translation. The nontrivial
nature of the point is seen by considering the motion of a polyatomic
ion like $Co(CO)_4^-$ or NH_4^+.

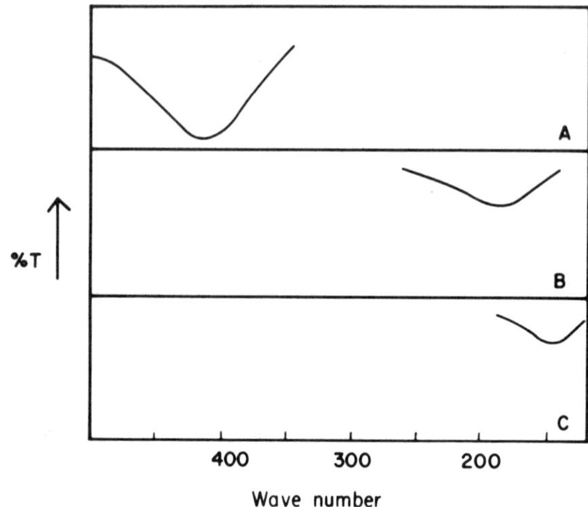

Fig. 1. The Li^+, Na^+, and K^+ ion translation bands in THF solutions. A: salt = $LiCo(CO)_4$.   B: salt = $NaCo(CO)_4$.   C: salt =
$KCo(CO)_4$.

TABLE 1

Alkali Ion Translation Bands in THF [1,2]

Salt	ν_{max}^{a}
$LiCo(CO)_4$	413
$LiBPh_4$	412
$LiNO_3$	407
$LiCl$	387
$LiBr$	378
LiI	373
$NaCo(CO)_4$	192
$NaBPh_4$	198
NaI	184
$KCo(CO)_4$	142

[a]Units in cm^{-1}.

TABLE 2

Solvent Influence on Translation Band

Salt	Solvent	ν_{max} [a]	Reference
NaCo(CO)$_4$	DMSO	199	1, 2
	DMF	197	1, 2
	THF	192	1, 2
	Piperidine	183	1, 2
	Pyridine	180	1, 2
LiCl	THF	387	1, 2
	M-Pyrrol	377(398)	4a
	Acetone	409	4c
	Pyridine	416	4b
	DMSO	425	1-3
LiI	THF	373	1-3
	M-Pyrrol	398	4a
	Pyridine	419	4b
	Acetone	423	4c
	DMSO	424	1-3
NaI	Pyridine	170	4b
	Acetone	192	4c
	DMSO	194	1-3
	M-Pyrrol	207	4a

[a]Units in cm^{-1}.

These data make it possible to draw some conclusions about these solutions. First, just the existence of the band itself indicates that one is dealing with "bound" quantum states. This means that structure exists in the solution at the cation whose lifetime is significant compared to the period of the ion motion, i.e., the lifetime is $>>10^{-13}$ sec [1,2]. Second, the dependence of the frequency of the cation band upon the anion shown in Table 1, means that the cation and anion must be associated in an intimate way in the THF solutions. The simplest form that this association could take is a

tight ion pair. But the cation must also have solvent molecules in
its near-neighbor environment on both geometric and chemical grounds.
The data of Table 2 show that the variation of the cation frequency
with solvent is as large as that found for the anion variation.
These data led Edgell and co-workers to conclude that the alkali ion
moves in a "cage" composed of anion and solvent molecules in THF
solutions and that both anion and solvent molecules contribute to
the factors which determine the band frequency [1a-c,2]. A similar
situation was expected to be obtained in other solvents of moderate
solvating power.

In contrast to the behavior in THF, the alkali ion frequency is
independent of the anion in DMSO (see, for example, the data in Table
2). The ion band in solvents of this kind has been extensively studied by Popov and co-workers [3,4]. In this work, they have examined
the infrared spectra of a variety of salts in several sulfoxides [3],
several 2-pyrrolidones [4a], pyridine [4b], and acetone [4c]. Their
data for N-methyl-2-pyrrolidone (M-pyrrol) are found in Table 3. In
addition, other workers have studied salts in solvents of this type.
Thus, Lassigne and Baine [5] report ion band data for salts in DMF
and Buxton and Caruso [6] for sulfalone solutions.

These data can also be understood in terms of the same basic
model proposed above. As the solvating power of the solvent increases, it competes more effectively with the anion for a place in
the near-neighbor environment of the cation. Thus, with solvents of
high solvating power, the cation moves in a cage formed only of solvent molecules. This is the situation which is obtained in these
latter solutions [1d,2-6].

Two other early studies of ion motion in solution which were
interpreted in terms of different models should be mentioned at this
point. Evans and Lo [7] observed two far infrared bands in benzene
solutions of tetra-alkyl ammonium chlorides and bromides which they
assigned to the halide ions moving in ring vibration modes of the
salt dimer in the solution. French and Wood [8] reported the alkali
ion mode for $NaBPh_4$ dissolved in pyridine, 1,4-dioxane, piperidine,

TABLE 3

Cation Ion Translation Band in M-Pyrrol [4a][a]

Salt	ν_{max}[b]	Salt	ν_{max}
LiI	398	NaI	204
LiBr	398	NaBr	204
LiCl	377 (398)	$NaClO_4$	204
$LiClO_4$	398	$NaNO_3$	206
$LiNO_3$	398	NaSCN	207
LiSCN	398	$NaBPh_4$	205
		$NaBF_4$	207
Ki	140		
$KClO_4$	140	RbI	106
KSCN	138	$RbClO_4$	106
$KBPh_4$	138	$RbBPh_4$	106

[a]M-Pyrrol = N-methyl-2-pyrrolidone.
[b]Units in cm^{-1}.

and THF; $LiBPh_4$ in THF; and NH_4BPh_4, ND_4BPh_4, and $KBPh_4$ in pyridine.
These authors assign these bands to the vibration of a "diatomic"
type of ion-pair species.

We return now to a consideration of the ion environment. One
can say something more about the cation environment in mixed solvent
systems in which one solvent component is highly solvating and the
other is weakly solvating. The basic technique is illustrated with
the infrared measurements of Wuepper and Popov [4a]. The basic
scheme is to add highly-solvating molecules to a solution of a salt
in a weakly-solvating solvent. One may describe what happens in an
idealized manner as follows. The highly solvating molecules displace
the weakly solvating molecules in the near neighborhood of the cation.
This replacement changes the factors (forces, geometry) which control
the cation frequency. As a result, there is a continuing change in
frequency until further additions of the highly-solvating molecule

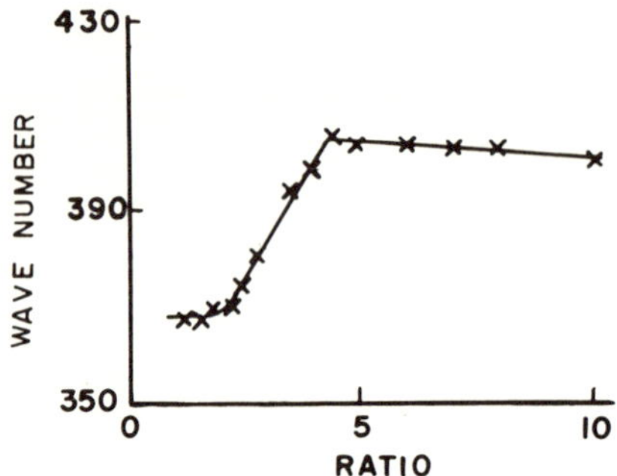

FIG. 2. The Li$^+$ ion band frequency in dioxane with added M-pyrrol. R = moles M-pyrrol/moles LiClO$_4$.

produce no further change at the cation--presumably because all the substrate solvent molecules at the cation have been replaced. Figure 2 shows the results when M-pyrrol is added to LiClO$_4$ in dioxane. The structure of the solution at the Li$^+$ stabilizes when there are 4 M-pyrrol molecules for each cation. The results show little change in this structure as the ratio of M-pyrrol/Li$^+$ increases from 4 to 10. It is customary to infer from this that there are 4 molecules of M-pyrrol around the Li$^+$ when the solvent is pure M-pyrrol. However, this is a long extrapolation whose validity has yet to be demonstrated--a factor which in no way detracts from the utility of the information in the experiment.

This type of experiment is, perhaps, easier carried out by nuclear magnetic resonance spectroscopy (nmr), as first demonstrated by Schaschel and Day [9]. Table 4 shows solvation numbers determined in this manner by Schaschel and Day [9], Popov and co-workers [4a, 10], Lassigne and Baine [5], and Buston and Caruso [6].

The terms "highly-solvating" and "weakly-solvating" must not be interpreted in an absolute manner in defining the conditions required for the above phenomena. The key to its occurrence is a "solvator" molecule with substantially greater solvating power than that of the "substrate" solvent molecule. This is neatly demonstrated in the recent work by Risen and co-workers [11] who used crown ether (dibenzo-18-crown-6) as the solvator molecule and DMSO or pyridine

TABLE 4

Solvation Number of Alkali Ions

Ion	S.N.	Salt	Solvent/System	Method	Reference
Na^+	4	$NaAlBu_4$	THF/C_6H_{12}	nmr	9
Na^+	4	$NaAlBu_4$	M-Pyrrol/dioxane	nmr	10b
Na^+	6	$NaAlBu_4$	DMSO/dioxane	nmr	10b
Li^+	4	$LiClO_4$	M-Pyrrol/dioxane	ir/nmr	4a
Li^+	2	LiI	$DMSO/C_5H_{11}OH$	nmr	10a
Li^+	2	LiI	$Sulfolane/C_5H_{11}OH$	nmr	6
Li^+	4	LiI	DMF/dioxane	nmr	6

as the substrate solvent. Their results are collected in Table 5. These data not only demonstrate the solvating action of the crown ether (ν_{max} is independent of both anion and substrate solvent) but also show that the alkali ions are more tightly bound in their crown cage than in the cage formed only by substrate solvent molecules (DMSO or pyridine).

It might be pointed out here that these mixed solvent measurements give at-the-ion properties of systems which are related to such diverse areas as solvent extraction, the use of solubilizers to promote the solubility of salts in organic solvents, the transport of ionic substances (antibiotics, drugs used in treating mental illness, etc.) in the body and across its membranes, etc.

III. THE NATURE OF THE ION MOTION

If the cation were free to move in a relatively unrestricted manner, its translational motion could be described by a continuum of very broad states whose effect would be to contribute a "background" type of absorption to the far-infrared region of the spectrum. The effect of restricting the motion of the ions and solvent molecules is to replace these states with bound states, with a general shift of the absorption to higher frequencies. The translation of the alkali

TABLE 5

Solvation of the Alkali Ion by Crown Ether [9]

Salt	Solvent	ν_{max}[a] Solvent	ν_{max} Crown + Solvent
NaSCN	DMSO	205	215
	Pyridine	183	217
NaBPh$_4$	DMSO	203	213
	Pyridine	180	212
NaPF$_6$	DMSO	198	210
	Pyridine	180	214
KSCN	DMSO	150	169
	Pyridine	139	170
KBPh$_4$	DMSO	152	167
	Pyridine	135	165
KPF$_6$	DMSO	147	170
	Pyridine	134	167

[a]Units in cm^{-1}.

ion will, of course, make its contribution to the general absorption
in this spectral region. The "gathering together" of the contribution of the alkali ion into a band characteristic of the ion is the
consequence of the existence of quantum states of the solution which
localize translational motion at the cation. The quantum aspect
will be considered in greater detail. Here we are concerned with a
more classical description of the motion.

The simplest model for the bound ion is to consider it moving
in a square-well potential. This leads to the expression

$$\nu = \frac{h}{8Mc\ell^2} \tag{1}$$

for the wave number of the ion band. Here M is the mass of the ion
and ℓ is the amplitude of the motion of the center of mass of the
ion. If one uses $\nu = 200$ for the Na$^+$ in solution, one may calculate

ℓ to be 0.19Å--a not unreasonable value. An interesting feature of such a model is the inverse dependence of ν on the mass of the ion. Figure 3 shows a plot of the alkali ion band frequency for salts in DMSO [1-3] against the reciprocal of the atomic weight of the ion. It is apparent that one can not understand the change in ν as a mass effect on this model. To bring it into harmony with these data requires a continuous adjustment of the size of the well, e.g., to 0.11Å for the Cs^+. A more stringent test of the model is to compare the frequencies of two isotopic ions with the expectations of Eq. (1). Since the size of the well is the same for two isotopic ions, one has by Eq. (1)

$$\nu' - \nu = \nu \frac{M - M'}{M'} \tag{2}$$

where ν' and M' refer to the isotopic species. Both Edgell et al. [1,2] and Maxey and Popov [3] have obtained data for isotopic ions. Comparison of their ν values with Eq. (2) yields

Isotopic ions	Salt/solvent	$\Delta\nu$ calculated	$\Delta\nu$ experimental
$^6Li^+/^7Li^+$	LiCl/THF	65 cm^{-1}	29 cm^{-1}
ND_4^+/NH_4^+	NH_4Br/DMSO	39 cm^{-1}	14 cm^{-1}

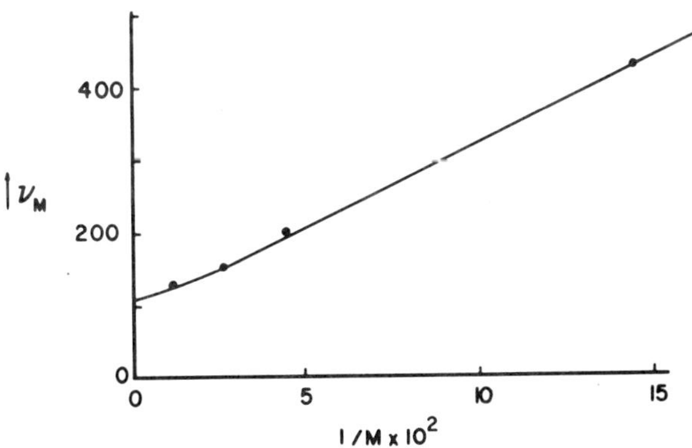

FIG. 3. The mass variation of the alkali ion band frequency for an ion-in-square well model. M = mass of alkali ion; salts are in DMSO.

It is clear that the ion-in-square-well model expresses little more
than the bound character of the ion and gives a poor representation
of its motion.

A more realistic model is to consider the ion moving in a potential well with a curved bottom. Since the considerations of the
last paragraph indicate the ion displacements are small, one might
approximate the potential by Hooke's law, in which case

$$\nu = \frac{\sqrt{K/M}}{2\pi c} \tag{3}$$

where K is the Hook'e law constant. Now, the frequency depends inversely on the square root of the ion mass. A plot of ν versus
$1/\sqrt{M}$ again shows a substantial positive intercept on the ν axis and
requires the value of K to vary with the ion. This would be expected
and the values of K are reasonable. Again, a more stringent test is
furnished by the data for the isotopic ions. The comparison of experiment with the consequences of Eq. (3) follows:

Isotopic ions	$\Delta\nu$ calculated	$\Delta\nu$ experimental
$^6Li^+/^7Li^+$	31 cm^{-1}	29 cm^{-1}
ND_4^+/NH_4^+	21 cm^{-1}	14 cm^{-1}

The agreement between experiment and theory is now much better
and one must conclude that the ions are subjected to forces as they
move in their solution environments.

The cage in the above model is rigid. The solution elements
(solvent, anion) determine the forces which apply, i.e., the value
of K through their individual character and geometry. However, the
existence of the force between the ion and its environment suggests
that the environment may also move in this state. The relationship
for ν depends upon the solution elements at the cation, their arrangement in space, the details of the motion, etc. Fortunately,
the results are not especially sensitive to the specifics of these
details (see Ref. 2) and we take here as a model the alkali ion surrounded by four solvent molecules in a tetrahedral arrangement (see
Fig. 4). It is shown in the next section that, when the solvent
molecules move as rigid entities,

S$_2$M MODEL

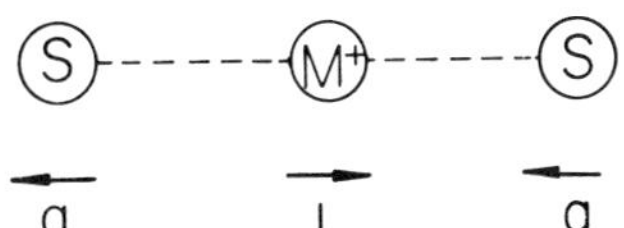

S$_4$M MODEL

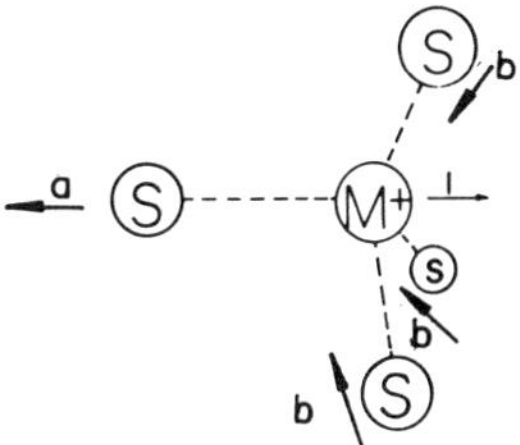

FIG. 4. Displacements in S$_2$M and S$_4$M models (at R = 1) for motion in cation translation band.

$$4\pi^2 c^2 \nu^2 = \frac{3M + 4S}{4MS} K_{net} \tag{4}$$

where K_{net} is the net force constant for the motion. The plot of ν for the alkali ions in DMSO against the mass factor of Eq. (4) is shown in Fig. 5. The ability to represent the data for the Na$^+$, K$^+$, Rb$^+$, and Cs$^+$ salts by a straight line passing through the origin means that these data are well fit by this model with a (nearly) constant value of K_{net}. The comparison of the experimental isotopic shift with the expectations for this model are

Isotopic ions	ν calculated	ν experimental
$^6Li^+/^7Li^+$	29 cm^{-1}	29 cm^{-1}
ND_4^+/NH_4^+	17 cm^{-1}	14 cm^{-1}

The agreement is excellent.

Considerations of the effect of mass on ν lead both Edgell [1, 2] and Popov [3,4] and their co-workers to conclude that the elements of the solution in the near neighborhood of the cation also move (in an opposed fashion) in these translation modes.

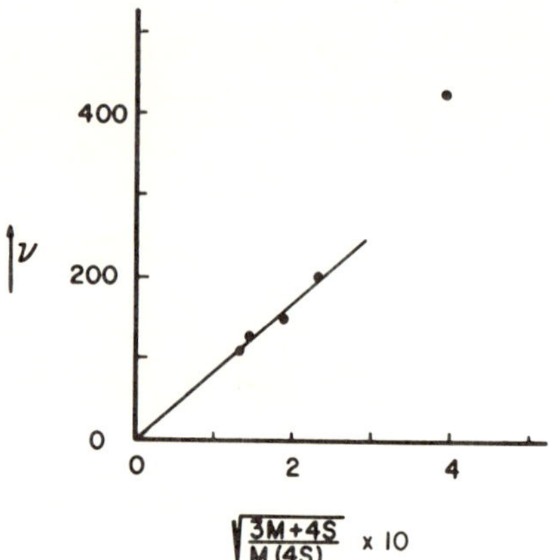

FIG. 5. The mass variation of the alkali ion bànd frequency
for the ion-in-local-lattice model: S_4M. S = solvent = DMSO.

Before leaving this topic it should be noted that the agreement
is also very good in terms of a linear S_2M^+ model or in terms of an
octahedral S_6M^+ model. Therefore, it is not possible to distinguish
between these models on this basis.

IV. FORCES ON THE IONS

To this point, we have learned something about the structure of the
solution at the ions and the character of the motion there. In this
section, the methods which may be used to estimate the forces acting
on the ions and the nature of these forces will be considered.

The alkali ion in solution is subject to forces from a number of
entities, but it is not possible at present to evaluate them separately. The central fact is that one observes one frequency, and this
dictates the methods which can be employed. However, it is possible
to evaluate the net force resisting the distortion of the solution
[1,2]. The net force constant expresses the composite of the individual forces and is a valid parameter for describing the ion environment in a solution. In particular, it gives primary information about
the short-range forces in electrolytic solutions. It is worthwhile
recalling at this point that it is just these short-range forces which
have proved so difficult to obtain from thermodynamic and conductance
measurements.

Edgell and co-workers [1,2] have attacked the problem with *descriptive coordinates*. This is a general method of treating oscillatory problems, which provides for the introduction of whatever information one has about the motion at the beginning of the computation [12]. Details of the general application of this method to oscillatory problems can be seen in the doctoral theses of Dr. Peter Moore and Dr. T. Seshan [13]. Examples of the application of this method to specific models for ion motion in solution may be found in the doctoral dissertations of Risen [14] and Lyford [15]. A more general treatment of ion motion in solution by this method is presented here.

It is possible to calculate the frequency of any mode of oscillation if one knows the displacements which occur and the forces which they bring into play. The vector displacement of the nth atom in the mode is expressed by

$$\rho_n = \underline{\tau}_n R$$

where R is the coordinate for the mode. The set of $\underline{\tau}_n$ vectors gives the vector displacement of each atom when R = 1 and, hence, defines the motion in the mode. The kinetic energy T for the mode is

$$2T = T_{net} R^2 \tag{5}$$

where

$$T_{net} = \sum_n M_n \, \underline{\tau}_n \cdot \underline{\tau}_n$$

The potential energy V of the system for its distortion in this mode is

$$2V = K_{net} R^2$$

in the harmonic approximation. Here K_{net} is the constant expressing the composite force which acts in this mode. Solving the equation of motion yields

$$K_{net} = 4\pi^2 c^2 \nu^2 T_{net} \tag{6}$$

Here ν is the wave number of the motion. Eq. (6) is used to calculate K_{net} from the experimental value of ν.

It is frequently desirable to compare the values of K_{net} with force constants for other types of distortions in molecular systems, e.g., the constants governing the stretching of chemical bonds in molecules, or to express K_{net} in alternate ways. This may be done in the following way. Suppose the potential energy of the ion system is expressed for any distortion by

$$2V = \sum_{qr} k_{qr}\, \Delta r_q\, \Delta r_r$$

where the k_{qr} are the force constants for the displacements expressed in terms of a convenient set r of "internal" coordinates. When the distortion of the system is that of the mode R,

$$\Delta r_q = \overline{\Delta r}_q\, R$$

and

$$K_{net} = \sum_{qr} k_{qr}\, \overline{\Delta r}_q\, \overline{\Delta r}_r \tag{7}$$

The change in r_q for the distortion of the ion mode (for R = 1) is given by

$$\overline{\Delta r}_q = \sum_n \underline{s}_{qn}\cdot \underline{\tau}_n$$

where the $\underline{s}_{qn}$ are the vectors introduced by Wilson [16] and the summation is over the n atoms whose displacement define Δr_q.

To calculate the net force constant from the observed frequency of motion requires a knowledge of the displacement of the various entities which move in the mode. This is not precisely known at present for these ion modes. The importance of having some knowledge of K_{net} encourages one to make approximations. Fortunately, as the isotope studies discussed above show, the ion modes being studied here are so dominated by the alkali ion displacement that the net force constants computed from a variety of models consistent with this localization do not differ much. Consequently, we limit the treatment here to rather simple models for the ion motion.

An interesting question is what response will occur in a solvent molecule as a cation moves toward it. Classically, this is determined by comparing the size of the forces between the ion and atoms of the solvent molecule with the forces between the several atoms comprising the solvent molecule. The latter forces are much larger, in general, than the former for the systems of these studies. As a result, the solvent molecule responds by moving as a whole with only modest internal distortion. In view of the small total contribution of the solvent molecules to the dynamics of the ion motion, the internal distortion is secondary, and our simple models treat the solvent molecules as rigid.

The simplest model for the motion of the solution entities in the ion mode of a solvent-surrounded cation has two solvent molecules moving linearly with the cation. The vector displacements in the S_2M model are shown in Fig. 4. Here $a = M/2S$ is determined by the conservation of momentum conditions. It is convenient to "normalize" the coordinate R by multiplying each displacement in Fig. 4 by $N = 1/(1 + a)$. When one uses these normalized displacements $(\underline{\tau}_n)$ in Eq. (5), one obtains for the S_2M Model:

$$N = \frac{2S}{M + 2S} \qquad T_{net} = \frac{2SM}{M + 2S} \tag{8}$$

where M and S are the masses of the ion and solvent molecule, respectively. If one would assume that all the forces resisting the distortion of the system in this mode depend only on the M-S distance shown, then the value of K_{net} obtained from the experiment by Eq. (6) may be identified as twice the force constant k_s for the distortion of this M-S distance by Eq. (7), that is,

$$K_{net} = 2k_s$$

There is a simple physical meaning which can be given to K_{net} in this case. It is the constant which defines the total force which acts on the ion when it alone is moved from its equilibrium position in the solution. This is readily verified by noting that when M (only) moves by an amount q, then $\Delta r_1 = q$, $\Delta r_2 = -q$ and

$$2V = k_s(\Delta r_1^2 + \Delta r_2^2) = 2k_s q^2$$

Now consider the model which arranges four solvent molecules tetrahedrally about the ion. Considerations of symmetry give the relative displacements shown in Fig. 4, where b = a/3. Momentum conservation in the mode requires a 3M/4S. Normalizing the displacements of Fig. 4 with N = 1/(1 + a) and inserting them into Eq. (5) yields the S_4M Model:

$$N = \frac{4S}{3M + 4S}$$
$$T_{net} = \frac{4SM}{3M + 4S} \tag{9}$$

If one can assume that the forces resisting the distortion of the solution in the ion mode can be approximated in terms of force constants for the M-S distances defined through terms like $k_s \Delta r_1^2$ and $k_{ss}\Delta r_1 \Delta r_2$ in 2V, then K_{net} is related to them by

$$K_{net} = \frac{4(k_s - k_{ss})}{3}$$

Take up now the case of six solvent molecules arranged octahedrally about the ion. The primary displacements in the ion mode are just those given in Fig. 4 for the linear S_2M Model. In the S_6M mode, symmetry also permits atoms of the four solvent molecules, off the S_2M axis, to move parallel to the S_2M axis. Only angles are changed as a consequence of this motion and the corresponding force constants are expected to be weak. Then, the displacements of these four solvent molecules are small and the equations for the S_2M model may be used for the S_6M model for the ion mode.

Some calculations of force constants have been made throughout these studies, reaching back to their very beginnings [1-3,8,11]. Maxey and Popov [3] demonstrated the similarity in the forces at the Na^+, K^+, Rb^+, and Cs^+ ions in DMSO solutions by calculating the frequency for the latter three ions from the Na^+ value under the assumption that the forces were the same. They used what is essentially the S_2M model and in addition calculated a value for a bond constant in Li^+ salts of 0.36 mdyn/Å. This is equivalent to $K_{net}/2$. We use their data here in the S_2M model to calculate the value of K_{net} for all the salts. The results are found in the second column of Table 6. The similarity in forces at the heavier ions referred to by Maxey and Popov is apparent.

Before proceeding, it is worthwhile to investigate the effect of the model choice on the force computed from the experiment. As pointed out above, the values obtained are identical in the S_2M and S_6M models. The values for the net force constant, which result from these data if one uses the S_4M model, are found in column three of Table 6. The values of K_{net} are essentially the same in the two models for the Li^+, Na^+, and K^+ ions while showing some difference for Rb^+ and Cs^+. Both models show the same trends. Thus, the value for the Li^+ salts is larger than those for the heavier alkali ions, while the K_{net} value for the latter are essentially the same. Risen and co-workers [11], while making force constant calculations for the crown-ether solvation study, were moved to consider a model for DMSO in which the SO unit moved rigidly in response to the cation, while the methyl groups remained fixed in space. At the same time,

TABLE 6

K_{net} for Salts in DMSO[a]

Salt	S_2M[b]	S_4M	$S_4'M$
Li^+	0.72	0.71	0.68
Na^+	0.47	0.44	0.40
K^+	0.43	0.39	0.33
Rb^+	0.51	0.43	0.34
Cs^+	0.51	0.42	0.31

[a]Units are mdyn/Å; S = 78.1; S' = 48.1.
[b]Also, valid for S_6M model.

the force arising from the resulting distortion of the DMSO molecule was ignored. As a further test of the effect of model choice
on the value of K_{net}, we apply the Risen model for DMSO motion to
these data. The values for K_{net} which result are listed in the last
column of Table 6. The values tend to be somewhat smaller than
those obtained with S_2M or S_4M models, but they also show the same
trends. One may conclude from these results that the values of K_{net}
calculated from the experimental ν by Eq. (6) will show significant
dependency on the model used only for the heavier alkali ions. Moreover, any substantial trends in the force acting at the ion in a
series of measurements will appear in K_{net}, whatever model is used.

The utility of resolving the total force at the ion into components between the ion and each solvent molecule to form bond constants, in analogy to the molecular case, may also be tested with these
data for salts in DMSO. The values of the bond constant k_s were
calculated by the expressions given above and are collected in Table
7. As can be seen, they are much more dependent upon the model used
in their calculation than the K_{net} values. Furthermore, as we shall
see below, the use of k_s can obscure important facts about ion binding. For these reasons, K_{net} values are generally used in the author's laboratory to investigate the force situation at ions in
solution.

TABLE 7

k_s for Salts in DMSO[a]

Salt	S_2M[b]	S_4M	$S_4'M$
Li^+	0.36	0.53	0.51
Na^+	0.24	0.33	0.30
K^+	0.22	0.30	0.25
Rb^+	0.26	0.33	0.25
Cs^+	0.26	0.31	0.23

[a]Units are mdyn/Å; S = 78.13; S' = 48.06.
[b]Also, valid for the S_6M model.

TABLE 8

Net Force on Li^+ in Several LiI Solutions[a]

Solvent	K_{net}, mdyn/Å
THF	0.54
M-pyrrol	0.63
Pyridine	0.69
Acetone	0.69
DMSO	0.70

[a]Calculated by S_2M model; virtually the same numbers are obtained with the S_4M (and S_6M) model.

Several examples of the calculation of the net force on the alkali ion will now be given. Table 8 shows the constant for the net force on the Li^+ in LiI solutions in several solvents from the data found in Table 2. One may conclude that the frequency difference shown in Table 2 does not arise from a mass effect of the solvent molecules but represents a real difference in the force field at the Li^+ ion in the several solutions. The source of this difference is not clearly understood at present.

When the anion is one of the near neighbors of the cation, it not only contributes to the force at the cation but also participates in the motion of the cation mode. Any substantial difference between cation-anion and cation-solvent force will be reflected in a difference between the solvent and anion displacements. No experimental means are presently available to measure any such difference. And fortunately, one does not expect appreciable differences here. Consequently, we make the simplest assumption in constructing a model for the motion at this kind of ion site: namely, that there is no difference between solvent and anion motion. This gives for the linear (or octahedral) model, the SAM model:

$$N = \frac{S + A}{M + S + A}$$

$$T_{net} = \frac{M(S + A)}{M + S + A}$$

(10)

TABLE 9

Net Force on Alkali Ion for some Salts in THF[a]

Salt	K_{net}, mdyn/Å	Salt	K_{net}, mdyn/Å
$LiCo(CO)_4$	0.68	$LiBPh_4$	0.68
$NaCo(CO)_4$	0.46	$LiNO_3$	0.64
$KCo(CO)_4$	0.40	$LiCl$	0.58
$NaBPh_4$	0.41	$LiBr$	0.56
NaI	0.50	LiI	0.55

[a]SAM model used; virtually the same values are obtained with the S_3AM and S_5AM models.

If one assumes that the forces resisting the motion in the ion mode may be approximated in terms of M-S and M-A force constants, then the net force constant can be equated to

$$K_{net} = k_S + k_A$$

In this way, the data of Table 1 were used to compute the net force on the alkali ions for salts in THF solutions; these are found in Table 9 [1.2]. One concludes, again, that the frequency differences seen in the spectra are reflected in force differences at the ion. The data show a dependence of the net force constant on the anion in these solutions, that the force on the Li^+ ion is greater than that on the Na^+, and that the force on the K^+ ion is about the same as that on the Na^+ ion.

The last example is from the study by Tsatsas et al. [11] of the effects of crown-ether solvation on the Na^+ and K^+ salts of SCN^-, BPh_4^-, and PF_6^- in DMSO and pyridine as substrate solvents. These authors calculated the bond force constant k_s^* for the solvent surrounded Na^+ and K^+ in DMSO and pyridine solutions and compared them to the value for these ions when solvated with crown-ether. The bond constant for each ion site was calculated for a variety of models for

---

*The interaction force constants included in the original study have been dropped here for simplicity of expression.

the ion motion. We abstract from their work the values for k_s in neat DMSO in the S_4M model (with rigid DMSO) and in the $S_4'M$ model (rigid SO) and the value in the crown solvate in the C_6M model (with C = rigid C_2O). These values are collected in column 4 of Table 10. The *drop* in the bond force constant k_s accompanying the crown-ether solvation is pronounced. The reader is referred to the original paper for the discussion of the energetics of the solvation.

Before leaving this case, it is interesting to use their data to calculate the net force at the alkali ions in the neat DMSO and in the crown-ether solvate. The values of K_{net} are found in the last column of Table 10. Note that the tighter binding of the ion to the equilibrium position which is expected when the crown-solvate is formed is readily apparent in the net force constant K_{net} but is obscure in the values of k_s. This seeming paradox results from the division of the total force on the ion into the several bond components of the crown-ether solvate.

The above paragraphs have shown how one may evaluate the force acting at ions in solution. We shall now see that it is also possible to get experimental information about the character of this force. The ion, of course, is subject to electrostatic forces from the other solution components. The question, really, is whether there is a co-valent component to the force at the ion. The existence of a compo-nent of this type will be reflected in a change in the polarizability

TABLE 10

Effect of Crown-ether Solvation on Alkali Ion Forces

Cation	System	Model[a]	k_s[b]	K_{net}[b]
Na^+	DMSO	S_4M	0.34	0.45
	DMSO	$S_4'M$	0.30	0.41
	Crown/DMSO	C_6M	0.17	0.52
K^+	DMSO	S_4M	0.28	0.38
	DMSO	$S_4'M$	0.24	0.32
	Crown/DMSO	C_6M	0.16	0.49

[a]$S = 78.1$; $S' = 48.1$; $C = 40.0$ is mass of C_2O part of crown ether.
[b]Units are mdyn/Å.

of the system with the ion motion. This will make the excitation
of the state associated with the ion motion active in the Raman effect. In the absence of a covalent component of force at the ion,
polarizability changes produced by the ion motion can be expected to
be small to nonexistent, a situation which would produce a weak Raman
band at best and more probably none at all.

Edgell and co-workers used this method of determining the nature
of the force at ions in a number of solutions [2]. They were unable
to find an alkali ion motion band in the Raman spectra of the
$NaCo(CO)_4$, $Na_2Cr_2(CO)_{10}$, and $NaHCr_2(CO)_{10}$ in THF and DMSO solutions;
of $NaCo(CO)_4$ in DMF; and of LiCl in THF. They concluded that covalent
forces on the alkali ion were small in these solutions and that the
ion motion in them takes place essentially under the influence of
electrostatic forces. Wong, McKinney, and Popov used the same technique to establish that the forces on the alkali ion of salts dissolved in acetone are also ionic [4c].

On the other hand, Tsatsas and Risen have observed a Raman band
from the Na^+ ion motion for $Na^+AlBu_4^-$ dissolved in cyclohexane [17].
This Raman band is not present for this salt in pure THF. These
authors attribute the observed spectral changes on addition of THF
to the cyclohexane solution to the formation of THF Na^+ and $(THF)_4Na^+$.
Application of the criterion above indicates that the Na^+ in one site
in cyclohexane is bound by forces with a substantial covalent component. In contrast, the force on the Na^+ for this salt in THF is
essentially electrostatic.

V.   THE COMPUTATION OF THE FORCE AT THE ION

In the preceding section we considered how one could convert the frequency of the ion translation band into a value for the force acting
at the ion. Further, the intensity of the ion band in the Raman
spectrum is a measure of the covalent component of this force. By
this criterion, the ion force is essentially electrostatic for all
salts examined so far in dipolar solvents (DMSO, THF, acetone, DMF).
The calculation of this force from electrostatic theory is considered
in this section.

The starting point for an electrostatic calculation is the potential energy U of the system. The force constant is the curvature of U at its minimum. There are many expressions which might be used for U, depending upon the approximations which one employs. A note of caution should be inserted at this point. Some potential energy expressions for the interaction between charges contain the dielectric constant in the denominator. They are valid only for dealing with long-range forces and can not be used to compute the short-range forces which control the dynamical behavior of the ion and which are our concern here.

Models of ion solvation have been formed by summing the interaction of the ion with the adjacent solvent molecules and considering the remaining molecules as a dielectric continuum [18-20]. These are "hard-shell" models and can not describe the force on the alkali ion near its equilibrium position because of the "discontinuity." The hard-shell model has been modified to remove this difficulty by adding appropriate repulsion terms [1,2].

As the first example of this model, consider each solvent molecule or polyatomic ion, adjacent to the central ion, as a small number of polarizable charge clouds. Then,

$$U = \sum_{ij} [A_{ij} \exp\frac{-r_{ij}}{\rho} + \frac{q_i q_j}{r_{ij}} - \frac{C_{ij}}{r_{ij}{}^6}] - \frac{1}{2} \sum_i (\alpha_i \underline{F}_{ij} \cdot \underline{F}_{ij}) + U_p$$

with   (11)

$$\underline{F}_{ij} = \sum_j q_j \frac{\underline{r}_{ij}}{r_{ij}{}^3}$$

The first time is the repulsion energy, the second term is the Coulombic interaction, the third is the London dispersion terms, the fourth is the energy resulting from the interaction of the induced dipole in each cloud with the field which creates it, and the last term results from the interaction of the ion and its neighbors with the remainder of the solution, i.e., the continuum. The first sum is over all cloud pairs such that the ith and jth clouds are in different entities (ion, molecule); each pair is included once. The

second sum is over all charge clouds, while the sum over j is over
all clouds but those of the ith entity. Here q_i is the charge of
the ith cloud, α_i is its polarizability, $\underline{r}_{ij}$ is a vector from the
center of the ith cloud to that of the jth cloud, and r_{ij} is the
distance between their centers.

Distorting the system in the ion mode changes the various r_a
which enter U. At the equilibrium position, U is a minimum and

$$0 = \frac{\partial U}{\partial R} = \sum_a \frac{\partial U}{\partial r_a} \frac{\partial r_a}{\partial R} \tag{12}$$

This equation marks the position of equilibrium if all the parameters
which enter Eq. (11) are known. The net force constant is obtained
from the second derivative,

$$K_{net} = N^2 \frac{\partial^2 U}{\partial R^2} = N^2 \sum_{ab} \frac{\partial^2 U}{\partial r_a \partial r_b} \frac{\partial r_a}{\partial R} \frac{\partial r_b}{\partial R} \tag{13}$$

where a and b cover the different coordinates which appear in U, and
R is the ion mode coordinate before normalization. N is the normalization constant as defined in the last section, and all derivatives
are evaluated at the equilibrium conformation.

In practice, we have used a digital computer to evaluate K_{net}
for the model of Eq. (11). The program computes U for a given configuration of the solution at the ion. Then the ion is moved a small
increment along its path in the ion mode. At the same time all other
elements of the solution are displaced, the proportionate amount specified by the ion mode and U is computed again. Another increment
is made in the positions, and U computed. This process is repeated
while the ion is swept through the minimum in U. The net force constant is obtained as

$$U = U_o + \frac{K_{net} R^2}{2N^2} \tag{14}$$

where U_o is the minimum and R is the displacement of the ion from it.
It is unlikely, of course, that the first choice of configuration
will be the one of lowest energy. The software steps the system
through successive configurations until it is found.

Calculations of this type have been made in this laboratory by Risen [14], Lyford [15], Spencer, and others. While the values obtained for K_{net} are dependent somewhat upon the values chosen for the parameters and the configuration assumed for the "cage," all are of the magnitude found to exist in the solution from the experimental data in the last section.

The electrostatic calculation of K_{net} will be illustrated here for the S_2M model of the ion motion with a somewhat less detailed treatment of the interactions. In it each solvent molecule is represented by a charge cloud with polarizability α, dipole moment μ, quadrupole moments θ_a and θ_b, and ionization potential I. When one reflects for a moment on the geometry of typical organic solvent molecules like acetone, acetonitrile, DMSO, etc., it is apparent that the center of the charge cloud which repulses an approaching ion is not located at the center for the dipole moment. We represent the distance between these two centers by the symbol a. The alkali ion is represented by the charge +e, the polarizability α_i, and the ionization potential I_i. The bulk dielectric constant of the solvent is ε. In these terms, Eq. (11) becomes

$$U = -\frac{e\mu}{r_1^2} + \frac{e(\theta_a + \theta_b)}{2r_1^3} - \frac{e^2\alpha}{2r_1^4} - \frac{C}{r_1^6} + A\,\exp\left[-\frac{(r_1 - a)}{\rho}\right]$$

$$-\frac{e\mu}{r_2^2} + \frac{e(\theta_a + \theta_b)}{2r_2^3} - \frac{e^2\alpha}{2r_2^4} - \frac{C}{r_2^6} + A\,\exp\left[-\frac{(r_2 - a)}{\rho}\right]$$

$$- 2\alpha_i\mu^2\left(\frac{1}{r_1^3} - \frac{1}{r_2^3}\right)^2 + \frac{e^2}{2\varepsilon r_o} + \cdots$$

with

$$C = \frac{3I_i\,I\alpha\alpha_i}{2(I + I_i)} \tag{15}$$

The first line expresses the interaction of M^+ ion with the first solvent molecule whose (dipole) center is a distance r_1 from it. The second line is the interaction of the ion with the second solvent

molecule. The first term in the third line is the energy of interaction of the solvent dipole with the induced dipole in M^+ and the
second term is the Born solvation energy of the ion. Here r_o is the
radius of the (spherical) hole in the continuum in which M^+ and the
solvent shell are contained. Among the terms following the last +
sign, but not explicitly expressed, are the energy of interaction of
the solvent molecules in the shell, the energy of interaction of the
solvent molecules in the shell with the continuum, the terms for
saturation effects in the continuum, etc.

The condition that U is a minimum at the equilibrium position

$$\frac{\partial U}{\partial r_1} = \frac{\partial U}{\partial r_2} = 0$$

yields

$$\frac{A}{\rho} \exp\left[-\frac{(r-a)}{\rho}\right] = \frac{2e\mu}{r^3} - \frac{3e(\theta_a + \theta_b)}{2r^4} + \frac{2\alpha e^2}{r^5} + \frac{6C}{r^7} - \frac{e^2}{4\varepsilon r_o^2} \tag{16}$$

This equation serves to determine $r = r_1 = r_2$ at equilibrium when
all the parameters are known. In carrying out the differentiation,
r_o was set to $r + b$, where b is the distance from the center of the
solvent molecule (dipole) to the center of the cavity, and all terms
not explicitly expressed were dropped.

Differentiating once more gives

$$K_{net} = N^2 \frac{\partial^2 U}{\partial R^2}$$

$$= 2\frac{A}{\rho^2} \exp\left[-\frac{(r-a)}{\rho}\right] - \frac{12e\mu}{r^4} + \frac{12e(\theta_a + \theta_b)}{r^5}$$

$$- \frac{20\alpha e^2}{r^6} - \frac{84C}{r^8} - \frac{144\alpha_i \mu^2}{r^8} \tag{17}$$

This equation has been applied to the computation of K_{net} for Li^+
salts in DMSO [2]. A was estimated from a plot of the repulsion
constants for gaseous alkali halides against the sum of the gaseous
ionic radii, which is shown in Fig. 6. The primary data for this

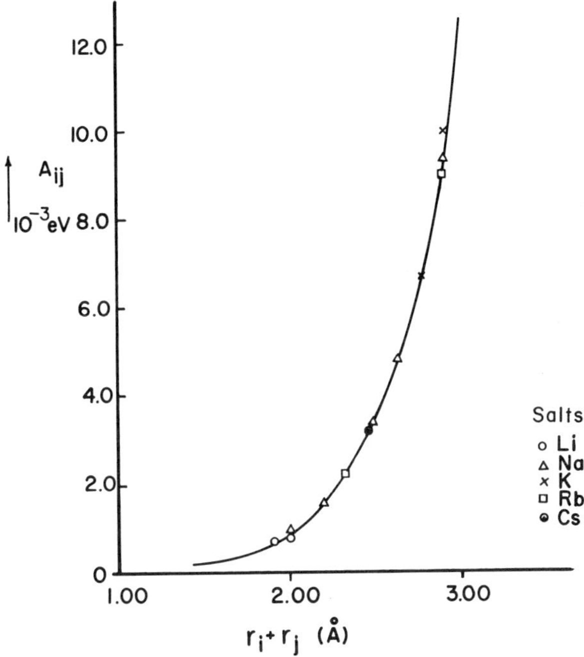

FIG. 6. Repulsion constant A_{ij} vs the sum of the ionic radii $r_i + r_j$.

figure is from the work of Rittner [21] and White et al. [22]. With a value of $r_i + r_j = 2.1$ Å, one obtains $A = 2.08 \times 10^{-9}$ ergs. Then ρ is given its nearly universal value of 0.32 Å. For DMSO, $\mu = 4.3$ D, $\alpha = 7 \times 10^{-24}$ cm^3, $\varepsilon = 43$ and a is taken as 0.9 Å; for Li$^+$, $\alpha_i = 0.03 \times 10^{-24}$ cm^3. No data are available for the quadrupole moments of DMSO, and C is set to zero because the contribution from the dispersion terms is small.

These data are used in Eq. (16) to obtain the value of r, which is then used in Eq. (17). The result is $K_{net} = 0.59$ mdyn/Å, which is to be compared with 0.72 mdyn/Å obtained from the experimental band (Table 6). This is a rather remarkable result when one considers the a priori character of the calculation. It is only necessary to change a to 0.8 Å in order to obtain agreement with the observed value. These results for the first calculations of K_{net} from theory encourages one to believe that it will be possible to construct reliable electrostatic models for the short range forces at an ion in solution. What is needed is an improved understanding of the geometry and character (symmetry, what's there) of the ion environment.

The above sample can be used to illustrate an important point about the energetics at an ion in solution [2]. Figure 7 shows a plot of the potential energy U (solid line) against R. The value of U at the minimum is the equilibrium value U_o of the potential energy. The contribution to U of the repulsion and electrostatic terms are shown separately in Fig. 7 by the dashed lines. It can be seen that while the contribution of repulsion to U_o is not negligible, the largest contribution comes from the electrostatic terms. This is well known.

Figure 8 shows a plot of the curvature of U (solid line) against R. Its value at the equilibrium position gives K_{net} by Eq. (13). The contribution to the curvature from the repulsion and electrostatic terms are again shown separately by the dashed lines. As can be seen, the major contribution to K_{net} comes from the *repulsion* terms. This point has not been well recognized for ionic solutions. Yet it is clearly in harmony with the major experimental findings. For example, translation band frequencies are clearly determined, in the main, by the fact that the ion has neighbors and only in a secondary way by

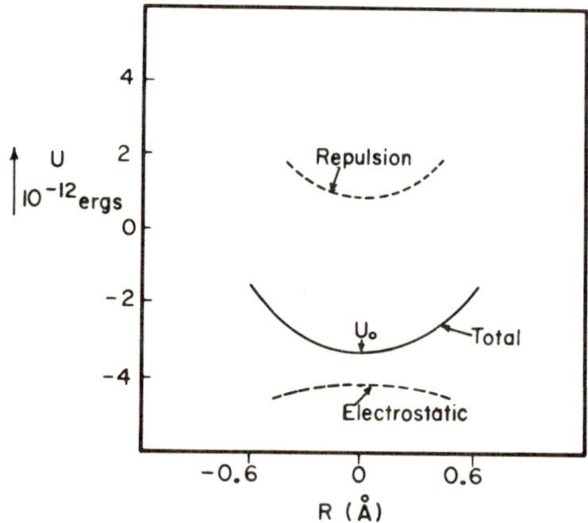

FIG. 7. Potential energy U vs cation motion coordinate R.

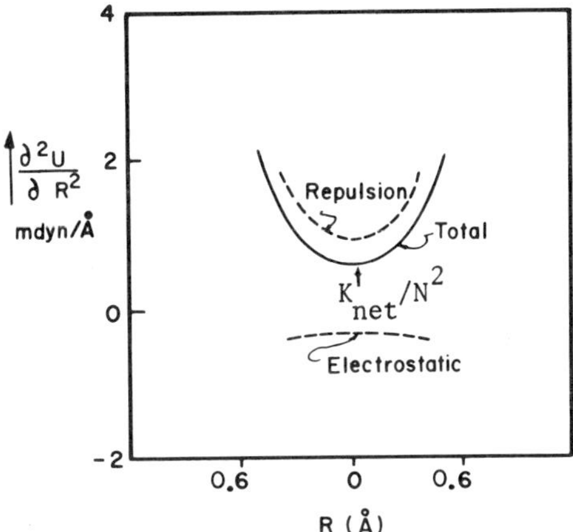

FIG. 8. Curvature in U vs cation motion coordinate R.

the specifics of the neighbor. Moreover, the localization of the
ion motion which produces the translation band results from the fact
that the ion has neighbors and is little dependent upon whether a
neighbor is bound to the ion by strong attractive interactions or
not. Thus, a realistic potential energy for an ion or ion pair in
solution must include the interaction of the ion with all its neighbors and not be limited just to those to which it is strongly (or
chemically) bound.

VI. SOLUTION STRUCTURE IN THE ION ENVIRONMENT

We return now to the question of structure in the solution. What,
in fact, does one mean by the term? Perhaps, the place to start is
to consider the absorption of infrared radiation by the ionic solution in the localized cation-translation band. The absorption experiment sums the interaction of the radiation field with all the
cations of the solution. Under the conditions of our experiment,
this instantaneous sum over all the cation centers can be replaced
by the sum over time of the behavior at a single cation. Thus, the
question reduces to the time behavior of the conformation of the
solution at the cation as determined by the instantaneous values of

the coordinates of all the relevant atoms. Will a (near) continuum
of conformations occur? Or will certain ones persist? And for how
long? The point probed by the last question can be represented in
the following way. Conformations which are passed through during a
time which is characteristic of the cation motion (~1/energy required
to excite the motion) are "averaged" to give the observable properties of that state. The word structure, then, refers to basic solution conformations, which persist for times which are long, depending
on the states which are being excited. For example, if all the possible conformations at an ion were passed through in this time, the
experiment would show only one structure, that of being " in solution." On the other hand, if the motion of the ion environment is
very slow compared to that of the ion, a large number of structures
occur. A situation intermediate between these extremes is also possible. In this event, several basic conformations persist for long
times, and one may say that an ion spends part of its time in each
of the structures or that the solution at any one time is described
as a mixture of ions in the several kinds of ion sites.

Two kinds of solutions have been seen in the ion band studies
reported above--one in which the ion band frequency is independent
of the anion and one in which it is anion dependent. Does this mean
that there is only one kind of ion site structure in each solution
within the meaning of the term? If this were the case, each cation
in the first kind of solution sees only solvent molecules as near
neighbors, while each cation in the second time of solution sees an
anion among the near neighbors. Intuitively, this seems too simple
a result.

One may look for evidence relative to this point. First, almost without exception, only one ion motion band is seen in the far
infrared. Second, the absorbance of this band is linear with concentration [1-4]. The simplest interpretation of these data involves
only one kind of solution structure in each solution. But they are
also consistent with the simultaneous presence of several kinds of
sites if the contribution from each site is contained in the envelope

of the single band. Early spectral evidence for the simultaneous
presence of more than one kind of ion site for a salt dissolved in
an organic solvent was presented by Edgell, Yang, and Koizumi [23],
and Hogen Esch and Smid [24]. The former authors arrived at this
conclusion from observations of the vibrational spectrum of the
$Co(CO)_4^-$ ion for $NaCo(CO)_4$ in THF, while the latter drew their conclusions from the electronic spectra of the fluorenyl anion for its
salts in THF and other solvents. In addition, two far-infrared ion
motion bands which come from different ion sites have been seen by
Tsatsas and Risen [17] for NaAlBu dissolved in cyclohexane. One
concludes from this evidence that more than one kind of ion site
may exist in many solutions. What one needs is a more sensitive
probe of solution structure than the anion variation of the frequency
of the intensity maximum of the alkali ion motion band.

Such a probe is suggested by the results for the anion vibrational spectra reported above [23]. The intramolecular vibrations
of a polyatomic ion reflect its immediate surroundings in the manner
discussed in the first paragraph of this section. Information about
the solution structure at the anion, of necessity, also leads to
some knowledge of the situation at the cation by direct connection
and by implication. The $Co(CO)_4^-$ ion is an excellent anionic probe
of structure because of the wide solubility of its salts in organic
solvents.

Edgell and co-workers [25] are carrying out a study of solution
structure in a wide variety of solvents by this method. Their technique consists of making a careful study of the contour of the infrared band arising from the C-O stretching modes of the $Co(CO)_4^-$
anion. The anion geometry is tetrahedral, as shown by both infrared
and Raman spectra [26] and x-ray studies [27]. An isolated ion with
this structure would have T_d symmetry and show a single, triply degenerate C-O stretching band in its infrared spectrum. What one will
actually see in the spectrum of this anion in solution will depend
upon the strength of its interaction with the environment and the
persistence of the solution conformations, as discussed above. If

the effective forces at an anion are tetrahedral, one band will appear. If, however, the effective forces are asymmetric, the degeneracy of the F_2 band is lifted. The observed spectrum is the result of the interaction of the radiation field with all the anion sites. If several kinds of solution conformation at the anion fall into the persistent category, i.e., if several kinds of anion sites exist in the solution, the observed band will be the sum of the spectra produced by each such kind of site. If only one ion site structure exists, however, all the features of the observed band arise from it.

Figure 9a shows the CO stretching band for $NaCo(CO)_4$ in THF at room temperature, taken from the work of Edgell, Barbetta and Heade [25e,f]. Three bands are readily seen in the absorption complex. Do all these bands come from the $Co(CO)_4^-$ in the same site? How many sites are present? These authors have developed three methods for dealing with this kind of question [25]. First, they remove one or more of the sites by the addition of small amounts of another solvent. Second, they vary the population of the several ion sites by changing the temperature. And third, they carry out a computer-aided analysis of the observed band contour. Each will be illustrated because of the fundamental importance of structure in solution.

Some solvents have little effect upon the contour of the band shown in Fig. 9a when they are added in modest amounts to the THF solution of $NaCo(CO)_4$. Others alter the band shape--some slowly, some more rapidly. Water is effective. With each addition, there is a decrease in the intensity of the 1857 and 1899 cm^{-1} bands and an increase in the 1886 cm^{-1} band. By the time the water content has risen to 5% only the latter band remains. This change can be seen in Fig. 9b.

The same change takes place in the spectrum when the temperature of the solution is dropped. This can be seen in Fig. 10, p. 318.

It follows, therefore, that there are two kinds of ion sites in this solution--one which produces the 1857 and 1899 cm^{-1} band components and the one which gives the 1886 cm^{-1} component. The former site has been identified as a contact ion pair with C_{3v} symmetry in

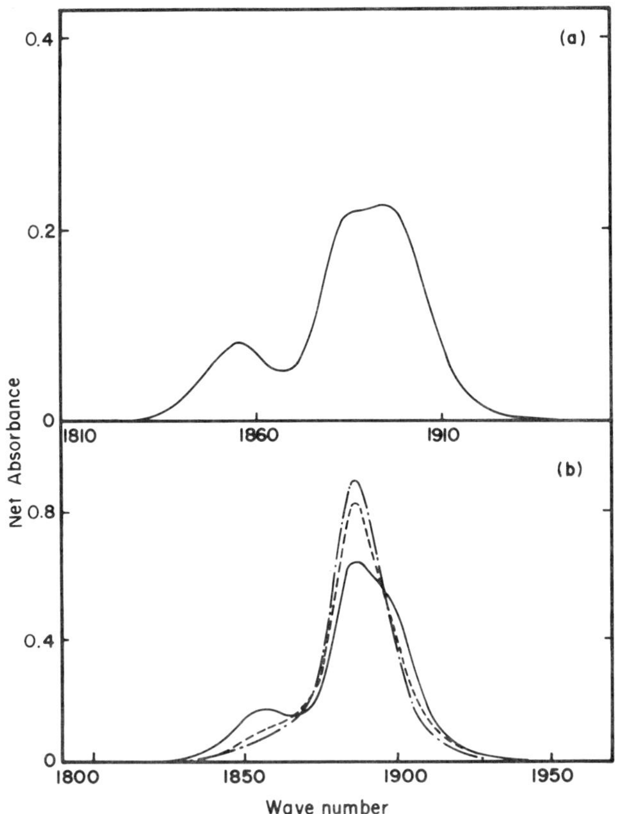

FIG. 9. The CO stretching band for NaCo(CO)$_4$ in THF. a: infrared spectrum at room temperature; b: infrared spectrum of solution laced with water; 0.11 M in H$_2$O (——) 0.55 M in H$_2$O (---) 1.09 M in H$_2$O (—•—).

the effective potential energy, while the latter has been assigned

to the solvent-surrounded ion site of T$_d$ symmetry [25b,e]. Solvent-

surrounded ions include both solvent-separated ion pairs and "free"

ions. While no spectroscopic distinction has been seen between these

last two kinds of sites, electrical conductivity measurements [23]

indicate that the solvent-surrounded ion site in this solution is

primarily a solvent-separated ion pair.* The assignment presented

---

*Tight and loose ion pairs and inner and outer complexes are
two other names given to contact- and solvent-separated ion pairs.

here is supported by the observations of the Raman spectrum [25b].

The third method developed for structure determination is a computer-aided analysis of the contour of the absorption complex [25b,d]. The net absorbance is calculated at each wave number ν in the complex band by the equation

$$A = \log\frac{T_o}{T}$$

Each component band is regarded as a sum of a Lorentz contribution of the form

$$\frac{H}{1 + D^2(\nu - \nu^o)^2}$$

and a Gauss contribution of the form

$$h \exp[-d^2(\nu - \nu^o)^2]$$

Here, ν^o is the center of the component band, H is the peak absorbance of the Lorentz component, $D = 1/LHW$, h is the peak absorbance of the Gauss component, and $d = \sqrt{\ln 2}/GHW$. LHW (or GHW) stands for half the width of the Lorentz (Gauss) contribution at half its peak absorbance. The net absorbance for a complex band is the sum of the absorbance of the component bands

$$A = \alpha + \sum_i \left\{ \frac{H_i}{1 + D_i^2(\nu - \nu_i^o)^2} + h_i \exp[-d_i^2(\nu - \nu_i^o)^2] \right\} \tag{18}$$

The quantity α is added to Eq. (18) to take care of small shifts in the baseline, if needed.

To analyze a complex band with this function, one selects the five parameters (ν^o, H, h, D, and d) which describe each of its component bands. The computer software uses a non linear, iterative process to do this which minimizes the difference between the experimental and the computed band contour. The quantity which is minimized is

$$\phi = \sum_n W_n^2(AX_n - AC_n)^2$$

where AX_n is the experimental absorbance value at the nth wave-number point, AC_n is the computed value there, and W_n is a weighting number which expresses the relative importance of a deviation at the point. This computation takes several forms in our laboratory. The software designed for direct interaction between the operator and the computer is given in the Appendix.

The computer program is able to describe the contour of a single band with accuracy. In general, this accuracy is greater than that to which the experimental band is defined by the chemical and spectroscopic methods. When corresponding care is used on the experimental side, a computer-aided band analysis is a powerful tool for solution structure determination.

The experimental methods used in this laboratory are described in Ref. 25d. In brief, on the spectroscopic side, it is necessary to expand the wave-number scale so that the band contour can be expressed with accuracy. The slits should be as wide as possible in order to enable the pen servo system to accurately track the changing spectrum. An upper limit to the slit widths is fixed by the point at which band distortion results from the finite wave-number spread of the radiation passing through the slits. The gain is set to produce a responsive pen and the intertia in the servo system is set to reduce the noise to a level consistent with the accuracy expected for the contour. Finally, attention must be paid to the accuracy of the wave-number scale of the spectrometer. The scale is readily corrected for a slow drift from "true" wave number by the usual calibration with standard reference lines. On the other hand, short-range excursions from true wave number are insidious in this application since they distort a band contour making it appear that an overlapped band is present where none occurs. Unfortunately, errors of this last type do occur in commercial spectrometers. They are detected and corrected for in this laboratory by recording the interference fringe pattern from an empty Irtran-2 cell with a pathlength of 0.5 mm. Fringe maxima and minima occur at equal spacing on a true wave-number scale. Their deviation from this behavior is

a measure of the error. A knowledge of the true wave number of one
fringe and the constant wave-number difference between fringes is
sufficient to calibrate any point on the wave-number scale. The
frequency of a fringe maximum or minimum at both ends of the region
of interest is determined by linear interpolation between two closely
spaced, known lines in the DCl and water vapor spectra. These data
give the wave number ν^o of the reference fringe and the spacing between adjacent maximum and minimum. The corrected wave number is
computed by the equation

$$\nu = \nu^o - (n + \frac{a}{b})s$$

where n is the number of maxima and minima between the reference
fringe and the point of interest, a is the scale wave-number distance
of the point from the last extremum, and b is the scale wave-number
spacing between the maximum and minimum which spans the point.

Not only is it necessary that the spectrometer furnish a faith-
ful reproduction of the spectrum of the sample but the sample must
be pure. Water is a solution contaminant which must be rigorously
excluded. Methods for drying solvents and measuring the residual
water have been described [25d].

The effectiveness of the computer-aided analysis is seen in the
results obtained for solutions of $NaCo(CO)_4$ in the six solvents:
dimethylsulfoxide, dimethylformamide, nitromethane, hexamethylphos-
phoramide, acetonitrile, and pyridine [25d]. While the CO stretching
band from these solutions appears to be a lone band, the computer
analysis shows the presence of a second weak band. The position and
intensity of the two bands selected by the computer program for each
solution is given in Table 11. The strong band has been assigned to
the F_2 modes of the $Co(CO)_4^-$ ion in a solvent-surrounded ion site
with effective potential energy of T_d symmetry. The weak band arises
from the A_1/F_2 mode of the isotopic $Co(CO)_3C^*O^-$ ion in the solvent-
surrounded site of T_d symmetry. Thus, there is only this one kind
of ion site in these solutions.

TABLE 11

Component Band Properties from Band Analysis

$NaCo(CO)_4$ in Several Solvents

Solvent	ν, cm^{-1}	$\nu - \nu'$, cm^{-1}	B' x 100/B
DMSO	1888.3	33.4	1.8
CH_3NO_2	1892.5	32.5	1.6
DMF	1889.3	32.3	1.7
Pyridine	1887.7	32.7	1.7
HMPA	1886.1	35.1	1.5
CH_3CN	1892.8	34.2	1.3

ν = wave number of the F_2 band of the normal anion

ν' = wave number of the A_1/F_2 band of isotopic anion

B, B' = integrated absorbances for ν and ν'.

Theory says that the isotopic band should lie 34 cm^{-1} below the main band and appear with an intensity which is 1.5% of that of the main band. The excellent agreement of the theoretical expectations with the computer resolution of the experimental data found in Table 11 is a measure of the accuracy obtainable with computer-aided analysis of accurate experimental data. Two important points can be made. Not only can this type of analysis find sites of small population, but also the inclusion of such sites will, in general, be required to obtain agreement between computed and experimental band contours when data are accurate.

Let us return to a consideration of the solution structure of $NaCo(CO)_4$ in THF. A computer-aided analysis of the band of Fig. 9a shows that it can not be satisfactorily understood in terms of two sites with three component bands, as appears to the eye. It is necessary to include a third site with a small population at room temperature which gives rise to weaker bands at 1906 and 1846 cm^{-1}. The excellent agreement between the experimental data and the band contour computed on this basis is shown in Fig. 11. Thus, it is concluded that three kinds of ion sites are found in THF solutions

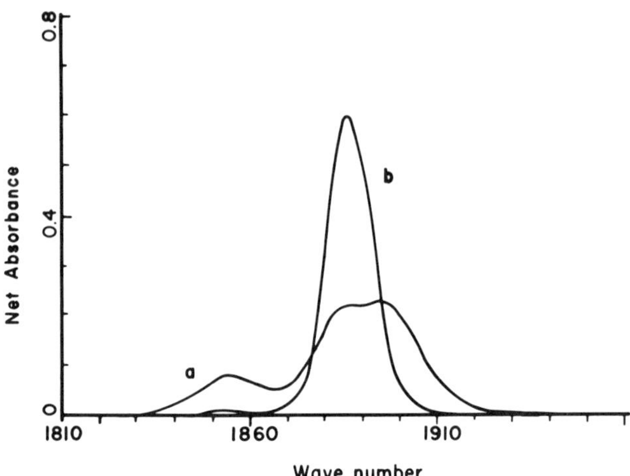

FIG. 10. The change in the CO stretching band with temperature for NaCo(CO)$_4$ in THF: a: 300°K, b: 175°K.

of NaCo(CO)$_4$ at room temperature [25e,f]. They are the contact ion pair, the solvent-surrounded ion, and a triple-ion Na[Co(CO)$_4$]$_2^-$. The frequency assignment of the band components is given in Table 12. The two very weak bands expected from a ^{13}C bearing anion in these sites would lie under the strong bands at 1857 and 1886 cm^{-1} and are not clearly resolvable.

The ion site mix has been studied for NaCo(CO)$_4$ in some 14 organic solvents [25]. In the six solvents listed above (DMSO, DMF, HMPA, CH$_3$CN, CH$_3$NO$_2$, and pyridine), only the solvent-surrounded ion is found. In others, the mix involves several sites in substantial quantity. A total of five different kinds of ion site has been identified to date in the whole range of the solvents.

Solution structure studies have been made for TlCo(CO)$_4$ in seven solvents [25c]. Three ion sites dominate these solutions. They are the solvent surrounded ion, the contact ion pair, and the triple ion. There is a difference in the solution structure at ions between NaCo(CO)$_4$ and TlCo(CO)$_4$ solutions. The contact ion pair in solutions of the first salt has the form

TABLE 12

Ion Site Band Assignments for $NaCO(CO)_4$ in THF

Ion site	Mode	ν, cm^{-1}
Solvent surrounded ion	F_2	1886
Contact ion pair	A_1/F_2	1857
Contact ion pair	E/F_2	1899
Triple ion	A_1F_2[a]	1846
Triple ion	E/F_2	1906

[a]Anion site symmetry.

while the corresponding structure for the thallous salt is represented by

where the solvent molecules are represented in these drawings as
ethers, for example, THF. The metal ion is opposite an oxygen atom
of the anion in the Na^+ salt and is subjected to forces which are
primarily electrostatic. On the other hand, the Tl^+ ion is opposite
an anion face and its interaction with the Co atom of the anion shows some covalent character. The infrared spectra for an
anion in these two different kinds of contact ion-pair sites are
characteristic and are shown in Fig. 12.

Structure is a fundamental starting point in the understanding
of the microscopic of any system. This applies to the chemical as
well as the physical properties of solutions. The chemistry of a salt
in solution is just the sum of the chemistry at the microscopic ion
sites. What makes this statement nontrivial is the building evidence

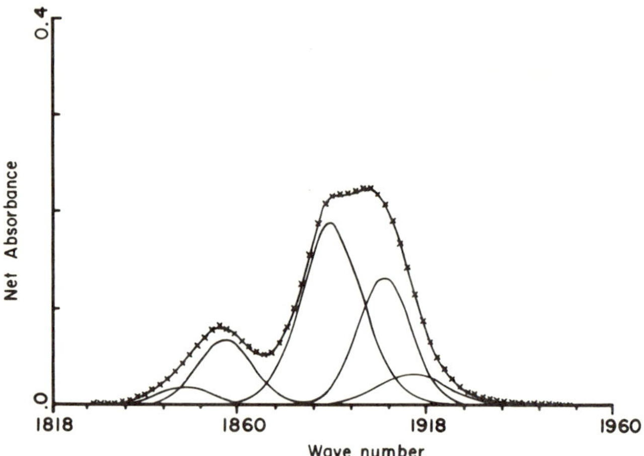

FIG. 11. Comparison of experimental (xxxx) band contour with computed (———) contour for NaCo(CO)$_4$ in THF. The separate computed bands are also shown.

of a salt in solution is just the sum of the chemistry at the microscopic ion sites. What makes this statement nontrivial is the building evidence that its chemistry is different at different ion sites. This situation is demonstrated for the Co(CO)$_4^-$ ion in thallous salt solutions. Thus, the reaction

$$3TlCo(CO)_4 + 3PPh_3 \rightarrow Tl[Co(CO)_3PPh_3]_3 + 2Tl + 3CO$$

has been found to take place in solutions with contact cation-anion sites and not in those with only a solvent-surrounded anion site [25c,27].

The role of contact ion pairs and solvent-surrounded ions has long been known in organic chemistry and their importance in metal-organic chemistry is now being elucidated. The spectroscopic work described above indicates that solution structure at ions is richer than this--for example, besides these two ion sites, at least three additional ones exist for the Co(CO)$_4^-$ ion; and there are two different kinds of contact pairs. The role of these other ion sites in the chemistry of ionic solutions has yet to be investigated.

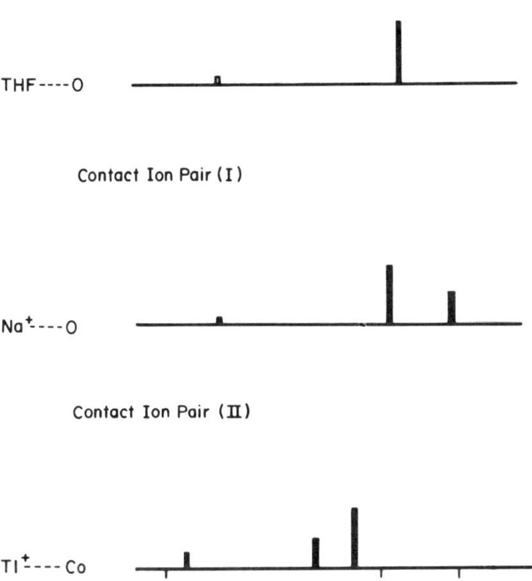

FIG. 12. The $Co(CO)_4^-$ anion stretching frequencies in several ion sites. The primary feature of the anion-environment interaction is represented on the left-hand side of the figure.

VII. A QUANTUM MODEL OF AN ORGANOMETALLIC ELECTROLYTIC SOLUTION

This section seeks to provide a basic quantum description for the solution phenomena described above. No effort will be made to obtain rigor. Instead, one seeks a zero order description which concentrates on the primary nature of the events.

In the absence of a radiation field, the state of the (whole) solution is described by the state function Ψ^o. The salt solution may be found in a number of states (Ψ_1^o, Ψ_2^o, Ψ_3^o, etc.) of energy (ε_1^o, ε_2^o, ε_3^o, etc.). These state functions satisfy the time-dependent Schrödinger equation

$$i\hbar \frac{\partial \Psi_n^O}{\partial t} = H^O \Psi_n^O$$

$$\Psi_n^O = \exp \frac{-i\varepsilon_n^O t}{\hbar} \; \psi_n^O(x\text{--}) \tag{19}$$

The character of the time-independent function $\psi_n^O(x\text{--})$ is form-ulated as follows:

1. The state function for the motion of the electrons $|e\rangle$ is treated by a Born-Oppenheimer type of approximation in which the solution for $|e\rangle$ is made while holding the position of the nuclei constant. The resulting energy eigenvalue W_e is a function of nuclear positions. Its minimum for the salt solution is the electronic energy ε_e of the system. Thus, one may write

$$W_e(rtv) = \varepsilon_e + V^e(rtv) \tag{20}$$

where the difference between W_e and ε_e is the potential energy under which the translation, rotation, and vibration of the ions and solvent molecules of the solution take place. At this level of approximation, we shall suppose that the electronic states are localized at the individual ions and solvent molecules. Then,

$$|e\rangle = |e_1\rangle |e_2\rangle \; \cdots$$

2. Solvent molecules in solution show vibrational spectra whose dominant character is its similarity to the gas phase spectra of the molecule. The same thing is true of neutral (solute) molecules when they are dissolved. Further, the vibrational spectra of a polyatomic ion (NO_3^-, ClO_4^-, $Co(CO)_4^-$, NH_4^+, etc.) in solution shows a basic similarity from one solvent to another. These facts lead one to transform the rectilinear coordinates of the nuclei of a solvent molecule or polyatomic ion into the coordinates of its center of mass, the Eulerian angles which give its orientation in space, and the internal coordinates which express its deformation, just as in the gas phase. These coordinates define, respectively, the translation, rotation, and vibration of a solution molecule or polyatomic

ion. In conformity with this nomenclature, the rectilinear coordinates which specify the position of a monatomic ion (Na^+, I^-, etc.) is said to define its *translation*.

In treating the vibrational states of the solution, another adiabatic (Born-Oppenheimer) approximation is made in which it is assumed that the vibration of a polyatomic species is fast, compared to the translation and rotational motion. The state function $|v\rangle$ for the vibration of the salt solution is defined by the relation

$$[\hat{T}_v + V(v)]\,|v\rangle = W_v(rt)\,|v\rangle \tag{21}$$

where $\hat{T}_v$ is the kinetic energy operator for the vibration and V is the quantity defined by Eq. (20). The rt coordinates are held fixed in Eq. (21), and the resulting eigenvalue W_v is a function of them. The localization of the vibration to the individual molecules and (polyatomic) ions is expressed by

$$|v\rangle = |v_1\rangle|v_2\rangle \cdots$$

3. W_v is expanded about its minimum for the solution

$$W_v(rt) = \varepsilon_v + U^v(rt) \tag{22}$$

to give the vibrational energy ε_v and the number U, which forms the potential for the translation and rotation of the solution elements. Not too much is known about this motion. It is expected to be coupled and to lead, in general, to collective modes which give rise to rather general absorption in the far region of the infrared spectrum and to low frequency scattering in the Raman effect. The rotation-translation states of the solution are obtained from

$$[\hat{T}_{rt} + U(rt)]\,|rt\rangle = \varepsilon_{rt}\,|rt\rangle \tag{23}$$

Thus, the energy ε and the state function Ψ^o has been partitioned as follows:

$$\varepsilon = \varepsilon_e + \varepsilon_v + \varepsilon_{rt} \tag{24}$$
$$\Psi^o = |e\rangle|v\rangle|rt\rangle$$

4. All the rotation-translation modes are not collective modes of the ionic solution. The presence of separate bands in the far-infrared spectrum which are associated with alkali ion motion imply the existence of localized motion at these ions--at least in dilute solution. If we let t_1 represent the coordinate (or coordinates) which defines the motion of all elements that move in the mode(s) localized at the first ion, one can write

$$|rt\rangle = |t_1\rangle|t_2\rangle \cdots |rt'\rangle \tag{25}$$

where the first terms on the right hand side of this equation define, in succession, the state of the localized motion at each ion, and the last term describes the state of each of the collective modes. This leads to

$$\varepsilon_{rt} = \sum_i \varepsilon_{t_i} + \varepsilon_{rt'} \tag{26}$$

where the first terms on the right hand side give the energy in the localized ion motion and the last term that in the collective motions.

One can now specify more precisely what quantum states of the solution are being excited in the two different types of infrared bands, which are being used above, to probe the nature of ionic solutions. The initial state of the solution ψ_i^o is here designated as $|i\rangle$ and the final state as $|f\rangle$. For the mid infrared band associated with the vibration of the $Co(CO)_4^-$ anion, which was used to probe solution structure, one has

$$|i\rangle = |t_1\rangle \cdots |rt'(i)\rangle \cdots |v_n(i)\rangle \cdots |e\rangle$$
$$|f\rangle = |t_1\rangle \cdots |rt'(f)\rangle \cdots |v_n(f)\rangle \cdots |e\rangle \tag{27}$$

One is exciting the vibration in the nth anion from the initial state $|v_n(i)\rangle$ to the final state $|v_n(f)\rangle$ while, at the same time, changing the collective translation-rotation state of the solution from $|rt'(i)\rangle$ to $|rt'(f)\rangle$. The wave number of the radiation being absorbed by this transition is given by the energy change in the solution

$$\nu = \frac{\varepsilon_f - \varepsilon_i}{hc}$$

$$= \frac{\varepsilon_{v_n}(f) - \varepsilon_{v_n}(i) + \varepsilon_{rt'}(f) - \varepsilon_{rt'}(i)}{hc}$$

At wave numbers larger than

$$\nu_c = \frac{\varepsilon_{v_n}(f) - \varepsilon_{v_n}(i)}{hc}$$

one is adding energy to the collective rotation-translation motions with the adsorption of the photon; at wave numbers less than ν_c, the converse is true. The center wave number ν_c lies near, although not necessarily precisely at, the maximum in the band intensity.

For the far-infrared band associated with the ion motion, the initial and final states of the solution are

$$|i\rangle = |t_1\rangle \cdots |t_n(i)\rangle \cdots |rt'(i)\rangle \cdots |e\rangle$$
$$|f\rangle = |t_1\rangle \cdots |t_n(f)\rangle \cdots |rt'(f)\rangle \cdots |e\rangle \tag{28}$$

Here one is exciting the localized translation modes at the nth ion from the state $|t_n(i)\rangle$ to $|t_n(f)\rangle$ while, at the same time, changing the collective translation-rotation state of the solution from $|rt'(i)\rangle$ to $|rt'(f)\rangle$. The wave number of the radiation being absorbed for this state change is

$$\nu = \frac{\varepsilon_{t_n}(f) - \varepsilon_{t_n}(i) + \varepsilon_{rt'}(f) - \varepsilon_{rt'}(i)}{hc}$$

The center wave number, defined by

$$\nu_c = \frac{\varepsilon_{t_n}(f) - \varepsilon_{t_n}(i)}{hc}$$

is expected to lie near, but not necessarily precisely at, the maximum in the band intensity.

VIII. BAND CONTOURS AND SOLUTION BEHAVIOR
IN TIME

There is much information about the nature of these ionic solutions
contained in the contours of the bands being excited. This applies
to both the far-infrared ion motion bands and the mid-infrared vibration bands of polyatomic ions. The shape of these bands is basically
connected with the character of the collective $|rt'\rangle$ states of the
solution. If we adopt the outlook of Heisenberg, the band shapes are
related to the behavior in time of basic system parameters. Consequently, the measurement of band contours can furnish dynamical information about the solution and, in particular, about the ions therein. This information forms a useful complement to that about states,
energy, and structure. Consequently, the knowledge of solutions
which lies in band shapes will be treated here. Only the detail required to define the quantities which appear and their relation to
the experiment will be given. Further, we will restrict the discussion to infrared bands.

In the last section, it was seen that the state of the solution
is described by the function Ψ^o. Now put the solution in a beam of
radiation at time zero. Specifically, we consider that part of the
solution in a differential volume element of unit cross section, perpendicular to the beam and thickness $d\ell$, which contains a large number of molecules and ions yet is small compared to the wavelength of
the radiation. The function which describes the state of this system
at time t is given the symbol Ψ; the Hamiltonian in the presence of
the radiation beam is H. These are connected by

$$i\hbar\,\frac{\partial\Psi}{\partial t} = H\Psi$$

H is related to the Hamiltonian operator for the solution in the absence of the radiation, that is, $\hat{H}^o$, by

$$H = H^o + H'$$

where H' is the operator which couples the salt solution to the radiation. The radiation beam is described by an electric field $\underline{E}$ passing through the solution. One can write

$$H' = \underline{M} \cdot \underline{E} \, d\ell \tag{29}$$

where

$$\underline{M} = \eta \sum_m (q_m \underline{R}_m + \underline{\mu}_m)$$

is the electric moment of the solution per unit volume. $\underline{R}_m$ is the position vector of the mth molecule or ion, q_m is its charge, $\underline{\mu}_m$ is its dipole moment referred to its center of mass, and the sum is over all the molecules and ions in the unit volume.

The consequence of the interaction of the radiation field with the solution is that energy is removed from the beam. On the experimental size, this is described by the Bouguer-Lambert law ($- dI = \alpha I d\ell$); it is computed from the theoretical side by conventional time-dependent perturbation theory. Equating the two permits one to write the experimental quantity $F(\omega)$ defined by

$$F(\omega) = \frac{3\hbar c}{4\pi^2} \frac{n(\alpha/\omega)}{1 - e^{-\beta\hbar\omega}} \tag{30}$$

as equal to

$$F(\omega) = 3 \sum_{i,f} \rho_i \langle i | \underline{M} \cdot \underline{w} | f \rangle \langle f | \underline{M} \cdot \underline{w} | i \rangle \delta(\omega_{fi} - \omega) \tag{31}$$

In these expressions, n is the real part of the index of refraction, $\beta = 1/kt$, and ω is the circular frequency of the radiation, i.e., $2\pi c$ times the wave number. Then, $|i\rangle$ and $|f\rangle$ are the initial and final states of the solution, ρ_i is the probability of finding it in the state $|i\rangle$, and $\underline{w}$ is a unit vector giving the electric field direction in the beam. Finally, $\delta(\omega_{fi} - \omega)$ is the Dirac delta function with

$$\omega_{fi} = \frac{\varepsilon_f - \varepsilon_i}{\hbar}$$

If one uses the Fourier expansion of the delta function

$$\delta(\omega_{fi} - \omega) = \frac{1}{2\pi} \int_\infty^\infty e^{i(\omega_{fi} - \omega)t} \, dt$$

and introduces the time dependent operators for the electric moment
given by

$$\underline{M}(t) = e^{iH^{o}t/\hbar} \, \underline{M} \, e^{-iH^{o}t/\hbar}$$
$$\underline{M}(o) = \underline{M}$$

then Eq. (31) will yield

$$\langle \underline{M}(o) \cdot \underline{M}(t) \rangle = \int_{\infty}^{\infty} F(\omega) e^{i\omega t} \, d\omega \tag{32}$$

where the ensemble average is defined by

$$\langle \underline{M}(o) \cdot \underline{M}(t) \rangle = 3 \sum_{i} \rho_{i} \langle i | \underline{M}(o) \cdot \underline{M}(t) | i \rangle \tag{33}$$

is equal to

$$\langle \underline{M}(o) \cdot \underline{M}(t) \rangle = 3 \sum_{i,f} \rho_{i} \langle i | \underline{M} \cdot \underline{w} | f \rangle \langle f | \underline{M} \cdot \underline{w} | e^{i\omega_{fi} t} \tag{34}$$

The left-hand side of the last three equations is just the time
correlation of the electric moment of the solution. It reflects its
dynamical behavior and may be related to the character of the ion
motion. While, strictly speaking, it is a quantum quantity, it has
a classical analogue (or near analogue) and may be more readily understood than some other quantum entities. It may be computed from
theory by Eq. (34). But what is of first importance, it may be evaluated from experiment by Eq. (32).

Before considering experimental results, let us see what kind of
dynamical information is contained in the shape of the ion motion
band. The time correlation function is related through Eq. (34) to

$$\langle i | \underline{M} \cdot \underline{w} | f \rangle = \langle i | \underline{M} | f \rangle \cdot \underline{w}$$

With the specification of the initial and final states for this band
given by Eq. (28) and the use of Eq. (29) for $\underline{M}$, the above expression
reduces to a term involving only the electric moment of the solution
at the alkali ion that is being excited. In the approximation that

the S_2M, S_4M, S_6M, or some similar model for the motion applies, then

$$\langle \underline{M}(o) \cdot \underline{M}(t) \rangle = \eta^2 q^2 \langle \underline{R}(o) \cdot \underline{R}(t) \rangle \tag{35}$$

and the time correlation of $\underline{M}$ for this far-infrared band just gives the time correlation of the position of the ion. This gives one a means of following the translational motion of an ion in solution as averaged over all the ions of the solution.

On the other hand, the transition which gives rise to a vibrational band of a polyatomic ion, Eq. (27), leads to

$$\langle i | \underline{M} | f \rangle = \langle rt'(i) | \underline{\mu}^V | rt'(f) \rangle$$

where

$$\underline{\mu}^V = \langle v(i) | \underline{\mu} | v(f) \rangle$$

Here $\underline{\mu}$ is the component of the dipole moment of the ion generated by its distortion in the vibrational mode of the band. Then one obtains

$$\langle \underline{M}(o) \cdot \underline{M}(t) \rangle = \eta^2 \langle \underline{\mu}(o) \cdot \underline{\mu}(t) \rangle \tag{36}$$

Thus, the shape of this kind of band of the ionic solution contains information about the time development of the dipole moment generated in the ion by the vibration.

To evaluate the time correlation function from the observed band, one computes the absorption coefficient α at each wave number in the band from the measured transmission of the sample. These values are then used to compute the experimental function $F(\omega)$ by Eq. (30). Then $\langle \underline{M}(o) \cdot \underline{M}(t) \rangle$ is obtained from the Fourier transform of $F(\omega)$ by Eq. (32). Figure 13 shows the results for the Li^+ band for $LiCo(CO)_4$ in THF, where the function has been normalized to unity at time zero.

The physical meaning of these results can be seen with the help of the following figure

$$\oplus \xleftarrow[\underline{R}(t)]{} \quad x \quad \xrightarrow[\underline{R}(o)]{} \oplus$$

where $\underline{R}(o)$ is the position vector of an ion at time zero and $\underline{R}(t)$ is the vector to its position at a later time t. The scalar product $\underline{R}(o) \cdot R(t)$ is just the length of $\underline{R}(o)$ multiplied by the projection of $\underline{R}$(t) on $\underline{R}$(o). Its value oscillates as the ion moves back and forth in its cage. But this motion of the ion is "disturbed" by its interaction with the collective translation-rotation modes of motion of the solution. The result is a change in the direction and/or amplitude of the ion motion and the value of the scalar product drops. The ensemble average $\langle \underline{R}$(o) $\cdot$ $\underline{R}$(t)$\rangle$ is the behavior at an individual ion averaged over all such ions in the solution.

The motion of the Li$^+$ ion back and forth in its cage is clearly seen in Fig. 13. The drop in amplitude shown is the averaged consequence of the disturbance of its motion by the environment. Thus,

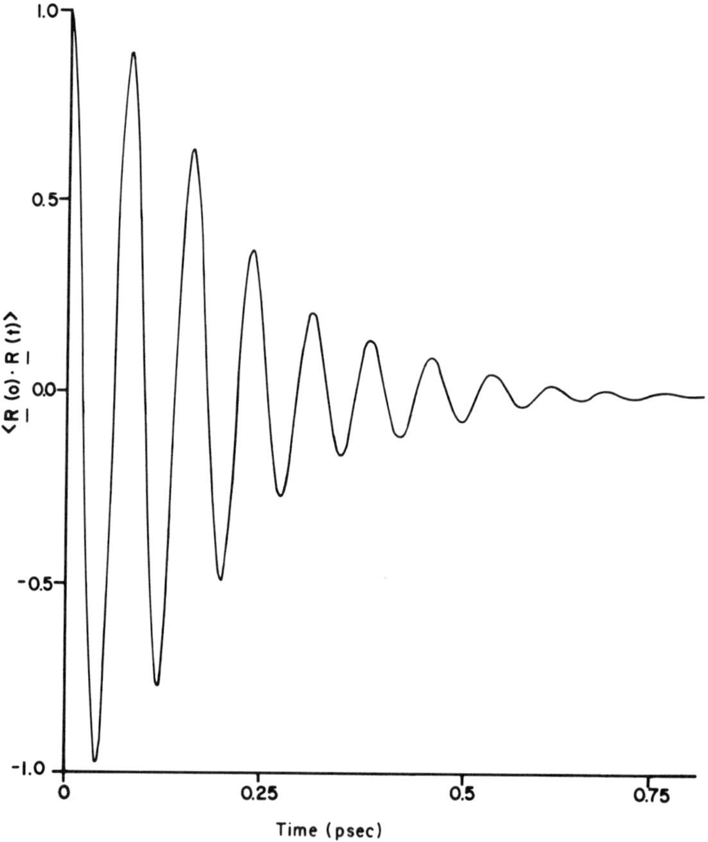

FIG. 13. Correlation on the motion of the Li$^+$ ion with time for the cation translation band.

the original direction of the motion persists for about 0.9 psec
(slightly more than 10 oscillations) before it is completely random-
ized. This dynamic result is not trivial. It is fundamental in
understanding the microscopic events in these solutions, and it is a
property of each solution. For example, the original direction of the
alkali ion motion in salt melts near the fusion temperature does not
persist for more than one oscillation [28]. Thus there is a fundamental difference in the behavior of the melt and these solutions.

   The basic similarity of Fig. 13 to the amplitude behavior of a
damped oscillator in classical mechanics is apparent. This is no
accident because both cases deal with dissipation in oscillatory
motion. Such a model, however, is only a crude representation of
events in solution. For example, it does not give a good representation of band contours to the extent that they are known today. Thus,
the mechanism of the decay of correlation in the motion in these
ionic solutions differs from that of the classic damped oscillator.

ACKNOWLEDGMENT

The author is grateful to Mr. Paul Lurix for help with the library
work for Chapters 3 and 4.

Appendix

DCKCLC: AN INTERACTIVE PROGRAM TO FIND THE BAND PARAMETERS
WHICH DESCRIBE A COMPLEX SPECTRAL BAND

NATURE OF DCKCLC

DICK CALC is a program which uses a nonlinear, weighted, least-squares
method to modify the parameters which describe the computed band
contour in order to bring it into closer agreement with the points
which define the experimental curve. An iterative procedure is used
with a damping factor which may be varied by the operator. Any set
of the parameters may be held constant in the iterative process.

The contour of the computed band may be the sum of as many as
15 individual bands. Each individual band is regarded as a sum of
Lorentz and Gauss components and requires five parameters to describe.
These are the band center plus the height and half-width-at-half-
height of each component. In practice, only four of these vary since
the half-width of the Gauss component is fixed by the half-width of
the Lorentz component in this computation.

The input to the calculation consists of the experimental absorbance values at equally spaced wave numbers throughout the spectral region, the weight to be assigned to each point, the number of
bands in the region, and the initial values for the five parameters
for each band.

The flow of the calculation is controlled by the operator sit-
ting at the teletype keyboard of a laboratory computer. This is
done by selecting a good damping factor for each iteration and by
deciding when sufficient iterations have been made. The calculation
proceeds as follows. The first iteration is made automatically with
a damping factor of P = 0.10, and the modified band parameters printed.
This is accompanied by PHI, the weighted sum of the squares of the
deviations. The operator selects a second value of the damping factor and the first iteration is repeated using the new value of P.
An examination of the new PHI and the new modification of the parameters enables the operator to judge if the second damping factor
did a better job on this iteration. A total of five different P
values (0.10 plus four to be selected) may be tried at each iteration. The operator selects which of these trials is best and, therefore, is to be taken as the first iteration. Then the calculation
proceeds to the second iteration, which begins by the first trial
with P = 0.10. The operator proceeds as before to select a damping
factor which gives a good second iteration. The calculation proceeds
in this way until the operator decides to terminate the computation.

A little reflection, and some experience, shows that the choice
of the values assigned as the weights for the experimental points
provides a powerful means of "guiding" the calculation. Their abuse
can produce weird results.

THE BASIC EQUATIONS

The problem of selecting a set of parameters to reproduce experimental data is an old one and has had many formulations. We use here the
particular form of the mathematics used by Pitha and Jones [29] in
their treatment of the band resolution problem.

DCKCLC computes the absorbance of an infrared band at the wave
number W, defined by the equation

$$ABSRB = \log_{10} \frac{T_o}{T}$$

The absorbance of a band is given by Eq. (18) of the text

$$ABSRB = \alpha + \sum_j \left\{ \frac{H_j}{1 + D_j^2(\nu - \nu_j^o)^2} + h_j \exp[-d_j^2(\nu - \nu_j^o)^2] \right\}$$

where the parameters H_j, ν_j^o, D_j, h_j, and d_j define the shape and
intensity of the jth component band. In the version of the program
presented here, the width of the Gauss contribution to a component
band is fixed by the width of the Lorentz contribution with a width
factor

$$d_j = WDF_j * D_j$$

The variable parameters for each component band are stored in the
program in the X array as follows: $X(1) = H_1$, $X(2) = \nu_1^o$, $X(3) = D_1$,
$X(4) = h$, $X(5) = H_2$, etc.

The absorbance at the Nth point, computed by the above equation,
is given the symbol AC(N) in the program. The weighted difference
between the experimental absorbance AX(N) and the computed value is
represented by F(N). Thus,

$$F(N) = WT(N) * (AX(N) - AC(N))$$

The sum of the squares of the weighted deviations is

$$PHI = \sum_N F(N) * F(N)$$

We wish to choose the X(I) so that PHI is a minimum. At this minimum,

$$\frac{\partial PHI}{\partial X(I)} = 0$$

For any other set of X(I),

$$\frac{\partial PHI}{\partial X(I)} = \sum_N 2F(N)\frac{\partial F(N)}{\partial X(I)}$$

We use an iterative procedure to push PHI toward a minimum. It yields
an equation

$$- \frac{1}{2} \frac{\partial PHI}{\partial X(I)} = \sum_N \frac{\partial F(N)}{\partial X(I)} \sum_J \frac{\partial F(N)}{\partial X(J)} (X(J)^{m + 1} - X(J)^m)$$

for each X(I). Here $X(J)^m$ is the value of this parameter produced
in the mth iteration and $X(J)^{m+1}$ that from the (m + 1)th iteration.
The solution of these equations permits one to compute the "corrected"
parameters $X(J)^{m + 1}$ from the previous set $X(J)^m$.

These equations are expressed in matrix notation by introducing
the quantities defined by

$$A(N,J) = \frac{\partial F(N)}{\partial X(J)}$$

$$G(I) = \frac{1}{2} \frac{\partial PHI}{\partial X(I)}$$

$$D(J) = X(J)^{m + 1} - X(J)^m$$

With $B = \tilde{A}A$, the equations become

$$- G = BD$$

In the mathematical procedures used in the computer program, it is
useful to "normalize" B so that the diagonal terms are unity. We
introduce the matrix $C^{1/2}(I,I) = \sqrt{B(I,I)}$; $C^{1/2}(K,J) = 0$. With this
matrix, these equations become

$$(C^{-1/2}BC^{-1/2})\, C^{1/2}D = -\, C^{-1/2}G$$

where the matrix in parenthesis now has unit diagonal terms. This
is a set of linear equations in the unknowns

$$C^{1/2}(I,I)(X(I)^{m + 1} - X(I)^m)$$

The equations actually solved by the computer software are

$$(C^{-1/2}BC^{-1/2} + P*E)(C^{1/2}\,D) = -\,C^{-1/2}G$$

where the damping factor P has been introduced to help obtain convergence. The program solves the equations by the successive elimination of the unknowns.

THE PROGRAM DETAILS

The computer used with this program in the author's laboratory is a PDP 11/45. The software is in FORTRAN and, hence, is readily transferable to other computers after some allowance is made for computer individuality. The program listing is found at the end of the Appendix.

The computer program consists of several main sections. The first section prepares the computer and reads the data which defines the specific calculation to be made into core. This is done by statements 0001 through 0036 (see Program Listing). The wave number of each point is calculated and certain quantities set to zero by statements 0037 through 0048. The absorbance at each point is computed by statements 0049 through 0066 while the elements of the A matrix are formed by 0067 through 0085. Note that all the 4*NB + 1 elements of a row of A are computed and then the terms corresponding to the parameters held constant are dropped and the row closed up. F(N) and PHI are calculated with 0086 and 0087. The B and G matrices are computed with 0088 through 0101 and the elements of $-C^{-1/2}$ G are added to B as an LIT1th column and the augmented matrix put on DEC tape (DT1) with 0102 through 0104. This is followed by adding the damping factor P to B and solving for the corrected band parameters with 0105 through 0118, while the vector of the corrected band parameters is formed with 0119 through 0131. These are printed together with the resulting least squares sum PHI and P with statements 0132 through 0168.

At this point the computer stops and asks the operator to specify the next cycle of the calculation. It indicates this stage by printing

INPUT REQUIRED TO CONTINUE

The statements 0171 through 0182 process the information about the new cycle of the computation and set it into motion. Statements 0183 through 0187 bring the computation to an end when the operator specifies this option with statement 0171.

The DCKCLC software as written here expects the experimental points which define the observed band to be in a file on magnetic tape unit DT0. The name of this file is read as input data to the program as the variable VAR. The magnetic tape on unit DT1 contains the set of weighting numbers to use in the specific calculation under the filename WTPTS and with FORMAT (11F6.2). This tape also serves as a "scratch" unit on which the augmented B matrix is stored between successive phases of the operation of the program. The subroutine DELETE, called in statement 183, removes this file from the "scratch" tape at the end of the calculation and thus restores the system to the condition to treat another band by the same program. It is not listed since it would be rather different for each model of computer used with the software.

The input to the program other than the two files on magnetic tape is made from the teletype keyboard. It takes the form

Line	Columns	Variable	Remarks
1	1-72	HEAD	Your printout heading; alphanumeric
2	1-6	VAR	Name of the experimental file; alphanumeric, ≤ 6 characters
3	1-3	NB	Integer; right justified
	4-18	WL	Use decimal point
	19-33	WH	Use decimal point
	34-48	WI	Use decimal point
4	1-14	H	Use decimal point
	15-28	ν^o	Use decimal point
	29-42	D	Use decimal point
	43-56	WDF	Use decimal point
	57-70	h	Use decimal point

Type a line as above for each of the NB individual bands of the complex band.

Line	Columns	Variable	Remarks
5	1-15	ALPHA	Use decimal point
6	1-3	NSKP	Integer; must be right justified
7	1-5	blank	
	6-8	JSKP(1)	Integer; must be right justified
	9-11	NJ(1)	Integer; must be right justified
	12-16	blank	
	17-19	JSKP(2)	Integer; must be right justified
	20-22	NJ(2)	Integer; must be right justified

Continue until each JSKP and its NJ have been entered. Note that only 5 JSKP (and NJ) values are to be entered on one line of input. The program starts when the RETURN key is pressed after the last NJ value has been typed. After an iteration, the program stops and the computer prints the results of the iteration and

    INPUT REQUIRED TO CONTINUE

Then type your choice of how to proceed on line 8.

Line	Columns	Variable	Remarks
8	1-3	NWP	Integer; must be right justified
	4-18	P	Use decimal point
	19-21	NWIT	Integer; must be right justified
	22-24	KK	Integer; must be right justified

A line like this is typed after each cycle. Note that one completes the computation with this line by typing NWP = 0 and NWIT = 0. Then the computer returns.

    $

The computer is now ready for another task under the DOS monitor.

---

    *The program currently operates under the RSX-11D monitor and all files and matrices are stored on the disk.

DICTIONARY OF SYMBOLS

AC(N)   Computed absorbance at the Nth point.

AX(N)   Experimental absorbance at the Nth point.

ALPHA   The constant offset of the baseline of the computed curve
             (in absorbance units) must be small (< 0.003).

AN(I)   The A(N,I) element of the A matrix (see text).

B(I,J)   See text for the definition of the B matrix. Note that
             after B is calculated, $C^{-1/2}BC^{-1/2}$ and $C^{-1/2}BC^{-1/2}$ + P
             is stored as B.

CC   The $C^{-1/2}$ matrix (see text).

F(N)   The weighted difference between the experimental and
             calcualted absorbance at the Nth point.

G(J)   See text for definition of the G matrix.

JSKP(NN)   The number of the band in which the NJ(NN) band parameter
             is to be held constant. The numbering is the same as
             used when the band parameters were entered as data at
             line 3. NN refers to the NNth parameter to be skipped.

KK,JJ   See NWIT.

LI   The last I.

LIT   The last IT.

NB   The number of individual bands in the complex band.

NJ(NN)   The parameters in a given component band which can change
             are listed in the program in the order H, ν^{o}, D, h. If
             D in the first band is to be held constant, NJ is 3. If
             h in the first band is to be held constant, NJ is 4. If
             D in the second band is to be held constant, NJ is again
             3. NN refers to the NNth parameter that is being skipped
             (held constant).

NP   The number of points in the experimental band.

NSKP   The number of band parameters to be held constant (or
             skipped) in the iterations.

NIT   The number of iterations which have been carried out.

NWP   If NWP = 1, another set of connected parameters is to be
             calculated with a new P value; if 0, no new P is to be
             entered.

NWIT   If NWIT = 1, a new iteration is to be made using the KKth value of P.

P   The damping factor.

PHI   The sum of the squares of the weighted differences between the experimental and computed absorbances.

VAR   The filename under which the experimental points are stored on DTØ:. Must be $\leq$ 6 characters--first one alphabetical.

W,WW(N)   The wave number of the Nth point.

WL   The smallest wave number for which experimental points exist.

WH   The largest wave number for which experimental points exist.

WI   The wave-number interval between the experimental points.

WT(N)   The weight which is assigned to the Nth experimental point for the least-squares calculation.

X(I)   The Ith parameter in the list of component band parameters which can change. The parameters which may change for the first two bands H_1, ν_1^o, D_1, h_1, H_2, ν_2^o, D_2, h_2 are placed in this order in the first eight positions of the X vector, $X(1) = H_1$, $X(2) = \nu_1^o$, $X(3) = D_1$, $X(4) = h_1$, $X(5) = H_2$, $X(6) = \nu_2^o$, $X(7) = D_2$, $X(8) = h_2$. Note that all X parameters to be held constant in the iterations are entered in their normal place in the X vector. They are "skipped" later in the program.

XC(JJ,I)   The new ("corrected") value of the parameter X(I) after the current iteration with the JJth value of P.

PROGRAM LISTING

FORTRAN VØØ4A

```
        C                         DCKCLC
ØØØ1              DIMENSION X(61),XC(5,61),AN(61),B(61,62),G(61),
                 1AX(3ØØ),AC(3ØØ),WW(3ØØ),F(3ØØ),CC(61),WDF(15),
                 2WT(3ØØ),HEAD(18),JSKP(61),NJ(61),VAR(2)
ØØØ2              READ(6,1Ø2) HEAD
ØØØ3         1Ø2 FORMAT(18A4)
ØØØ4              READ(6,1Ø3) VAR
ØØØ5         1Ø3 FORMAT(2A4)
ØØØ6              READ(6,1Ø4) NB,WL,WH,WI
ØØØ7         1Ø4 FORMAT(I3,3F15.5)
ØØØ8              CALL SETFIL(4,VAR,-1,'DT',Ø)
ØØØ9              NP=(WH-WL)/WI+1.Ø
ØØ1Ø         4ØØ FORMAT(15)
ØØ11              IF(NP.GT.3ØØ) GO TO 33Ø
ØØ12              READ(4,1Ø6) ( AX(N), N=1,NP )
ØØ13         1Ø6 FORMAT(F5.3)
ØØ14              ENDFILE 4
ØØ15              CALL SETFIL(2,'WTPTS',-1,'DT',1)
ØØ16              READ(2,1Ø7) ( WT(N), N=1,NP )
ØØ17         1Ø7 FORMAT(11F6.2)
ØØ18              ENDFILE 2
ØØ19              CALL SETFIL(4,'MTRXBA.DCK',-1,'DT',1)
ØØ2Ø              LI=4*NB+1
ØØ21              LI1=LI+1
ØØ22              NW=2*LI1
ØØ23              DEFINE FILE 4(LI,NW,U,INDX)
ØØ24                DO 1Ø8 J=1,NB
ØØ25                K=4*J-3
ØØ26                L=4*J-2
ØØ27                M=4*J-1
ØØ28                I=4*J
ØØ29         1Ø8 READ(6,1Ø9) X(K),X(L),X(M),WDF(J),X(I)
ØØ3Ø         1Ø9 FORMAT(5F14.4)
ØØ31              READ(6,11Ø) X(LI)
ØØ32         11Ø FORMAT(F15.5)
ØØ33              READ(6,114) NSKP
ØØ34         114    FORMAT(I3)
ØØ35              READ(6,116) ((JSKP(I),NJ(I)),I=1,NSKP)
ØØ36         116    FORMAT(5(5X,2I3))
ØØ37              DO 12Ø N=1,NP
ØØ38              CN=N-1
ØØ39         12Ø WW(N)=WL + CN*WI
ØØ4Ø              NIT=Ø
ØØ41         2ØØ   PHI=Ø.Ø
ØØ42              DO 2Ø2 I=1,LI
ØØ43         2Ø2   G(I) =Ø.Ø
ØØ44              DO 2Ø4 I=1,LI
```

```
ØØ45                 DO 2Ø4 J=I,LI
ØØ46      2Ø4        B(I,J)=Ø.Ø
ØØ47                 LIT=LI-NSKP
ØØ48                 LIT1=LIT+1
          C          COMPUTES THE NTH ROW OF THE A MATRIX FOR ALL MEMBERS OF
          C          THE X VECTOR.  DELETES THE TERMS CORRESPONDING TO THE
          C          NUMBERS OF THE X VECTOR TO BE HELD CONSTANT IN THE
          C          ITERATIONS.  COMPUTES F(N) AND PHI.
ØØ49                 DO 232 N=1,NP
ØØ5Ø                 AC(N)=X(LI)
ØØ51                 DO 22Ø J=1,NB
ØØ52                 K=4*J-3
ØØ53                 L=4*J-2
ØØ54                 M=4*J-1
ØØ55                 I=4*J
ØØ56                 DIF=WW(N)-X(L)
ØØ57                 R=DIF*X(M)
ØØ58                 D=R*R+1.Ø
ØØ59                 DG=WDF(J)*X(M)
ØØ6Ø                 R1=DIF*DG
ØØ61                 D1=R1*R1
ØØ62                 IF(D1.GT.86.8) GO TO 21Ø
ØØ63                 E=EXP(-D1)
ØØ64                 GO TO 214
ØØ65      21Ø        E=Ø.Ø
ØØ66      214        AC(N)=AC(N) + (X(K)/D) + X(I)*E
ØØ67                 S=2.Ø*R*X(K)/D*D
ØØ68                 T=2.Ø*R1*X(I)*E
ØØ69                 AN(K)=-WT(N)/D
ØØ7Ø                 AN(L)=-WT(N)*(S*X(M) +T*DG )
ØØ71                 AN(M)= WT(N)*DIF*( S + T*WDF(J) )
ØØ72      22Ø        AN(I)=-WT(N)*E
ØØ73                 AN(LI)=-WT(N)
ØØ74                 IF(NSKP.EQ.Ø) GO TO 225
ØØ75                 NN=1
ØØ76                 ISKP=4*(JSKP(NN)-1) + NJ(NN)
ØØ77                 IT=Ø
ØØ78                 DO 224 I=1,LI
ØØ79                 IT=IT+1
ØØ8Ø                 IF(I.EQ.ISKP) GO TO 222
ØØ81                 AN(IT)=AN(I)
ØØ82                 GO TO 224
ØØ83      222        IF(NN.LT.NSKP) NN=NN+1
ØØ84                 ISKP=4*(JSKP(NN)-1) +NJ(NN)
ØØ85                 IT=IT-1
ØØ86      224        CONTINUE
ØØ87      225        F(N)=WT(N)*( AX(N) - AC(N) )
ØØ88                 PHI=PHI + F(N)*F(N)
          C          CALCULATES THE UPPER HALF OF THE B MATRIX.  THE NTH
          C          PASS THROUGH STATEMENT 23Ø CONTRIBUTES THE NTH TERM
          C          TO ALL ELEMENTS OF UPPER HALF OF B.  COMPUTES THE G
```

```
       C         VECTOR, THE CC=C**-Ø.5 MATRIX, AND  (C**-Ø.5)*B*
       C         C**-Ø.5 WHICH IT STORES IN THE CORE AS B.  THE COLUMN
       C         VECTOR -(C**-Ø.5)*G IS ADDED TO B AS A LAST COLUMN
       C         AND THIS AUGMENTED B MATRIX IS STORED ON DT1 UNDER
       C         THE FILE NAME MTRXBA.
ØØ89             DO 232 I=1,LIT
ØØ9Ø             DO 23Ø J=I,LIT
ØØ91   23Ø       B(I,J)=AN(I)*AN(J) + B(I,J)
ØØ92   232       G(I)=F(N)*AN(I) + G(I)
ØØ93             NIT=NIT+1
ØØ94             WRITE(6,234) PHI
ØØ95   234       FORMAT(8X,4HPHI=,E12.4//)
ØØ96             DO 24Ø I=1,LIT
ØØ97           BD=B(I,I)
ØØ98       24Ø CC(I)=1/SQRT(BD)
ØØ99             DO 252 I=1,LIT
Ø1ØØ             DO 25Ø J=I,LIT
Ø1Ø1           B(I,J)=CC(I)*B(I,J)*CC(J)
Ø1Ø2       25Ø B(J,I)=B(I,J)
Ø1Ø3       252 B(I,LIT1)=-CC(I)*G(I)
Ø1Ø4             DO 254 I=1,LIT
Ø1Ø5        254  WRITE(4'I) ( B(I,J), J=1,LIT1 )
       C         P IS ADDED TO THE AUGMENTED "B" MATRIX IN CORE
       C         AND THE ELEMENTS OF D ARE SOLVED FOR BY SUCCESSIVE
       C         ELIMINATION (GAUSS-JORDAN METHOD).  THE CORRECTED
       C         X VECTOR = XC IS FORMED.
Ø1Ø6             JJ=1
Ø1Ø7             P=Ø.1Ø
Ø1Ø8   26Ø       DO 262 I=1,LIT
Ø1Ø9   262       B(I,I)=B(I,I) + P
Ø11Ø             DO 28Ø I=1,LIT
Ø111             PVOT=1.Ø/B(I,I)
Ø112             DO 272 J=1,LIT1
Ø113   272       B(I,J)=B(I,J)*PVOT
Ø114             DO 28Ø K=1,LIT
Ø115             IF(K.EQ.I) GO TO 28Ø
Ø116             U=B(K,I)
Ø117             DO 274 J=I,LIT1
Ø118   274       B(K,J)=B(K,J) - B(I,J)*U
Ø119   28Ø       CONTINUE
Ø12Ø           NN=1
Ø121           ISKP=4*( JSKP(NN) - 1 ) + NJ(NN)
Ø122           IT=Ø
Ø123           DO 286 I=1,LI
Ø124           IT=IT+1
Ø125           IF(I.EQ.ISKP)GO TO 284
Ø126           XC(JJ,I)=X(I) + B(IT,LIT1)*CC(IT)
Ø127           GO TO 286
Ø128       284 XC(JJ,I)=X(I)
Ø129           IF(NN.LT.NSKP) NN=NN+1
Ø13Ø           IT=IT-1
```

```
Ø131                ISKP=4*( JSKP(NN)-1 ) + NJ(NN)
Ø132            286 CONTINUE
            C       THE CORRECTED BAND PARAMETERS ARE WRITTEN AND THE
            C       CORRESPONDING VALUE OF PHI COMPUTED AND WRITTEN.
Ø133                PHI=Ø.Ø
Ø134                WRITE(6,291)
Ø135        291     FORMAT(8X,'THE BAND PARAMETERS ARE :')
Ø136                DO 298 N=1,NP
Ø137                AC(N)=XC(JJ,LI)
Ø138                DO 294 J=1,NB
Ø139                K=4*J-3
Ø14Ø                L=4*J-2
Ø141                M=4*J-1
Ø142                I=4*J
Ø143                DIF=WW(N)-XC(JJ,L)
Ø144                R=DIF*XC(JJ,M)
Ø145                D=R*R+1.Ø
Ø146                DG=WDF(J)*XC(JJ,M)
Ø147                R1=DIF*DG
Ø148                D1=R1*R1
Ø149                IF(D1.GT.86.8) GO TO 292
Ø15Ø                E=EXP(-D1)
Ø151                GO TO 294
Ø152        292     E=Ø.Ø
Ø153        294     AC(N)=AC(N) + (XC(JJ,K)/D) + XC(JJ,I)*E
Ø154                F(N)=WT(N)*( AX(N) - AC(N) )
Ø155        298     PHI=PHI + F(N)*F(N)
Ø156                DO 296 J=1,NB
Ø157                K=4*J-3
Ø158                L=4*J-2
Ø159                M=4*J-1
Ø16Ø                I=4*J
Ø161                WRITE(6,295)
Ø162        295 FORMAT(1ØX,2HHL,11X,2HNU,13X,2HDL,1ØX,3HWDF,1OX,2HHG)
Ø163        296     WRITE(6,297) XC(JJ,K),XC(JJ,L),XC(JJ,M),WDF(J),XC(JJ,I)
Ø164        297     FORMAT(F14.4,F14,2,F14.4,F11.2,F14.4)
Ø165            WRITE(6,299) XC(JJ,LI)
Ø166        299     FORMAT(8X,'ALPHA=',F8.4)
Ø167                WRITE(6,3ØØ) NIT,P,PHI
Ø168        3ØØ     FORMAT(3X,15HAFTER ITERATION,I3,6X,7HWITH P=,F10.3,
                16X,8HAND PHI=,E11.4//)
Ø169                JJ=JJ+1
            C       THE COMPUTER STOPS AT THIS POINT AND ASKS FOR INPUT
            C       REGARDING THE NEXT STEP OF THE ITERATION OR TERMINATION.
Ø17Ø                WRITE(6,3Ø5)
Ø171        3Ø5     FORMAT(8X,'INPUT REQUIRED TO CONTINUE')
Ø172            READ(6,31Ø) NWP,P,NWIT,KK
Ø173        31Ø     FORMAT(I3,F15.5,2I3)
Ø174                IF(NWP.EQ.Ø) GO TO 316
Ø175                DO 312 I=1,LIT
Ø176            312 READ(4'I) ( B(I,J), J=1,LIT1 )
```

```
Ø177              GO TO 26Ø
Ø178    316       IF(NWIT.EQ.Ø) GO TO 326
Ø179          JJ=KK
Ø18Ø          DO 32Ø I=1,LI
Ø181    32Ø       X(I)=XC(JJ,I)
Ø182              GO TO 2ØØ
Ø183      326 ENDFILE 4
Ø184              CALL DELETE
Ø185              CALL EXIT
Ø186    33Ø       WRITE(6,332)
Ø187    332       FORMAT(3ØX,'TOO MANY POINTS')
Ø188              STOP
Ø189              END


ROUTINES CALLED:
SETFIL,EXP     , SQRT   , DELETE, EXIT
```

REFERENCES

1. (a) W. F. Edgell and A. T. Watts, *Abstracts,* Symposium on Molecular Structure and Spectroscopy, Ohio State University, June 1965, p. 85; (b) W. F. Edgell, A. T. Watts, J. Lyford, IV, and W. Risen, Jr., *J. Amer. Chem. Soc., 88,* 1815 (1966); (c) W. F. Edgell, *Abstracts,* 153rd National Meeting of the American Chemical Society, Miami Beach, Florida, April 1967, No. R-149; (d) W. F. Edgell, J. Lyford, and J. Fisher, *Abstracts,* 155th National Meeting of the American Chemical Society, San Francisco, California, April 1968, No. S-136.

2. W. F. Edgell, J. Lyford, IV, R. Wright, W. Risen, Jr., and R. Watts, *J. Amer. Chem. Soc., 92,* 2240 (1970).

3. (a) B. W. Maxey and A. I. Popov, *J. Amer. Chem. Soc., 89,* 2230 (1967); (b) *ibid., 91,* 20 (1969).

4. (a) J. L. Wuepper and A. I. Popov, *J. Amer. Chem. Soc., 91,* 4352 (1969); (b) W. J. McKinney and A. I. Popov, *J. Phys. Chem., 74,* 535 (1970); (c) M. K. Wong, W. J. McKinney, and A. I. Popov, *ibid., 75,* 56 (1971).

5. C. Lassigne and P. Baine, *J. Phys. Chem., 75,* 3188 (1971).

6. T. L. Buxton and J. A. Caruso, *J. Phys. Chem., 77,* 1882 (1973).

7. J. C. Evans and G. Y.-S. Lo, *J. Phys. Chem., 69,* 3223 (1965).

8. M. J. French and J. L. Wood, *J. Chem. Phys., 49,* 2358 (1968).

9. E. Schaschel and M. C. Day, *J. Amer. Chem. Soc., 90,* 503 (1968).

10. (a) B. W. Maxey and A. I. Popov, *J. Amer. Chem. Soc., 90,* 4470 (1968); (b) J. L. Wuepper and A. I. Popov, *J. Amer. Chem. Soc., 92,* 1493 (1970).

11. A. T. Tsatsas, R. W. Stearns, and W. M. Risen, Jr., *J. Amer. Chem. Soc., 94,* 5247 (1972).

12. W. F. Edgell, unpublished.

13. (a) Dr. Peter Moore, "Studies in Molecular Spectroscopy," Ph.D. Thesis, Purdue University, June, 1961; (b) Dr. T. Seshan, "Studies in Molecular Spectroscopy," Ph.D. Thesis, Purdue University, January, 1969.

14. Dr. William M. Risen, Jr., "Studies in Molecular Spectroscopy," Ph.D. Thesis, Purdue University, January, 1967.

15. Dr. John Lyford, IV, "Studies in Molecular Spectroscopy," Ph.D. Thesis, Purdue University, August, 1969.

16. (a) E. B. Wilson, Jr., *J. Chem. Phys.*, *9*, 76 (1941); (b) E. B. Wilson, Jr., J. C. Decus, and P. C. Cross, "Molecular Vibrations," McGraw-Hill, New York, 1955, p. 54.

17. A. T. Tsatsas and W. R. Risen, Jr., *J. Amer. Chem. Soc.*, *92*, 1789 (1970).

18. J. D. Bernal and R. H. Fowler, *J. Chem. Phys.*, *1*, 515 (1933).

19. D. D. Eley and M. G. Evans, *Trans. Faraday Soc.*, *34*, 1093 (1938).

20. A. D. Buckingham, *Disc. Faraday Soc.*, *24*, 15 (1957).

21. E. S. Rittner, *J. Chem. Phys.*, *19*, 1030 (1951).

22. D. White, K. S. Seshadri, D. F. Dever, D. E. Mann, and M. J. Linevsky, *J. Chem. Phys.*, *39*, 2463 (1963).

23. W. F. Edgell, M. T. Yang, and M. Koizumi, *J. Amer. Chem. Soc.*, *87*, 2563 (1965).

24. (a) T. E. Hogen Esch and J. Smid., *J. Amer. Chem. Soc.*, *87*, 669 (1965); (b) *ibid.*, *88*, 307, 318 (1966).

25. (a) W. F. Edgell, J. Lyford, IV, A. Barbetta, and C. I. Jose, *J. Amer. Chem. Soc.*, *93*, 6403 (1971); (b) W. F. Edgell and J. Lyford, IV, *J. Amer. Chem. Soc.*, *93*, 6407 (1971); (c) W. F. Edgell, W. R. Robinson, A. Barbetta, and D. P. Schussler, unpublished; (d) W. F. Edgell and A. Barbetta, *J. Amer. Chem. Soc.*, *96*, 415(1974); (e) W. F. Edgell and A. Barbetta, unpublished; (f) W. F. Edgell and S. Hegde, unpublished.

26. W. F. Edgell and J. Lyford, IV, *J. Chem. Phys.*, *52*, 4329 (1970).

27. D. P. Schussler, W. R. Robinson, and W. F. Edgell, *Inorg. Chem.*, *13*, 153 (1974).

28. G. H. Wegdam, J. B. te Beck, H. van der Linden, and J. van der Elsken, *J. Chem. Phys.*, *55*, 5207 (1971).

29. J. Pitha and R. N. Jones, *National Research Council of Canada Bulletin No. 12*, Ottawa, Canada (1968).